GOING INSIDE

Going Inside

A Tour round a Single Moment of Consciousness

JOHN MCCRONE

faber and faber

First published in 1999
by Faber and Faber Limited
3 Queen Square London WC1N 3AU

Photoset by Parker Typesetting Service, Leicester
Printed in England by Clays Ltd, St Ives plc

A CIP record for this book
is available from the British Library

ISBN 0—571—17319—5

PICTURE ACKNOWLEDGEMENTS

Plate 1: Alfred Pasieka/Science Photo Library. 2: John Bavosi/Science Photo Library. 3, 4: Wellcome Department of Neurology. 5: Wellcome Trust Medical Photographic Library. 6–9: Stephen M. Kosslyn. 10: Marcus E. Raichle, MD, Washington University School of Medicine, St Louis, Missouri. 11: Alex Martin, Laboratory of Psychology and Psychopathology, National Institute of Mental Health, Bethesda, Marlyland. 12: Leslie Ungerleider. 13: Alan Gilliland and Richard Burgess/*Daily Telegraph.*

2 4 6 8 10 9 7 5 3 1

Contents

CHAPTER ONE

Reluctant Beginnings

I remember the cut of sunlight across the court, a ragged oblong burning the varnished floor, far down below. Fuzzily, out of the corner of my eye, a pair of hands strained to block a path. But they may as well have been on another planet. The ball was somewhere, not yet visible but sure to mushroom before my hand as I swung. It was 1975, the New Zealand university volleyball championships, and I was hanging high above the net. What I remember most clearly is the upturned face of the opposing player who was crouched right in close, waiting exactly at the point where I wanted to hit the ball.

No matter. I had exploded into the air with such venom that it seemed to have stopped time. The world was moving with treacly slowness, illuminated with a giddy brilliance. I dangled above the net, the only one free to think and act. Gleefully, I bounced the ball right off his head and away into the stands.

It only happened once. We did not even win the game. But it is the kind of moment that makes you wish you understood consciousness a little better. Why were things usually such a hurried blur when I went up to smash a ball? Often you would do something good – jam the ball down the front of some tall guy who was blocking too far back from the net, or snap the wrist to cut the ball down the sideline – and it would seem that it was your body making the decisions. You had only a fleeting glimpse, a flash of memory to be salvaged as you crashed back to the ground. Did time always freeze for really good players? Did they always have the luxury of slow-motion action? Or was my moment just an illusion? Did I bang it off that head by happy accident and find the instant so hilarious that every sensory detail was etched into my memory, leaving me with a false feeling of being in complete control at the time?

Back on the hard benches of the lecture hall after the Easter break, there was little likelihood of getting answers. A junior teacher over from the medical school was inexpertly juggling a plastic brain in her hands. A purple bit – perhaps the hippocampus – dropped to the floor as she held half the organ aloft to show the room. The hour had hardly begun, yet already a blanket of deep boredom weighed down upon the introductory neurology class. Lists of names scratched up on the blackboard were dutifully copied onto pads. Caudate nucleus, pons, medulla oblongata, anterior commissure, mammillary bodies, posterior this, superior that, the lateral something or other. Meaningless fact piled upon meaningless fact. No one seemed too sure what any of these bits did, but by God, every dent and bulge had a label. It was like learning the street map of a city you would never visit. This was science without plot or theme. So dull was it that it was not unknown for some students to give up, rest their cheeks on folded arms, and fall gently asleep.

The thunk of a swing door and an apologetic patter of feet announced a latecomer. Heads turned in mild interest to watch as the person slipped into a seat halfway down the room, tucked an army surplus gas mask bag behind his legs and meticulously set out paper and pen before him. Just a few seconds passed before the student realised he had made a horrid mistake and rushed into the wrong lecture hall. With ducked head, he packed his bag again and made his escape. The door slapped shut. Then from somewhere up at the back, a leaden voice drawled: 'Couldn't take the pace, huh?' The answering howl of laughter was so wild, so heart-felt, it chilled the mind as much as it warmed the body. It was true, then: this was the pits. Just when science should be at its most exciting, it could not get more dusty or stale.

There was probably never a worse time than the mid-1970s to be at university studying the mind. Neurology was an anatomy lesson tailored for the fact-cramming needs of medical students. Psychology was little better. Your choice lay between rats pressing bars in boxes or unconvincing computer models. Philosophy was just ancient history. Psychophysics was actually rather good – you got to measure the speed of your reflexes or to check your ability to tell apart two shades of green

– but it was obvious even to a first-year student that it was a field that engendered little respect. Within the academic hierarchy, its professors seemed no more than glorified lab technicians.

Disillusioned, I gravitated towards biology, where at least everyone seemed to be having fun. You were out at dawn with binoculars to watch blackbirds squabble over territories or having beach parties on marine ecology weekends. Black rugby shorts, flip-flops and thick beards were the uniform, even for many of the lecturers. It was an informal science, but confident and organised. A potential career, if not exactly a challenge.

'Gee, I wish I had your job,' said the professor with touching sincerity. We were sitting having lunch during a break in a conference at which he had been speaking. I had spent about three-quarters of an hour grilling him in flatteringly careful detail about his work. Now, just out of politeness really, the questioning was going the other way a little.

I, of course, had been envying him his position as a top neuroscientist. He had had the doggedness to stick the long years at medical school, then even more years as a poorly paid post-doc. And here he was in the 1990s, in a good lab at a good university, part of the vast army at least doing battle with the brain, if not quite yet in a position to proclaim victory.

Yet the academic life was a treadmill, he confessed. The flood of new facts, new ideas, was disorienting. Stuff kept popping up all over the place. Then there was the unexpected pettiness – the nit-picking, jealousies and hatreds which showed through in anonymous peer reviews of papers or grant applications. And students, he sighed. Even the brightest often seemed so helpless. You were lucky if more than one in ten could be steered towards a posting that would do your own career some good in the future, stealthily expanding your circle of influence.

Worst of all, science had a way of pushing you into a tiny niche. As soon as you became known for something, you found yourself spending all your time defending it. Forever in the thick of the fray, there was precious little chance to look up, take bearings and start making sense of things. It certainly was not how, as a child, he had imagined science to be.

I, on the other hand, had taken the lazy route to this conference. After sleep-walking a first degree, I had gone off to cycle round Europe, searching for a reason to go back. But then I got stuck in London having a good time and took a job on a local paper to pay the way. Life could hardly have been less demanding. Typically there was only laughter from the rest of the news-desk when for the second time in a fortnight I stumbled into work at midday, having slept through the alarm clock.

Yet while I was lazy, I was not dumb. I could see there were few, if any, journalists whose chosen beat was the science of consciousness. Plenty of chemistry and biology writers. A ton of good physics writers. Even quite a few neuroscience and psychology writers. But not consciousness. Head that way and there would be pleasantly little competition. Of course, there was a reason for the lack of consciousness specialists. For a start, there was not a lot of breaking news to report. Nor were there the kind of bread-and-butter trade magazines and specialist journals that supplement the livelihoods of freelancers in other fields such as computing, pharmaceuticals or the physical sciences. Still, push gently for enough years and it is surprising how far you can travel.

I was trying to cushion the blow while describing my basic working day to the professor. Well, I sit around at home reading books and research papers, making notes and then notes about notes. If I seem to spend a lot of time taking breaks, watching the bumble-bees and lolloping frogs at the end of the garden, it is because I have learnt it is vital to let all this information settle into place. Then doing the occasional article for a serious newspaper or magazine like the *New Scientist* gives me the licence to get out and go to the conferences which seem interesting, or phone the researchers with something interesting to say. I have access without the ties.

There is a downside, I protest. Not just the uncertainty about paying the bills. As a writer, your mistakes are always going to be public ones. And even when you do a good job, no one remembers your by-line. Yes, said the professor, but just being able to get around and hear what everyone is thinking. To be able to ask the awkward or obvious question. To have the luxury of time.

Around the room, chairs were scraping as people headed back for the afternoon session of talks. I snapped my notebook shut and we rose to join them. These days, you would not want to miss a minute of a neuroscience lecture.

The 1990s were when science got serious about consciousness again. It had been a respectable enough subject originally. Turn-of-the-century psychologists like William James wrote in great detail about the nature of conscious experience – what else would a psychologist want to study?

But the problem for the researcher was that there was no hard data. Mental states could not be recorded, only reported. Because many early psychologists had started out as philosophers, at first this did not seem such an issue. However, pretty soon real scientists – physicists, chemists and biologists – were scoffing at their lack of rigour. Stung, the psychologists switched tracks to study what could be objectively recorded: behaviour. They concentrated on sensory inputs and motor outputs, carefully ignoring any thinking or aware states that might come in between. It helped that they adopted experimental subjects, like rats and pigeons, which did not have a lot of complicated thoughts going on in the first place. Then, to guarantee that the patterns of behaviour seemed as mindless and robotic as possible, the animals were stuck in featureless boxes that allowed just a single action like a peck on a button or a push on a bar. The level of rigour was matched only by the tedium of the results.

If psychology backed off from consciousness as a matter of principle, neurology was never really there to start with. Neurology was a branch of medical science – only doctors could have ready access to human brains, dead or alive – and so was always infected by the starchy, sober-sides attitude that appears necessary to the practice of medicine. Of course, the field had its mavericks happy to plunge in and speculate about how brains might make minds. But it was a case of good luck to them so long as they did it in their own time and did not distract the rest from their official business of amassing a detailed anatomical description of one of the body's more substantial organs.

Science's failure to get to grips with consciousness gradually became an embarrassment. But in the late 1980s, something crucial happened. Brain scanners arrived. These medical imaging systems – monster machines with acronyms like PET and MRI – had been designed for more ordinary purposes like investigating heart defects or locating cancers. But it turned out they could also be used to map minds. The active brain demands nourishment. Every thought, feeling and impulse produces a minute localised surge in the flow of blood. By doping the brain with radioactive tracers, or pulsing it with massive magnetic fields, it was possible to track tell-tale patterns of consumption. All a researcher had to do was slide a volunteer headfirst into the jaws of one of these metal marvels, ask the person to think a particular kind of thought, then take snapshots of the brain as it twinkled with activity. It seemed that at last a good and honest science of the mind was possible because there was a method of measuring the invisible. Such was the hope that sprang from this and other less publicised advances in recording techniques that the 1990s were quickly dubbed the decade of the brain. It was said that particle physics had had its golden age in the 1960s, computing in the 1970s and genetics in the 1980s. Now it was going to be the turn of the greatest mystery of them all: the human mind.

There was the confidence, and probably more importantly, the funding, to set to work. By the mid-1990s, the field of mind science was buzzing. Conferences, on-line discussion groups and the pages of august journals like *Science* bubbled with talk about consciousness. Quickly, two camps emerged.

The official neuroscience approach – centred around the veteran figure of Francis Crick, already famous for cracking the secret of life with his work on DNA – went looking for the clever neural trick, the brain gimmick, that must turn consciousness on. It seemed plain that the brain was similar to a computer in that its neurons did processing. The human brain has about a hundred billion nerve cells and some hundred trillion connections (that's one followed by fourteen noughts!), so there was no shortage of wiring for representing states of information. But something special about the way the brain did its job must turn these neural transmissions into a web of signals aglow with the added property of inner experience.

For most of the decade, with Crick's musings frequently acting as the spur, packs of neuroscientists chased one candidate mechanism after the other. There was the great excitement over oscillations – the discovery that brain cells tend to synchronise their firing when representing some perceptual experience. Then there was the idea that an entraining forty-Hertz rhythm drummed out by some central part of the brain might serve to force its scattered patterns of neural activity into a state of conscious coherence. Another suggestion was that a special layer of output cells within the grey rind of the cortex might hold the key. Or that the anterior cingulate cortex – an obscure bit of the frontal lobes which always seemed to light up during brain scans whenever a person was asked to be mindful of their actions – could be the brain's secret consciousness centre. Nearly every month in *Science* or *Nature* there would be another report of an experiment attempting to shed light on whatever happened to be the most fashionable hypothesis at the time.

While the neuroscientists rode a wave of cautious optimism, there was a second camp – largely made up of philosophers, physicists and other interested onlookers – who felt that consciousness was such an extraordinary phenomenon that ordinary science was never going to be of much help. Hunting for some processing gimmick was a little pathetic as there seemed no reason why any form of computation, neural or otherwise, should come alight with the extra property of conscious experience. There looked to be an explanatory gap that no amount of neurological detail could ever fill. For many, this feeling was crystallised by the brash young philosopher David Chalmers, who said existing science could answer all the easy questions, the questions of how, but not the hard one of why. Why should the activity of the brain result in a self-aware state rather than just an equally effective, but blank and zombie-like, handling of informa-tion? The processing story simply did not demand a brain with inner experiences.

A few brave souls suggested that a suitable bridge might be found in the further mysteries of quantum physics. The brain could be a kind of biological antenna capable of harnessing the strange powers of the quantum world. Nerve pathways would defy the usual prohibitions against room temperature, macro-scale, and states of quantum indeterminacy to feel out the future, literally probing an infinity of alternative space–time realities before collapsing around whichever offered the most creative answer to a problem or the most logical next focus of attention.

Hardliners like Chalmers, however, merely scoffed. Nice try, they said, but even quantum trickery was nowhere near weird enough to fit the bill. Whether the brain's activity was computational or quantum, the hard question about why a something rather than a nothing remained. Quantum gropes in the dark seemed no more to require a subjective aspect than the spit and crackle of a neuron. So if the much-trumpeted decade of the brain was proving anything, it was only how far science still had to go. To date, there was not even a sniff of the kind of mechanism that might be up to the job.

The often grumpy face-off between Chalmers's hard questionites and Crick's neural correlate seekers was good theatre. It certainly created a pleasing sense of polarisation – of a spread of opinions and things getting done – in a field which had finally gone overground again. For the consciousness journalist, there was at last some colourful controversy to report. However, as it turned out, these public debates masked a much deeper story. Behind the scenes, seismic shifts in thought were taking place.

Psychologists and neuroscientists had gone into the 1990s feeling that they sort of knew what was going on. Naturally they did not have the answers yet, but they had a grip on the questions and methods. Computer science had bred a fair understanding of the principles of information processing systems, a century or so of neurology had uncovered quite a lot of facts about the brain, and science's success in other areas had proved that its basic intellectual tools were in pretty good shape. A framework for doing research already existed. If science had been cautiously holding itself back, it was only because of the lack of a solid experimental technique. Now, with scanners to snap pictures of living brains, it was time to crack on and unpick the mechanisms of consciousness.

Instead, almost from the first, what brain-imaging experiments brought home was just how unprepared science actually was. Researchers found that they had been making all sorts of glib assumptions about the kinds of questions they should be asking and the likely form the answers would take. Even their approach to measuring the activity of the brain suddenly looked suspect. They discovered that just because a neuron was firing noisily, or a corner of the cortex glowed hot, did not automatically make it central to what was going on. The properties that caused activity to count could be a lot more subtle and so, at the very least, their clever machines would need recalibrating.

It was a chastening period. Some things needed a complete rethink. Others were a matter of rediscovering things that science already knew – pockets of findings that had been ignored largely because no one could see how they fitted in before. Gradually, however, out of the upheaval started to come real advance.

The big change has been a move towards a more dynamic view of the brain and brain processing, one rooted in the new sciences of chaos and complexity. The classical approach to science is reductionist. A system is understood through its parts. For the psychologist, this meant it was natural to tackle the mind faculty by faculty, dealing with perception, memory, cognition, speech, motor output, and so on in turn until the complete list of mental sub-systems was exhausted. The neurologist would do something very similar, proceeding perhaps level by level – from synapses to neurons and then the behaviour of whole networks – or else structure by structure, ticking off the parts of the brain in turn. Such a methodical route seems sound; indeed, it appears the very essence of being scientific. But as many a first-year student can vouch, it is ploddingly dull. Worse still, it somehow does not fit the character of what is being studied. The brain is above all an organ that is lively, responsive, and acts as a whole. A science of the brain ought to be equally light on its feet.

What a dynamic account of the brain means in practice can have many interpretations, but this book tries to tell the story by looking at what the brain has to do to create a single moment of consciousness.

The evidence is that it takes the brain about half a second to produce a fully realised state of awareness. It follows an arc of activity that begins with the establishing of plans and expectations, then passes through a pre-conscious stage before eventually flowering as an organised, tightly focused state of response. In other words, each moment of consciousness has a hidden structure. To begin to understand what consciousness is all about, you have first to get inside a single instant.

And if this book seems a little long to be all about half a second in the life of the brain, it is only because the brain is turning out to be the most complex system that science is ever likely to encounter. It has not just complexity, but complexity of an almost alien form. It has rococo dimensions of structure and process unimagined even in other complex systems like economies, ecosystems and living cells. There is a lot to talk about. And while, so far, science can offer only a glimpse of the brain's depths, already the glimpse is dazzling enough.

CHAPTER TWO

Disturbing the Surface

Standing by a pond at London Zoo on what was supposed to be a holiday outing with their families, two researchers were grabbing a moment to talk shop. Karl Friston, a theoretical neurobiologist at London's Institute of Neurology, was trying to describe a new vision of the brain to his friend, the Harvard University psychologist Stephen Kosslyn. Look, said Friston, traditional thinking holds that the brain is some kind of computer, crunching its way through billions of inputs each second to output a state of consciousness. But really, the brain acts more as if the arrival of input provokes a widespread disturbance in some already existing state.

Nodding down at the pond, Friston said that gives you a better way of thinking about it. The brain is like a surface, its circuits drawn tight in a certain state of tension. You toss in a pebble – that's your sensory input – and you immediately get ripples of activity. Sure, the patterns say something about the way the pebble hit the surface, but they are mixed with the lingering patterns of earlier pebbles of input. And then everything begins echoing off the sides of the pond. The overall shape of the system has an effect on the patterns you see. Nothing is being calculated. The response of the pond evolves organically. Or to use the proper term, dynamically.

Right, agreed Kosslyn. And as we throw in more pebbles – or rather, experiences – into this particular pond, we change its shape, and thus the kinds of patterns it tends to produce. This is a system that learns. It has a memory!

Early in the 1990s, in hundreds of private conversations like this, mind scientists everywhere were groping for a clearer view of what brains were all about. It quickly became the issue of the decade. If the brain did not work like a computer as so many had assumed, then what sort of information-processing system was it? For Friston and Kosslyn, the stakes were especially high. Friston was the methods guru of the brain scanner community, charged with the job of making sure the new multi-million-pound brain-imaging machines really measured what they claimed to be measuring. Kosslyn, an equally respected figure for his research on mental imagery, had been one of the great champions of the computer model but was now being forced by his own results to think again. Dragging along in the wake of their wives and children, oblivious to the bustle of the August crowds, the two continued to wrestle with what was turning out to be a startlingly different conception of how the brain might work.

Tanned, balding, shortish and dapper in his Ivy League uniform of cream trousers and soft blue denim shirt, Stephen Kosslyn speaks with calm precision, showing the same relentless logic in his person as he does in his work. As a professor of psychology at Harvard, Kosslyn had been one of the key figures in the cognitive science movement – science's attempt to use the computer metaphor to get back into the study of the mind after the long, fallow years of behaviourism. The behaviourists had banned all talk of mental events because there did not seem any way that serious science could get at them. But computer modelling appeared to offer a way around the problem of objective data. What psychologists could do was continue exactly like the behaviourists in recording patterns of stimulus and response – observing what inputs led to what outputs – but then add the extra step of making a computer model of the processing routines, the supposed mental events, that might come in between.

This seemed quite brilliant. Psychologists would be making no strong claim about studying consciousness itself; they would merely be treating the mind as a black box system and asking what a machine would have to look like to be able to do the same outward job. The flowcharts and the paper designs would be the field's objective data, the stuff that got tested and debated.

As a further precaution, cognitive scientists would start small. The ultimate aim would be to get an insight into conscious minds. But consciousness was clearly too big a phenomenon to attempt to reverse-engineer in one go, so psychologists would make models of the mind's many individual processes, dealing with it component by component, until gradually they began to see the logic by which all its parts slotted together.

It was not a hard plan to sell. Not only was it modest and impeccably empirical, but computers and minds seemed naturally connected. Both operated with rules and symbols. Both used memories and processed information. In fact, when you got down to it, a computer program was really just the automation of some poor forgotten human's thought patterns. A person would have sat down to work out a foolproof sequence of steps for carrying out a task, like handling a company's payroll or booking airline seats, then dumped this procedure on a machine that could repeat it endlessly. Obviously, computers and humans had their differences. For a start, computers did not get bored with their programs, while you could never trust a human to complete a task the same way twice. But computer technology had been born out of an idealisation of human thought patterns and so it seemed only right that psychologists should turn things round and start treating computer programs as their surrogate subject of research.

Right from the dawn of computing, comparisons were being drawn. In the 1940s, as part of the war effort, a group of British psychologists had produced engineering-style flowcharts of human mental processes such as perception and attention in the hope of better understanding what it took to make a good fighter pilot or gun aimer. However, it was only during the 1970s that cognitive science really took off. The push came from the computer side. Computer scientists had done the mundane things like payroll and word-processing systems. Now they were stepping up their ambitions and wondering about artificial intelligence – machines that would show the same kind of creativity and perhaps even awareness as humans. More to the point, there was money to bankroll this line of research. The idea of intelligent hardware was so tantalising that governments and military research agencies such as DARPA, the US Defense Department's advanced research projects agency, were willing to invest enormous sums in the most unlikely projects. Not everyone believed a computer could be conscious, but even a computer that was just a little bit smarter, a little more like a human in its operation, would bring a huge return.

Suddenly, cognitive science was a pot of gold. There was the cash for new labs, new careers. It did not matter whether you were leaping into the field to use computers as a way of modelling the mind, or applying what was known about the mind to build better computers. The belief was that in the end, a program was a program whether it ran on a brain or a silicon chip. Everyone was now essentially in the same business – that of understanding information-processing systems.

By the 1980s, the cognitive approach had taken over from behaviourism as the mainstream of psychology. There was real conviction that at last mind science had found its rightful path and could start to deliver the goods. But despite the confidence, consciousness itself was not yet truly back on the agenda. It was the implicit goal, of course. However, for a number of reasons, it could not be an explicit one.

The situation was complicated because it was only cognitive psychologists who had to be careful of uttering the c-word. Those who came from the computer science side of things were free to say more or less what they liked about consciousness, or the potential consciousness of their machines, because ultimately they would be judged on the systems they built. Who cared if an inventor had some crackpot ideas so long as something useful was emerging from his or her lab? Cognitive psychologists, on the other hand, had to take a tighter line with themselves.

Firstly, they were going forward on the understanding that they were dealing in computer models that might – only might – do things the way human brains did them, and it would be wrong to let this fact become blurred in the public's mind. But there was also the more uncomfortable truth that too much talk about consciousness would raise the problem of the many ways in which minds did not seem like computers. What would it mean for a computer to have subjective experiences, or even to represent the many different shades of awareness that a brain seems to support, from bright focal attention to the vague subterranean urgings and stirrings which actually make up so much of our mental lives? The belief was that cognitive psychology would eventually get there. It would start with crude models and end up with realistic ones. But to make too much fuss about consciousness in the early days would be a bad tactic. For computer science, even the slightest advance in machine intelligence would be much admired. But cognitive psychology would only ever be judged on how far it still had to go if it let consciousness become its official target.

So cognitive science forged ahead during the 1970s and 1980s, promising much without making those promises too explicit. Not only did psychologists avoid the general issue of subjective experience, they also tended to avoid any forms of mental activity with an overtly subjective element to them. Processes which appeared rule-based, and so computer-like, were favoured. There were many models exploring how brains might compute the recognition of an object seen from an unfamiliar angle, or apply the laws of grammar to decode the meaning of a sentence. But cognitive psychologists shied away from anything to do with the emotions, the imagination, the intuitive – things which it seemed could only make sense to a system which had awareness.

All this helps explain why Kosslyn's chosen line of research created such mixed feelings within the field. Kosslyn wanted to uncover what goes on inside the head when a person seems to be looking at an imaginary scene or having some other form of imaginative experience. Mental imagery had fascinated early psychologists, like William James, who had taken the apparently natural view that having an image was much like a weak form of seeing (or hearing, or feeling). The brain was using the same machinery that it used for perception, but instead of experiencing something real – a flow of messages from the senses – it was constructing an experience out of memories. Snatches of stored knowledge and scenery were being artfully projected onto the perceptual apparatus to create an artificial sensory pattern.

However, while this seemed a fair explanation, there is plainly a little more to a mental image than the idea of a weakly remembered sensation might suggest. For example, if we try to imagine a child's plastic beach bucket or a rhinoceros charging out of the African scrub, some sharp impression may form before our mind's eye. But most people report that the picture is fleeting, just a glimpse that dissolves as soon as they try to inspect it too closely. Furthermore, an important part of the experience of having an image is the feeling that other potential views of the object lie close at hand. When imagining a charging rhino, we would have both the flash of a specific rhinoceros image and also a background sense of preparedness to see other rhino-related views. Indeed, part of the instability of the initial image seems to come from a desire of our minds to move on. The pictures mutate of their own accord. So, while our first view may have been of a rhino charging out of a clump of thorn bushes, it might quickly have switched to an image of a dust-caked rhino standing side-on, snorting and tossing its head. Or we may have pulled back to a wider view in which the charging rhino was now seen to be headed at a safari Land-Rover, its fearful occupants clutching at guns and hats.

The Oxford philosopher Gilbert Ryle tried to capture the elusive nature of imagery by saying that it was not so much a case of being a spectator to a resemblance as of resembling being a spectator. That is, we do not view unyielding objects summoned from the vaults of memory. Rather, what we are doing when we are imagining is adopting a frame of mind in which we stand ready to see something happen. This prepares us in a general way for what we wish to see – we feel oriented to the idea of rhino-ness, kids' buckets, or whatever – but the images flare pretty much of their own accord. Also, the feeling of being oriented is as much a positive part of the experience as any of the passing parade of sensations.

Imagery was clearly a tricky subject for the cognitive psychologist. The way information graded from being in consciousness to near-consciousness was a much more difficult issue to duck than was the case with straight perception. More seriously, the act of imagining took place entirely out of sight. It was not bracketed by an input and an output – an overt chain of stimulus and response – so there seemed little hope of getting a handle on whatever might be going on inside a subject's head. However, Kosslyn spotted an experiment that gave him a way to begin a systematic study of imagery.

In the early 1970s, a group at Stanford University in California had come up with the clever idea of timing people to see how long it took them to rotate an image in their heads. Subjects would be shown two shapes at different orientations and asked to judge whether they were identical or mirror images of each other. The researchers found that the time it took to give an answer was proportional to the amount of rotation required, as if the subjects had been physically turning the pictures around inside their heads. If the images – usually something familiar like a letter of the alphabet or a silhouette of the state of Texas – were set at the same angle, then the reply came within half a second; if a 180-degree rotation was demanded, answering took over a second.

The task created its own input–output trail, with the drawings being the stimulus and the time to reply the objectively measurable response. Kosslyn realised that by playing about with different tasks, he could use this basic experimental design to get a real feel for the capacities and limitations of the brain's imagery display.

Throughout the 1970s and into the 1980s, Kosslyn and his students carried out numerous tests. In a typical experiment, subjects were asked to memorise a map of a treasure island dotted with various landmarks. Having done that, they were then asked to think about a named spot, like a particular sandy cove, then scan across, tracing their path using an imaginary black dot, to a second landmark, such as a lagoon, a hut, or a clump of palm trees. As with the mental rotations, there turned out to be a precise relationship between the time taken and the actual distance travelled, as if the subjects were physically moving a cursor inside their heads and so were being limited by the manipulation characteristics of their mental hardware.

Another kind of experiment measured the resolution and field of view of this display space. Kosslyn first asked his subjects to imagine a rabbit crouched next to an elephant, then the same rabbit standing alongside a fly. By asking questions about the imaginary rabbit, such as whether it had a pointy nose or a white tail, and timing how long it took a person to respond, Kosslyn could deduce something about the scale at which the rabbit was being represented. What he found was that when subjects imagined a rabbit alongside an elephant, the rabbit appeared small and therefore lacked detail. It took them measurably longer to reply because they had to pause for a closer look. But when the rabbit was imagined next to the fly, its picture was large and the answers came much faster.

As Kosslyn discovered more and more about the production of images, he felt he could begin to build a computer model of the process. In fact, taking the cognitive approach to its logical extreme, Kosslyn compared what the brain did directly to a desktop terminal or PC. He argued that just like the pictures appearing on the screen of a personal computer, the brain must generate its images from information stored in memory and then project them across a neural display space of fixed size and resolution. The system was free to combine the available information in any way it wanted to, but the qualities of the hardware – such as the resolution of the display area or the speed of the manipulations – were constant.

Kosslyn also believed that the images were constructed systematically. To build a mental image of something like a duck, our brains would first visit its duck files to grab a general outline sketch – a duck as seen from one typical point of view. Then if we stuck with this view, the image would quickly intensify as the brain fetched more and more details to flesh out the display, painting in the gleaming feathers of the duck's back or the glint of the water on which it floats. This would explain why sometimes images were fleetingly vague and at other times richly vivid.

Furthermore, the memory files would hold more than simple sensory details. Sometimes the brain might access files with knowledge about how ducks typically move or behave. When this happened we would get an animated picture of a duck flapping its wings or quacking. Or we might deliberately delve into our files to make our duck do things ducks normally do not do, like wear hats or flutter long eyelashes. Experiments showed that it took subjects longer to make odd conjunctions like this, backing up the idea that the information was having to be extracted from a number of physically separate mental files.

With timing tests to support each part of his story, by the mid-

1980s Kosslyn was famous for having the most computer-like of computer models of a brain process. He had even adopted the jargon of computers in postulating a set of programming instructions – specific commands like look for, fetch, put, scan, pan, zoom, and rotate – which the brain must somehow implement to control the display of mental images. It seemed that he had taken one of psychology's most awkward subjects and produced the most thorough theory. Yet despite all his efforts, it turned out that Kosslyn had run up a dead end. No sooner had his work become a textbook classic than critics emerged to point out a basic flaw in his experiments for which there seemed no answer.

A number of cognitive science's hardliners – the most prominent being Zenon Pylyshyn, a computer scientist at the University of Western Ontario – noted that Kosslyn could not be sure that his subjects were not faking the delays in their responses to make it seem as if manipulating mental images required time. The subjects might be pretending to take longer to check the colour of a rabbit's tail or track a dot across a map either out of sheer perversity or because of a misguided desire to give an experimenter the results they guessed he wanted. However unlikely, such a possibility could never be ruled out. In fact, when it really came down to it, the subjects may have been unsensing zombies with no form of mental experience being involved in their computing of map co-ordinates or retrieval of factual information. There was no way of distinguishing between Kosslyn's mental display theory and a theory in which subjects simply were pretending to have images.

As an argument, it was technically correct but also ludicrous. Kosslyn's critics were pushing for black and white certainty when it was quite normal for science to judge theories on their plausibility. Faced with two possible interpretations of the same data, the simplest answer wins – at least until new evidence or an even better theory comes along. And the same argument could be used against almost any psychology experiment. By the same token, even something so straightforward as checking a person's hearing would be open to the accusation that the subject might be pretending not to hear faint sounds so as to delude the tester.

Such comments should have been treated as minor nit-picking, a gentle reminder to control for some potential experimental biases. Indeed, to counter the possibility that the subjects had been faking delays to please him, Kosslyn went ahead and re-ran a few experiments using testers who had been briefed to expect the opposite results and so could not be accused of influencing their subjects. The fact that the criticisms were still seen as fatal to Kosslyn's approach says a lot about psychology's lingering nervousness when it came to anything touching on the question of conscious experience. When dealing with language, rational thought, or other aspects of the human mind that looked reassuringly computational, the attitude towards the experimental data was much more relaxed. No one worried that the subjects might be behaving deviously. But with imagery, the general instinct was still to don full protective gear.

Bruised by what became one of cognitive psychology's great set-piece battles, Kosslyn was left wondering where to go next. With his timing experiments, it had seemed almost as if he had had his hands in his subjects' heads, feeling the physical outlines of the machinery of their minds. Despite the fact that he was using computer jargon and analogies, he had been dealing not in the usual abstract flowchart models but in the real thing: how brains made particular mental states. However, while flicking through the pages of *Science* one day, Kosslyn stumbled upon an experiment which completely threw him, making him realise, as a psychologist, just how little neurology he really knew.

Many people were floored by the experiment reported by Roger Tootell and his colleagues at the University of California-Berkeley in 1982. Quite simply, Tootell had stained the brain of a monkey with a radioactive tracer as the animal watched a flickering dartboard pattern of lights. On dissecting the brain of the monkey afterwards, the spokes of the pattern remained imprinted on its visual cortex, the part of the brain which maps visual sensations. What the monkey had been seeing had been frozen for all time and could now be printed in the pages of a magazine! The method Tootell used was called autoradiography, a technique originally developed by medical researchers in the early 1970s to study the metabolism of organs like the pancreas or liver. The experimenter would inject a radioactively tagged form of the sugar glucose into the bloodstream of a living animal. The tracer would circulate around the body to the target organ where it would be taken up in greatest concentration by the most active cells. After giving the organ time to soak up the tracer, the animal would be killed (or sacrificed, to use the preferred euphemism) and the organ sliced up, then spread out across sheets of photographic film. As the radioactivity blackened the film, the organ would create an image of its own pattern of metabolic activity.

It was an obvious idea to apply autoradiography to the brain, although Tootell had a few technical problems to overcome. For a start, the monkey had to be kept staring at the light pattern for nearly half an hour to allow enough time for the glucose to be absorbed. This meant the animal was in fact anaesthetised. The monkey had had its eyelids pinned back and head propped to look at the screen, so it was not actually conscious of what its brain was seeing. However, the results were still astounding. The way a pattern was etched across a thumbnail-sized patch of visual cortex seemed to say something fundamental about how the brain handled sensations. The representation of experience was topographic; brain cells fired in a way that preserved the essential outline of the images falling upon the eyes. Computers might traffic in abstract bits and bytes, but there was something tellingly literal about the way the brain handled information.

Looking more closely at Tootell's results, it was apparent that the shrunken neural map was even distorted in a way that matched the variable acuity of the visual field. Our eyes take in a panorama which spans nearly a full 180 degrees. But only a tiny spot right at the centre – an area of about a thousandth of the visual field, or the size of a fingertip held out at arm's length – is seen with real sharpness. A look at the anatomy of the retina, the back of the eyeball, soon tells us why. Most of the retina is only thinly covered with light-detecting nerve cells. The cells are also rather coarsely tuned so that they will react to light of any colour. But a small pit in the middle of each eyeball, known as the fovea, is so jam-packed with tightly tuned receptors that it is hundreds of times more sensitive. We actually see the world with a kind of tunnel vision. The reason why we do not notice the dramatic variations in acuity is that we always keep our eyes on the move. We shift them every third of a second or so to take in whatever is our latest point of interest. We can get by with a weak awareness for what lies on the periphery because we are really only bothered by the bits we are putting at the centre.

Tootell's autoradiography pictures showed that this psychological bias was faithfully reflected in the mapping that appeared in the brain. Whereas the fovea accounted for only a pin-point fraction of the visual field, the mapping of what it was seeing took up most of the visual cortex. The periphery of the field of view had to be crammed into the margins. There was a perfect match between the intensity of the experiencing and the amount of display space given over to the mapping.

In truth, no one should have been surprised by Tootell's experiment. Neuroanatomists had come to much the same conclusions back in the 1940s by laboriously tracing thousands of individual nerve fibres back from the eye to the brain. But there was something tremendously immediate and haunting about seeing an actual brain pattern that had once flickered in the mind of a monkey. For Kosslyn, it was an especial shock because it seemed at a stroke to prove his claim that the brain had a physical display. There really was a place where mental images could dance. Kosslyn recalls: 'I was blown away because here was a picture of the brain and you couldn't argue with it. The evidence was there. I thought if you could do similar kinds of experiments with humans, you could really prove something about imagery – but, of course, you can hardly go around slicing up human brains, so there didn't seem to be anything for me there.'

However, while the technique was not of much practical use, its results certainly convinced Kosslyn it was time he got his nose stuck into the neurology textbooks. If there was to be a future for his imagery theories, they would have to be grounded in a thorough understanding of the workings of the brain.

With hindsight, it seems odder and odder that mainstream

psychologists were so intent on studying the mind without also studying the brain. Even if they could not do actual experiments, there was already enough known about neurology to help frame their theories. However, a generation of researchers had grown up in the belief that information processing was all about programs. It did not really matter what kind of hardware a program was running on – whether it was flesh and blood or silicon – so long as the underlying logic was preserved. The brain was simply a particular implementation of something more general. So how the brain might choose to arrange its circuits was of marginal interest at best.

The justification for this view was put forward in 1936 by the spiritual father of computing, the British mathematician Alan Turing. Turing's proof was famously simple. He created an imaginary device, later dubbed the Turing machine, which was nothing more than a long strip of paper tape and a processing gate which could carry out four operations. The gate could move the tape a step to the left or the right and then it could either print or erase a mark. Given an infinite amount of time and an infinite length of tape, Turing demonstrated that this most rudimentary of computers could crunch its way through any problem that could be reduced to a string of 0s and 1s. Using a binary code to represent both the data and the instructions which told the gate how to manipulate the data, a Turing machine had all it needed to get the job done.

For computer science, this proof was enormously important because it said that all computers were basically the same and so there was no limit to their future improvement. Whether a machine used a single gate, or millions, or trillions; whether it was built of paper tape, silicon chips, or something really exotic like beams of laser light, the principles of its operation would be identical. The computer revolution had rock-solid foundations.

In 1960, one of the founders of cognitive science, the Princeton philosopher Hilary Putnam, seized on Turing's proof to argue that it meant brains did not matter. If the processes of the human mind could be expressed in computational form, then any old contraption could be used to recreate what brains did. The brain might be an incredibly complicated system, involving billions of nerve cells all chattering at once – not to mention the biochemical reactions taking place within each cell – but, in the end, everything boiled down to a shifting pattern of information. It was the logic of what the brain was trying to do that counted. So given enough time, even the simplest Turing machine could recreate these flows. A few doubters – notably the philosopher John Searle of the University of California-Berkeley – tried to lampoon this belief. Searle asked whether a simulation of the brain using tin cans and string would also be conscious? The cognitive scientists came straight back and said, sure, why not? If the cans and string really captured the programs in all their detail, then the system should have some form of subjective life.

Again there was a noticeable difference between the way computer scientists and psychologists talked about the issue. Those on the computer side of the divide could be as bullish as they liked. Many seemed convinced their creations were practically conscious already; certainly, artificial intelligence was only a matter of decades away. The psychologists had to choose their words more carefully. Yet what Turing's proof did mean was that they never need feel guilty about failing to take a neuroscience class or open a volume on neuroanatomy. During the 1970s and for most of the 1980s, it was information theory which was the future of mind science. So while a psychologist might be embarrassed by not being up to date with the latest programming tricks or computer jargon, a complete ignorance of the brain was no bar to a successful career.

Kosslyn, however, had done the computer bit and found it wanting. He was still a believer, but it had become clear that cognitive models alone could not settle arguments and so neurology would have to be part of the mix as well. He would at least be guided by what he could glean from current neuroscience. But what would be even better was if he could find some way of doing actual experiments to find out how brains handled images. 'I knew I had to pin something down. Once things were anatomically rooted, then everyone would be working off something concrete. There'd be limits to the kinds of objections people could make about a theory,' says Kosslyn, although he was not holding out high hopes. Catching the brain having something so ephemeral as a mental image seemed a rather unlikely prospect.

In fact, the odds were better than Kosslyn realised. Being conscious seems such an effortless affair – we simply open our eyes and ears and let the world flood in – that it is tempting to view the brain as a silent organ, its soft grey folds closed round the gentle dance of awareness. But actually mental activity causes our brains to seethe. Generating a mind is a huge drain on the body's resources. The brain accounts for just 2 per cent of our body weight, yet it runs so hot that it burns up about a fifth of the food we eat and the oxygen we breathe. This means that every passing thought, urge or feeling must create its own small rustle of activity. There will be a flashing discharge of neurons followed almost immediately by a local surge in blood flow and glucose consumption as the neurons top themselves up again.

The idea of reading something into such changes is not new. In the 1870s, the French surgeon Paul Broca put thermometers on the skulls of his junior staff and found that changes in their mental activity created hot spots of greater blood flow. Then there was the famous case of a patient who complained to the Boston neurologist John Fulton that every time he opened his eyes he could hear a bruit, or blowing sound. The patient said this gurgling got worse when he did something difficult like trying to read in poor light. Fulton found that the man had a bulging knot of blood vessels in the visual area of his brain. The noise he had been hearing was the turbulent rush of blood that accompanied the act of seeing. By rigging up a stethoscope so that it drove a recording needle across a smoke-blackened drum, Fulton was able to capture the change as the man opened and shut his eyes. In another case, the Italian physiologist Angelo Mosso did something very similar with a peasant who had a square of bone missing from the front of his skull. Using a pressure pad to measure increases of blood flow into this area of the brain, Mosso found he could produce differences simply by asking the man to multiply two numbers together.

Attempts to decode the electrical activity of the brain started just as early. The first electrode recordings were made in the 1870s by the British physiologist Richard Caton, who sawed the tops off the skulls of rabbits and monkeys to expose their brains and then recorded what happened when he flashed lights into their eyes. Even with his crude methods, Caton was able to catch the jagged changes in potential that accompanied each visual stimulus. Obviously, Caton's technique was not much use with humans. But by the 1930s, electrode equipment had become sensitive enough to record right through the bone of the scalp, and electroencephalography (EEG) was born. An array of electrodes would be glued to the skull with conductive paste and the noisy chatter of the brain recorded on long rolls of chart paper.

The EEG machine seemed to have great promise. It was cheap, easy, and safe to use. However, it also had some equally serious drawbacks. The brain's electrical signals are bent as they pass through the skull and so it is hard to be sure which part of the brain they are coming from. Worse still, EEG electrodes are unselective. They pick up everything, and in the clamour of activity it is difficult to pick out the crackle of an individual thought or reaction. The EEG catches the roar of the crowd, but it cannot track a particular conversation.

A small, dedicated band of researchers set to work, collecting roll upon roll of squiggly recordings and trying to classify the wave patterns that seemed to be associated with different states of consciousness. They found that a busy brain would produce a quick, shallow rhythm of thirteen to thirty peaks a second, which they called the beta wave. A relaxed brain would show a slower and deeper rhythm of between eight and a dozen peaks a second – the alpha wave. In sleep, the peaks would drop as low as one or two a second. But the only thing these traces meant was that a busy brain breaks into many local pockets of activity while a quiet brain has just a few. In the public mind, the alpha and beta rhythms took on something of a mystical aura, especially with the popularisation of biofeedback machines. But to most scientists, such results were virtually meaningless.

There was a crucial change in the 1960s with the development of the evoked response potential (ERP) recording technique, an advance made possible by the newly available calculating power of computers. An ordinary EEG system would be used to record a subject doing a simple task, such as listening to the ring of a bell or feeling a swift tap on the back of the hand, many hundreds of times. Then by lumping the sessions together and averaging the results, it was possible to filter out the background noise and just be left with the brain activity associated with the processing of the sensation.

There were still problems with recording this way. The bending of the signals by the scalp meant that a researcher remained unable to say much about which bits of the brain produced which bumps in a trace. And the fact that so many trials were needed raised its own problems. The mental task had to be one that would bear constant repetition. But in skilled hands, ERP studies could produce some intriguing results. For example, there was the famous P300 wave or oddball response, a positively charged surge of activity which showed about a third of a second after some surprise. The stimulus might be something so slight as a louder tone following a string of softer ones, but anything that made a person sit up and take notice – to have a sense of aha! – would produce this characteristic blip of extra activity. It seemed that if only the EEG machine could reveal which parts of the brain produced this rather late reaction, then it might tell researchers something important about the parts of the brain that dealt with shifts in attention, or perhaps even the feeling of being strongly conscious.

Unfortunately, just as EEG techniques were beginning to show true promise, the mainstream of psychology was too busy scrambling aboard the cognitive science bandwagon to take notice. The EEG continued to be used in a few specialist areas, such as sleep research or the study of epilepsy, but there was little further talk about using it as a window on the processes behind consciousness.

It so happens that in his search for some route into the brain, Kosslyn dabbled briefly in EEG recordings. In 1986, he took out a year to set up his own lab at Harvard in the hope that the patterns of activity might say something definite about the parts of the brain which were active in generating images. However, the work did not bear fruit. 'It was a dismal failure,' admits Kosslyn. 'There were so many degrees of freedom in interpreting EEG recordings that the whole thing was just unreliable. In the end, I never published anything from the work.'

Then in 1988, Kosslyn was giving a talk about how the imagery debate seemed to have ended in a stalemate. After the lecture, a couple of medical physicists came up, tapped him on the shoulder, and said they wanted to tell him about something new called PET.

PET, or positron emission tomography, is a true space-age technology. The idea is simple enough, but the equipment necessary to make PET happen is practically science fiction. A fast-decaying radioactive tracer is injected into the bloodstream of a volunteer and then the brain is scanned while the person lies down having thoughts. Concentrations of the tracer show which parts of the brain have to work hardest. As the name suggests, PET depends on the emission of positrons, the positively charged electrons released by certain isotopes, like oxygen-15, upon their decay. The positrons themselves are never actually seen. Each one travels only a short distance – a couple of millimetres at most – before bumping into one of the brain's many electrons and combusting in a puff of pure energy. It is this quick anti-particle annihilation that produces the gamma rays picked up by the system's head-encircling detector ring. Vitally, the gamma rays come in pairs which fly off in exactly opposite directions. By catching both, the scanner's computers can draw a line running straight back through the original positron event in the brain. After a few million such readings, the scanner has an accurate, three-dimensional image of any metabolic hot spots in the brain.

Something like PET had actually been tried in the 1970s. A team of Scandinavian researchers from two medical schools a short boat ride across the water from each other in Lund, Sweden and Copenhagen, Denmark injected a radioactive version of the inert gas xenon into the bloodstreams of subjects and then used a bank of several hundred Geiger counters to map its passage through the brain. It was not a technique for the faint-hearted, for to get a crisp enough trace the dissolved gas had to be injected straight into a neck artery using a hand-held syringe – the local fire station in Lund provided most of the volunteers for the experiments. More importantly, the results were not a three-dimensional reconstruction like PET. The Geiger counter array simply registered a flat, two-dimensional picture, so it was impossible to tell if the signal was coming from the surface of the brain or somewhere deep inside. Nor was the resolution anything like PET. But despite the many shortcomings, the Scandinavian team was able to produce some stunning pictures of the brain's blood flow as it performed tasks as varied as tapping a finger, listening to speech, or even imagining making a sudden movement. The xenon studies certainly proved what might be possible if medical researchers with deeper pockets could come up with a more precise method of measurement.

A number of teams set off in pursuit of this prize. A group from the radiology department at the Washington University School of Medicine in St Louis stole an early lead, but a rival team based at the Hammersmith Hospital in London was always close behind. The first requirement even to be in this technological race was to have an on-site cyclotron – to be a hospital with its own private atom smasher! The isotopes used in PET are especially short-lived, their decay being measured in just minutes, so the accelerator ring which makes them has to be no more than a corridor's walk away. Fortunately, a number of the world's leading teaching hospitals installed cyclotrons during the 1950s when researchers were exploring novel treatments for cancer such as neutron bombardment. Nothing much came of the work and the machines had become something of an expensive embarrassment. Hospital authorities at places like Washington and Hammersmith were rather grateful that there might turn out to be another use for them.

The PET pioneers were also helped by the fact that much of the technology for the scanner's detector rings was tried and tested. The bismuth germanate oxide crystals used to trap the gamma rays were the same as those used by particle colliders at high-energy physics laboratories like CERN in Switzerland and Fermilab in Illinois to catch the debris from nuclear smashes. The expensive electronics needed to match pairs of events also came from these labs. A PET machine has to be able to sift through some ten million potential pairings a second, and the cost of developing such hardware from scratch would have been prohibitive.

But even with these savings, PET scanning was big science. It took brain research into a new league. Psychology and neurology had always been reasonably small-scale in their use of technology. Following a rat through a maze or examining a stained slice of brain tissue might require some specialist equipment, such as computerised recording devices or electron microscopes, but by and large the cost of such kit ran into the tens of thousands rather than the tens of millions. More importantly, a researcher could operate the equipment without needing any great technical assistance. PET was another world. An army of Ph.D.s was needed just to switch on a machine. At the Hammersmith unit, for instance, it took thirteen physicists and engineers to run the cyclotron; another seventeen chemists and nuclear medicine specialists prepared the isotopes for use. There were eight computer professionals to process the gigabytes of data produced during each scanning session. Doctors and nurses were needed to take care of the subjects, putting lines into their veins to deliver the tracer and then monitoring their progress. Only after all these people came the actual scientists there to do the experiment.

The very scale of PET research had a direct impact on the way science began to behave towards the question of consciousness at the tail-end of the 1980s. Of course it was important that scanners gave scientists a research method where there had been none. They provided a credible way to tie mental activity to actual events in the brain. But like the space race, the human genome project, and other examples of big science, the cost of PET, and the size of the teams involved, simply demanded a level of ambition to match. Governments and research foundations were hardly going to put so much money into anything less than a hunt for the secrets of the mind. So after decades of talking down the subject, suddenly people were talking it up.

The irony for the PET sites, of course, is that the researchers who built the machines and ran the labs were not equipped to do the actual experiments. Most were radiologists, or were drawn from other fields such as nuclear physics and haematology. They did not know much about the brain, let alone psychology. So as soon as they had got their systems operating, they had to rush out and find someone with a decent hypothesis to test.

Although he did not realise it, Kosslyn was in exactly the right place at the right time. The two medical physicists tapping him on the shoulder were from the Massachusetts General Hospital in nearby Boston. As one of the most famous medical schools in the US, Massachusetts was not about to be left out of the PET revolution and the radiology department had just bought its own system. Kosslyn had been pondering how he could possibly begin to tackle the problem of how brains handle mental images. Now he turned around to discover that not only was there a way of doing his absolute dream experiment – simply asking a subject to have a mental image and then printing out the results like Tootell's autoradiography experiment – but his local hospital was desperate for him to come and do it. Kosslyn did not have to think too long: 'Oh, I was pretty keen. I was in there the next week.'

In principle, the experiment should have taken a month at most. A subject could be scanned in under an hour and, even in the early days of PET, getting the results out of the computers took just a few days. But due to the novelty of the method, it was never going to be quite that easy. In fact, Kosslyn needed a year just to get to grips with the system. 'There were a lot of technical problems to solve – really trivial things like working out how to mount a computer monitor at an angle so subjects could see it with their heads stuck inside the scanner. We also found we had to build a foot pedal for subjects to respond to questions with, rather than a hand button, because moving the hand created brain activity in motor areas too close to some of the imagery areas we were trying to look at,' says Kosslyn. Once his experiment was done, it took even longer to get the results published. The paper did not appear in the *Journal of Cognitive Science* – a journal Kosslyn helped found – until 1993 because reviewers kept demanding he double-check his findings. However, the results were well worth the wait.

Kosslyn had two things he wanted to prove about mental images. The first was the simple fact that imagery took place in the same areas the brain used to map perceptions. The second was that this mapping had a topographical logic. Like Tootell's monkeys, there would be some sort of physical reality to the way the brain represented information and the brain was not the abstract machine many cognitive scientists seemed to assume.

The resolution of the Massachusetts PET scanner was not good enough just to ask subjects to imagine something like a wheel pattern and then take a snapshot of it. Such sharpness may come in the scanners of the next century, but first-generation PET systems could only really show which bits of a brain area were active. Indeed, Kosslyn had to make quite a few compromises to get any result out of the machine. Because of the weakness of the signal, he could not test just one person but had to average the readings from a dozen or more subjects doing the same task. This caused a problem because every person's brain is a slightly different size and shape; so while the averaging would increase the certainty about the existence of any hot spot, its actual shape would become even fuzzier. Another handicap was that PET scanning still took a reasonable time – several minutes, in fact – to accumulate enough positron events for a reading. Mental images might come and go in a flash, but somehow Kosslyn had to find a way of keeping a subject's brain active over the 250 trials of each half-hour scanning session.

The various restrictions meant that the subjects were not asked to do anything too fancy. Kosslyn got them to visualise letters of the alphabet. The letter was given at the beginning of the trial and then a subject would have to imagine it in a controlled fashion by projecting it onto a grid of squares showing on a computer screen hanging over them. As a guide to what they should be seeing, the subjects had earlier been shown examples of the same letters actually filling in the grid block by block. By limiting the subjects in this way, Kosslyn was ensuring that the brain activity would be pretty much the same from trial to trial, and from person to person. Kosslyn even added an extra check to make certain the subjects were doing the task as he had laid down. During each trial, one of the squares would be marked by a cross and the subjects would have to kick the foot pedal to say whether or not it fell on their imagined letter. Kosslyn knew that any hesitation would be a sign that the person had not properly visualised that trial's target.

A still further problem was that all the extra brain activity caused by such things as making decisions, kicking pedals, and even simply seeing a computer screen had to be distinguished from the blood flow rises due to the actual act of imagining a letter. This meant that the subjects had to be given two control conditions on top of the imagery trials. In one, the subjects merely watched a cross on the computer screen and kicked the foot pedal when it disappeared. By subtracting the activity needed to accomplish this simple task from the final result, Kosslyn could filter out the brain's decision-making and foot-moving processes. In the next control, the subjects were scanned as they looked at real letters and reported whether or not a cross lay on them. This removed the extra activity caused by the brain's processing of the sight of the computer display with all its grids and crosses, so that by the end, Kosslyn was left with a picture of just the effort needed mentally to project a letter onto the grid.

There were a lot of steps, a lot of considerations, but in the end no doubt about the result. The experiment showed that the visual cortex did indeed light up when people had mental images. Kosslyn's next move was to demonstrate that this activity was topographically organised. This was done simply by asking the subjects to shut their eyes and imagine the letters as either very large or very small. Kosslyn reasoned that a small letter should create a concentrated patch of metabolic activity right in the middle of the visual cortex, while a large one would spread the activation right to the edges.

Again, he had to use a few tricks to make sure the subjects were actually doing the task as specified. To get a strong enough reading out of the PET scanner, the subjects had been asked to do fifteen letters of each size and hold them for about four seconds, if they could. It was a demanding job. So to check they were managing to change the scale, after each letter the subjects had to answer a question about its shape, such as whether it contained a curve or had a long upright. It was not the answer that mattered but the speed of the reply. As Kosslyn had shown with his earlier experiments in which people imagined rabbits alongside elephants then flies, the need to stop and zoom in to see detail should slow the responses of subjects when they were making the letters very small. In fact, three of the subjects had to be dropped because they did not seem to be managing the task. However, the result from the rest was exactly as Kosslyn had predicted. The PET scans showed that small images took up a small area of the visual cortex while big images filled it to the edges.

The crowning moment for Kosslyn came when he first presented these findings to a roomful of fellow cognitive psychologists. To the untutored eye, the garishly bright scanner pictures were a little hard to take in. The marking of hot spots of blood flow in red, and quieter regions of the brain in cool greens and blues, was plain enough. But following the conventions of radiology, the PET images were presented as a series of cross-section slices through the brain. Even for someone expert in neuroanatomy, it could be difficult to work out exactly which bits of the brain were being shown. However, as Kosslyn walked them through the experiment, the results began to hit home. At a stroke, there could be no more arguments about whether the brain might do imagery this way or that way. 'It's hard to argue with the evidence of your own eyes,' says Kosslyn. When the talk ended, he remembers there was a stunned silence. 'That was kind of fun.'

Kosslyn's experiment was significant as more than just a victory within a small, vitriolic group of researchers. It was also the first high-profile use of PET to explore a question with obvious bearing on the problem of consciousness itself. The earliest PET experiments had played safe by looking at reasonably straightforward brain processes like perception, speech, or hand movements. But mental images were another matter. If PET was ready to take on something completely subjective, then the era of real discovery seemed set to begin.

There were certainly plenty of reasons for optimism. For a start, the performance of PET scanners was only going to get better and better. More and smaller detector crystals, new kinds of isotopes, quicker computers to handle greater quantities of data – the prospect was that the improvements would come so thick and fast that the latest systems would be obsolete after just three or four years' service. And then in the early 1990s, there was the surprise arrival of a completely new scanning method which offered all the power of PET without its many disadvantages.

Scanning with radioactive tracers has obvious drawbacks. Although the doses are low, it is hard to get permission to use the technique on children or women of child-bearing age. And even with the male volunteers who are the subjects of most PET experiments, the maximum exposure is reached after just a couple of scanning sessions. This means it is impossible to use PET to carry out a really in-depth study of a single person's brain or to play around simply to see what the system might achieve. However, as soon as the vast potential of PET became apparent, medical physicists began looking for other methods without such handicaps. What they came up with – functional magnetic resonance imaging, or f-MRI – was still more extraordinary.

The MRI scanner itself was already one of the great inventions of the 1980s. For the same reason that radioactivity caused problems for PET imaging, hospital radiology departments had been looking for a safe alternative to their X-ray machines. With an MRI scanner, a person is slid inside the coil of a gigantic magnet. So strong is the magnet that when the current is switched on, every proton – the positively charged part of an atom – in the body becomes aligned with the field. In this artificially ordered state, the particles can then be probed by firing a tuned pulse of radio energy at them. Different kinds of atom resonate at different frequencies, so it is possible to measure with fantastic accuracy the concentrations of various elements like iron, oxygen, or hydrogen. By taking these readings from many angles, just as with PET, a computer can be used to turn the information into a three-dimensional reconstruction.

Compared to X-ray machines, the images from an MRI system have an almost photographic quality. They are not only sharper but show the soft tissues of the body, like organs, muscles, and, of course, brains, that X-rays pass straight through. The improvement was so great that despite the multi-million-pound price tag of an MRI scanner, by the 1990s a machine had been installed in the radiography department of nearly every major hospital in the world.

Having established that MRI could do anatomical imaging, its developers then began looking for ways to do functional work – to use the scanners to catch organs like the heart and brain in action. At first, there were experiments which involved injecting magnetically visible dyes into the bloodstream. But then it was realised the machines could simply measure changes in the actual oxygen level of the blood as it flowed through the brain. There was no need to add anything. The tiny spurts in circulation that followed a thought or action would reveal themselves. Even better, unlike PET, it would be a true snapshot reading. There was no waiting around for a tracer to be pumped up the arm and around the body, so the timeframe of the activity could be narrowed down from minutes to just seconds.

The list of advantages was impressive: no radioactivity; no expensive cyclotrons or costly armies of nuclear medicine technicians; the pictures were going to be even sharper; and, above all, just about every hospital already owned an MRI machine which could be adapted for functional imaging for a few million extra. PET was something that was always going to be an exclusive club. The prediction had been that only a dozen or so well-equipped laboratories might ever be built. But with f-MRI, it was clear that there would eventually be hundreds, if not thousands, of sites. So instead of imaging experiments being the preserve of a lucky few, any scientist with a decent idea could expect to get access to a system. It was brain scanning for the masses.

On top of all this, researchers could see yet another imaging technology – magnetoencephalography, or MEG – lurking in the wings. A MEG system is much like EEG in that it measures the electrical activity of the brain, picking up actual cell firing and not just the metabolic surges that follow a burst of activity. But unlike EEG, MEG records the magnetic fields created by this firing rather than the changes in electrical potential. These magnetic fields are extraordinarily weak – billions of times fainter than the background magnetism produced by the Earth – but they have the advantage that they pass straight through the bone of the skull without distortion. This means that by using a helmet of several hundred sensors to record the brain from many angles at once, it is possible to build up a three-dimensional view that reveals not just what the brain is doing, but where it is doing it.

36

MEG has its own drawbacks that have hampered its introduction. It relies on the use of SQUIDS – super-conducting quantum interference devices – as its sensors. These delicate electronic circuits use quantum effects to pick up the brain's magnetic fields, but they only work when they are cooled to almost absolute zero in a bath of liquid helium. Finding ways of lowering something so chilly until it nearly touches a subject's head is a challenge in itself. A more serious problem is that the brain's magnetic fields are so faint that experiments have to take place inside expensively shielded rooms. And even then, a car passing in the street outside or the movement of a lift at the far end of the building can cause enough interference to force researchers into resorting to working in the dead of night, when all is quiet.

However, like all the imaging techniques, MEG equipment can onlyl continue to improve by leaps and bounds, and an advance like room temperature super-conductors might suddenly make the technology look surprisingly practical. But the real reason for suspecting that MEG may turn out to be the technique of choice for the twenty-first century is that while its spatial resolution will probably never match that of PET or f-MRI, MEG already wins hands down on temporal resolution. There is no delay between a cell's firing and the emission of bursts of magnetism. So whereas PET measures brain activity over minutes, and f-MRI can show its changes second by second, MEG deals in thousandths of a second. If getting inside an individual moment of processing is what counts in making sense of consciousness, then MEG will have the edge.

Yet the issue was not about which imaging modality would eventually win the race. Instead, the point was that from having nothing, suddenly the mind scientist had an abundance of techniques from which to choose. Scanning was just going to keep on getting better, cheaper, safer, and easier to use. So a feat of science that a few years earlier had been unimaginable, even to someone like Kosslyn, would soon be routine.

This meant that there would have to be a radical shift in the outlook of psychologists. There could no longer be any question about ducking the issue of how the brain did things. Researchers who did not ground their work in an understanding of neurology would simply appear ridiculous. It was just too easy to go and do the necessary experiments. In the 1970s and 1980s, it was psychology and computer science that had been getting hitched; now, in the 1990s, there needed to be a further – or perhaps alternative – marriage with neuroscience.

Kosslyn was riding at the front of this intellectual switchback. He had pushed the cognitive approach further and got into neuroscience earlier than just about any of his peers. But even in his hour of triumph, as he was presenting his PET findings to a roomful of incredulous colleagues, he was aware that the rollercoaster had not levelled out yet. The very same set of slides was saying that there needed to be one more revolution. If science was going to make sense of the brain processes behind consciousness, there was a marriage to be made beyond the one with neurology.

The fact troubling Kosslyn was that even after all the foot-moving, decision-making and display-watching activity had been subtracted away from the PET readings, there still seemed too much going on in the brain of his subjects as they were forming their mental images. The cognitive model he had spent so many years developing suggested that he should find just two or three regions of the brain lighting up to produce an image. There would be a display area, a place where the memory traces were stored, and perhaps some sort of executive centre that organised the actual business of projection. Instead, what the PET pictures showed was a brain sprinkled with hot spots. There was the activation of the visual cortex that Kosslyn had been looking for to prove his ideas about imaging making use of the brain's perceptual apparatus and having a topographic logic. There also happened to be some firing in the temporal lobes, an area of the cortex known to play a part in handling memories. Some of the activation towards the front of the brain might mark the location of the executive areas needed to control the display. But there was still more – too much more to square with a view of the brain as being made up of a collection of simple processing modules.

This same horrible finding was bothering nearly everyone doing scanner experiments at the time. People were designing their tests so as to reveal the parts of the brain that did some action, such as make a hand movement, or felt some sensation, like anxiety or fear. But no matter how many control conditions they used to try to eliminate unwanted background activity, they could not produce a PET image cleaned up to show just one or two isolated patches of activity. The expected areas would usually light up, but then so would a whole lot of others.

At first, it was hoped that this problem would disappear as the resolution of PET systems improved, or researchers got better at setting up their experiments. It was easy to blame some of the extra activity on the fact that subjects were being recorded over many minutes and so had plenty of time to think about things not strictly connected with performing the experimental task. The difficult-to-wash-away activation might be just the brain firing with feelings of boredom, or wandering off on a day-dream. Gradually, however, PET researchers came to realise that the widespread activity was telling them something basic about the brain. It was the expectation that the brain would be divided into neat little compartments of processing that must be wrong. As Kosslyn and his friend Karl Friston discussed by the pond at London Zoo, the brain seemed to respond in a much more holistic way. Even the smallest mental effort appeared to send reverberations right across all its circuits. If this were true, then the attempt in PET experiments to subtract away the scattered echoes of a moment of activity would be completely misguided. The machines might have to be used in an entirely different way.

CHAPTER THREE

Ugly Questions about Chaos

For science, computers and minds seem to make such natural bedfellows. After all, both are systems that do information processing. Anything learnt about the workings of one seems almost bound to help in the understanding of the other. Research into the mechanisms of the mind will lead to better ideas of how to build computers, while computer science should throw up some good models for theories about consciousness. The history of the two disciplines meant they would always be somewhat different in character, computer science being brash and mind science rather proper. But the closer the fields grew, the more both would profit.

The cognitive science movement which so dominated the 1970s and 1980s was just one possible face of this relationship. Cognitive science tried to unify the fields by dealing with the idea of information processing at an abstract level. The messy detail of whether a system was made out of flesh or silicon would be ignored. Instead, researchers would talk about the general architectural plan. Their theory of a process like mental imagery would be a blueprint that listed a set of components, then spelt out the flow of logic connecting these components, but paid no attention to what material was used actually to implement the underlying hardware.

This seemed like a more than reasonable approach at the time. Psychology needed a credible way forward while computer researchers were keen to start injecting some intelligence into their programs. But even before the dramatic rise of cognitive science, there had always been another school of thought that felt the link between the two fields ought to be made at a much lower level. Maybe brains were not so uninteresting as people made out. Perhaps there was something distinctive about the circuitry of the brain – about the way its cells were connected – which made it unique as an information-processing system. There could be a trick of processing stamped into its wiring which accounted for its particularly responsive computing style. Now this would not contradict Turing's argument that any kind of material could be used to implement an information-processing system; it just said that brain cells might make certain jobs, such as producing conscious states, an awful lot easier. The secret of the brain's power might lie in the low-level logic of its wiring rather than the high-level logic of its global information flows.

While cognitive science ruled, only a handful of researchers bothered to pick away at this possibility. But as the abstract approach began to run into increasing difficulties during the 1980s, suddenly computer scientists as well as psychologists began to wonder what lessons they might learn by spending a few years with their noses stuck in the neurology textbooks.

Neuroscientists had a few ideas about how networks of neurons might process information. The basic fact that brains are made of a mass of individually connected cells was discovered relatively late in the nineteenth century. Under the microscopes of the era, it looked to many researchers that brain tissue was just a continuous mesh of wiring. Electrical signals would simply flow cell to cell. But using a silver stain to show up the cell walls, it became possible to see that there was a tiny gap – a synaptic junction – between the output fibres of one neuron and the input fibres of the next. They did not touch. Instead, like a baton changing hands in a relay race, a cell had to release a chemical messenger to carry its message across the gap and set a fresh electrical impulse racing down the line.

The most obvious reason for such an interruption being built into the brain's circuitry seemed to be so that cells could act as switches – tiny stop-lights controlling patterns of information. And researchers were amazed at just how many connections the cells formed with each other. A typical neuron could be busy exchanging messages with over 10,000 of its fellows. Most of these connections would be to cells just a millimetre or two away, but some might reach right across the brain. By the 1930s, there was general agreement that the brain was some kind of information-processing network. In a famous public lecture, the neurologist Sir Charles Sherrington likened it to an enchanted loom, 'where millions of flashing shuttles weave a dissolving pattern, always a meaningful pattern though never an abiding one'. The question was, what were the actual rules governing the behaviour of brain cells? What was the code or the principles that allowed them to weave patterns tight enough to produce a conscious state?

In the 1940s, a Canadian psychologist, Donald O. Hebb, latched on to a further recent discovery about brain cells to offer some answers. Research was suggesting that many of the connections between neurons were feedback connections. When tracking nerve fibres to see where they might lead, neuroanatomists were often startled to find that a cell receiving input would send a line back again. This made a loop that could be used to amplify, or even switch off, the firing of the original cell. If this were true, then information in the brain did not flow in simple straight lines of stimulus and response. Each output would produce its own flood of echoes. Signals would come bouncing back down the chain, reflecting the impact a cell was having on its many targets and perhaps even forcing a change in the message it was putting out.

This seemed a big clue. But it took a bold individual to make something of it. Hebb was unusual for a psychologist in that he was trained by a famous physiologist and then worked alongside an equally eminent neurosurgeon. He felt it was clear that a system of feedback connections would bring a network of cells alive. It would turn a network into a kind of processing surface on which patterns of firing ebbed and flowed, driven by the dynamics of their own internal feedback. Only a spark of input would be needed to start a chain reaction of firing that would spread rapidly through a group of cells. But then through the feedback connections between them, the firestorm of activity would begin to settle. A broadly organised pattern would emerge as each cell fell into some new balance of tension with its neighbours. So input did not create the patterns; it was how the network responded as a whole that shaped the final state of firing. In other words, Hebb was saying that brain processing was a competition rather than a calculation. Networks evolved their way to stable solutions.

The second part of Hebb's story concerned what happened after each wave of firing, as the embers were dying down. He suggested that a sudden burst of activity would be followed by changes in all the cells that had taken part in the fleeting pattern. The cells would strengthen their connections with each other, growing extra synapses or increasing their stocks of neurotransmitters so that it became easier for them to fire together again in the future. In this way, every pattern could leave behind a memory trace. A faint ghost of what had just happened would become etched into the brain's circuitry where it would affect the course of all future processing competitions. So a nerve network could learn. Indeed, if the new connections were strong enough, it might be possible to fan the whole of the original pattern to life again. Through the amplifying power of feedback, simply tugging on one corner of a set of connections might be enough to stir the rest and so return an earlier experience or thought to bright consciousness. Nerve networks would be able to remember.

This was an immensely impressive theory. By combining the idea of feedback-driven competitions with that of connection strength changes, Hebb was talking about the brain as a living landscape of processing. It was not at all like a conventional computer, which keeps its memories and processing separate. A computer is basically a set of logic gates, a storage medium, and a ticking clock. At each tick of the clock, the computer pulls either a program instruction or a bit of data out of its memory bank. The logic gates carry out the selected operation and then throw the results back into memory, clearing themselves for the next step of the program. So everything is organised to take place in strict sequence – there is no room for error – and change only happens to the active bit of data. Once back in memory, any information is frozen again.

Hebb was saying that processing in the brain was something quite different. There was no clock. Competitions took place in their own time. They could start as soon as there was fresh input and they finished whenever some new balance of firing had been reached. More importantly, memory connections and processing connections were the same thing. Experience carved out a pattern of bumps and hollows in a pliant surface, creating a landscape that would channel input down certain established paths. But in processing the input, the landscape would itself be still further changed. The new experiences would wear either new or deeper grooves in the surface, so nothing was ever static. The machine made the moment and the moments made the machine.

Drawing from some of his own research into the development of baby rats and chimpanzees, Hebb argued that at birth the infant brain would present a fairly flat and even landscape of connections. Then, over the course of a lifetime, its circuits would become richly textured with experiences. Of course, the brain would not start as a completely blank slate. Genes might be responsible for some basic patterning, and certainly a baby would have experienced sounds, tastes, touches, and even movement while still in the womb. There would be some level of stimulation from the moment the nervous system began to form. But still the brain would be a machine that built itself. A baby would come into the world and have to learn how to make sense of things.

On first seeing a complicated three-dimensional scene, a baby might lack the processing structures to put all the parts in their place. It may see hazy patches of colour and shape, but not realise that one jumble of lines was the arms and legs of a teddy bear poking out near its face, while another was the curtain across the far side of the room. But through trial and error learning, strengthening the connections that worked and eroding the ones that did not, a baby's brain would swiftly learn how to place objects. Instead of sensations being left to slop around a rather featureless landscape, and so the patterns of one moment looking much like the patterns of the next, there would be a geography to guide each neural competition to a more well-defined conclusion. The brain would react to life through the eyes of experience.

Hebb's theory was by no means entirely original – many other researchers of the period were saying some of the same things. But Hebb, who had once intended to be a novelist, was able to express his ideas with particular clarity. His book, *The Organization of Behavior*, became an instant classic when it was published in 1949. Yet there was a problem. Scientists felt there was no obvious way of following up on its insights. Neuroscience lacked the equipment to test Hebb's ideas. The microscope might be able to show that brain cells had connections back to each other, but there was no means of telling whether these connections carried any signals, let alone whether they really supported flows of controlling feedback. As Hebb admitted, it was all just speculation. And while many might like his theories, there were others who favoured a radically different interpretation of what might be going on in the brain.

Hebb saw information processing in the brain as something biological. Like life itself, mental states and neural circuitry evolved. But the 1940s were also the decade that saw the birth of the first electronic computers – room-filling, vacuum tube monsters used for calculating artillery ranging tables or cracking secret codes – and their arrival prompted a number of theorists to take a much more mechanistic view of the brain. Two scientists in particular – Warren McCulloch, a psychiatrist at the University of Illinois, and a young mathematician, Walter Pitts – made much of the idea that brain circuits might do their job exactly like one of the new-fangled number-crunching machines.

McCulloch and Pitts started by assuming that neurons represented information using a simple binary code, firing fast to signal 'on' and slow, or not at all, to signal 'off'. Just like a Turing machine, data would be carried as a pattern of marks and blank spaces. Then again like a computer, groups of these two-way switches would be wired together to form logic gates – little circuits that could execute the various operations, such as AND, OR, and NOT, which were the basic building blocks of a digital computer program. For example, an OR gate would be a cell that let through a signal when stimulated by either of its two input cells, while an AND gate would be one that required the firing of both its input cells. So the wild-looking tangles of the brain would actually conceal a highly regular and formal organisation. Neurons would line up to form tidy logic arrays shunting data down pre-set paths.

45

Few people felt the McCulloch–Pitts story was realistic. It was already known that neurons formed thousands of connections, not just the two or three of a simple logic gate, so as an explanation of brain processing, the model fell at the first hurdle. But despite this the idea caused quite a sensation, because while it did not tell neuroscientists much about brains, the suggestion that networks of neuron-like switches could perform logical operations gave computer scientists some exciting new thoughts about ways to build computers.

The problem for the early designers was that the strict step-by-step logic of a computer created a crippling processing bottleneck. Because a machine had to execute each step of a program in turn, it was essentially a single-gate system. Shifting a reasonable amount of data was like trying to move a sandpile armed only with a teaspoon. Making a computer bigger by giving it more memory did not help as it merely made the pile higher. But McCulloch and Pitts seemed to be describing a network system in which there were multiple processing gates and data could flow through any that were free. It was like giving a machine an unlimited number of teaspoons to do its job.

The 1950s saw many attempts to build such a network – or parallel processing – system. There were even some minor successes. But researchers soon found that it was almost impossible to get data to find its own way across a grid of processing units. The problem was that there was no intelligence built into the gates to help channel the flow. Each gate still had to be told what to do by a central program; so a network design simply moved the bottleneck from the hardware to the software. The effort of co-ordinating the work cancelled out any advantage gained in distributing the processing. Meanwhile, computer technology was improving so fast that the fact a machine might only advance a single step at a time was ceasing to matter. Transistors, then silicon chips, meant that systems could zip along, executing millions of processing operations each second. The computer revolution could go ahead without needing the network computer.

Research never stopped, but the field went quiet for many years. Then suddenly, in the 1980s, there was a sharp revival of interest following the discovery of some computer learning rules – a distributed form of intelligence – which could take the place of a centralised program. Rather than each processing unit being instructed about how to handle the next step in a computation, it was made to work it out for itself through trial and error. A node might make the wrong choice the first few times, but with feedback about the network's overall performance being used to adjust its behaviour during each run-through, eventually it would start doing the right thing. In short, computer scientists had rediscovered Hebb's idea that competition and adaptation among a network of connected elements could create a self-organising processing landscape.

It was one learning rule in particular – the 'back propagation of errors' algorithm – that was almost single-handedly responsible for the new breed of neural computer. A backprop network consisted of just three rows of neurons, or computing cells. The first row was for the input – a pattern of values representing the data to be fed through the system; the second row of neurons did some processing; and the third did a little more processing before displaying the results. The connections between the nodes were equally rudimentary. There were no links between cells on the same level. Each node simply sent its output signal to every one on the next. Cells made a decision about whether or not to fire by summing the combined input reaching them from the tier below, then checking to see if it exceeded some stored threshold value or weight.

Obviously, if all the processing nodes used the same threshold value and all received the same input, the network would never do anything particularly exciting. But by seeding the network with a random set of weights, and then using feedback to tune the strength of chance connections, some sort of processing terrain could be produced. At the start of training, the network would be given a set of input values – perhaps a string of numbers representing the current state of the stock market or the digitised image of a warplane. These values would ripple forward from level to level, becoming trans-formed in the process into a slightly rearranged set of numbers that represented the answer – perhaps a buy or sell command with a stock market system, or a friend or foe identification in an aircraft recognition application. In the first few trials the answers would be a long way off beam. But by calculating the degree of error and then using it to suggest a small adjustment in the firing threshold of each processing node, the network could begin to home in on the correct answer. After enough trials, it would arrive at the balance of weights needed to turn a given input into the right output.

Backprop networks were not a lot of use for conventional computing jobs, but they proved to be terrific at anything that involved some form of recognition or pattern matching. One of their great strengths was that if trained on a wide enough variety of stimuli – the stock market in many different conditions, or aircraft seen from many angles – they could start giving answers about input patterns they had never come across before. They did not have to be programmed to deal with each individual case. Quite automatically, they began to generalise.

Plainly, a backprop network was still a long way from being biologically realistic. For a start, while it developed some sort of internal processing landscape, feedback played no part during the actual moment of processing. Adjustments were made only at the end of each training trial. And then, once a balance of weights had been established, the network did not employ feedback at all. It simply emitted its output. By contrast, brain cells are always being pushed by feedback. A balance of activity is being negotiated during every moment of awareness. Nor should it be forgotten how human intelligence gets smuggled in through the back door during the training of a backprop network. There has to be a programmer standing over the machine, telling it whether its answers are right or wrong. The network simply translates one set of numbers into another. A conscious human is needed to define the task and judge the results.

Yet having said this, what excited computer scientists was that something just a little bit like a brain circuit performed an awful lot better than might be expected. It was almost as if the connectionist design had tapped into some secret well of computational energy. The power of the backprop network was all the more surprising because throughout the 1970s and 1980s, computer scientists had failed so dismally in their efforts to build intelligent systems the cognitive science way. Billions had been sunk into an attempt to bring conventional computers alive through the use of fancy coding languages or complex reasoning programs. But one after another, big government-sponsored programs like Japan's fifth-generation computer project and Britain's Alvey initiative turned into embarrassing flops. The humble backprop network seemed to spell out just where computer science had been going wrong. Instead of trying to write smart programs for dumb hardware, people should have been making the hardware smart and getting it to evolve its own processing routines. Intelligence – or at least an ability to learn and organise – needed to be built in at the level of the circuit.

Computer science seemed to be on to something. At the tail-end of the 1980s, in an explosion of conferences, new labs, and new journals, the field of artificial neural networks was born. Almost overnight, computer academics and neuroscientists started getting together in their thousands with the aim of extracting some basic principles about brain processing that might be useful to both sides. Computer scientists wanted to know what it was about brain circuits that they should be putting into their machines. Brain researchers, in turn, felt they could do with some of the theoretical clarity of computer models. Neuroscientists had spent many years learning about how neurons fired and what sort of circuits they formed without really getting anywhere. They had a mass of detail, but no broad story. Now the mathematical rigour of computer modelling seemed to offer a way of discovering what it was about the brain's design that counted. The principles of information processing could actually be tested.

So there was still a marriage to be made between the study of computers and the mind, only at a different level. However, just as the neural network bandwagon was beginning to roll, promising to take over where cognitive science had left off, doubts began to grow about whether the two sides were such natural soulmates. It was clear that to go any further, neural networks would have to become more truly dynamic. Processing and feedback could not remain separate in the way they were with backprop networks. If brain cells used active competition to evolve their way to answers, then neural networks would have to follow suit. But one of the hallmarks of a genuine competition is that there is an element of unpredictability about its outcome. A certain result may be most probable, but how and when an answer will be reached, as well as the exact nature of the conclusion, may vary each time the competition is run.

Yet the success of the digital computer is founded on precisely its ability to squeeze out any uncertainty in its behaviour. Computers are built to follow the same steps and come to the same answers every time they run a program. The advantage of this is that once the design of the hardware or software has been perfected, endless copies can be churned out, all with identical characteristics and performance. By freezing the logic, the logic can be mass-produced. But the flipside is that the smallest bug or mis-step can bring the whole system crashing down. For a computer to work, chance events have to be ruled out right down to the level of the electrical components from which the hardware is built. A transistor is engineered so that it can put up with things like slight fluctuations in the power supply or changes in temperature without changing state. Some kinds of electronic devices are analogue – they produce a continuously varying output – but a transistor is a binary switch. That is the meaning of digital. It is either on or off, a 1 or a 0. There is no room for shades of grey, only black and white. A bit of information either exists, or it does not.

The assumption was that brain cells were also basically digital devices. The brain might be a pink handful of gloopy mush; brain cells themselves might be rather unsightly tangles of protoplasm, no two ever shaped the same; but it was believed that information processing in the brain must somehow rise above this organic squalor. There might be no engineer to draw neat circuit diagrams, but something about neurons had to allow them to act together with logic and precision.

Brain cells certainly had a few suggestive features. To start with, the very fact that they have a separate input and output end says there is a direction in which information flows. Signals arrive at a root-like bush of fibres known as the dendrites. Then, sprouting from the other end of the neuron, is the axon, the long fibre which carries its output message. It is true that a few synapses are also usually found on the cell body, and sometimes even on the axon itself, but generally speaking dendrites collect the information and axons deliver the response. Even more obviously, all the messages come from somewhere and go somewhere. Whether two cells are connected is a black and white issue. Under the microscope, dendrites and axons may look as though they are forming unruly tangles, but there is an unbiological precision in the way that a signal can be sent to a fixed destination – and only that destination – almost instantly anywhere in the brain.

There is a physical logic in the wiring patterns of the brain. Then, on top of this, there is something quite plainly binary about the all-or-nothing nature of a neuron's decision to fire. The simple story about how a cell fires is that incoming messages pool as a series of small charges in the dendrites. These individual charges creep up the branches and over the surface of the cell body to converge on a trigger zone at the base of the axon, known as the axon hillock. The hillock is delicately balanced so that it will only spark an output signal if the accumulation of charge exceeds some threshold value. The pooling of charge can take many different courses. Sometimes a cell might be triggered by just a few strong impulses arriving at almost the same instant; at other times, the threshold might be reached more gradually by the slow addition of many weaker or fading impulses. But the decision to fire is black and white. The cell convulses and a message is sent flying down the line to all the other cells to which it is connected. A bit of information has either been created, or it hasn't.

So, despite the brain being made of flesh and blood, the propagation of signals looks to have a digital clarity. But the question is whether brains are exclusively digital in their operation. A computer knows only a world of blacks and whites. It relies on its circuits being completely insulated from any source of noise which might interfere with the clockwork progression of os and 1s. But it is not so clear that brain cells are designed to be shielded from the messy details of their biology. Indeed, a closer look at a neuron soon suggests the exact opposite: it is alive to every small fluctuation or nuance in its internal environment. Where a transistor is engineered for stability, a brain cell trades in the almost overwhelming sensitivity of its response.

It could hardly be any other way, because the firing of a neuron is actually an electro-chemical process – and more chemical than electrical. Nerves do not conduct impulses like wires. Their electrical activity is based on moving charge-carrying ions, such as sodium, potassium, chlorine, and calcium, across the cell wall. The membrane of a neuron is finely covered with pores. Some of these pores are like pumps which can force ions either in or out of a cell to set up an imbalance in the concentration of charge. Then other pores are simply valves which open to let the ions flood back through again, swiftly righting the balance.

The principle of pushing ions back and forth across a membrane to create a voltage drop is simple enough, but the control of the channels is an immensely complex business, being both electrical and chemical. It is electrical because a change in membrane potential can itself cause a pore to open or shut. Not only does this mean that a pore can influence its own level of activity, any changes feeding back either to amplify or stabilise whatever it happens to be doing, but a drop in voltage in one region of the membrane will tend to spread. The opening and shutting of a group of pores will create a creeping electrical potential drop that causes neighbouring pores to follow suit, setting up a chain reaction that propagates across the surface of the neuron.

The electrical response is complex enough because there are many classes of pores, each handling a different kind of ion and reacting to different voltage levels in different ways. But pores can also be controlled by a whole range of chemical messengers – neurotransmitters and neuromodulators – which either bind directly to a channel to change its shape, or cause it to alter its activity through some more subtle chain of events. There are hundreds of different signalling substances that the brain uses to open and shut pores, from simple amino acids like glutamate right up to hefty proteins similar in chemical structure to a drug like morphine. Some cause an instant change, others work over minutes or even days; some affect just one kind of pore, others affect all. So, depending on what mix of pores is built into an area of membrane – something which itself can be changed in minutes or hours – the cell wall of a neuron can show a tremendous variety of responses. A computer is made of standardised components. One transistor is exactly like the next. But every bit of membrane in the brain is individual. The blend of pores can be tailored to do a particular job, and that blend can be fine-tuned at any time. There is a plasticity that makes the outside of a neuron itself seem like a learning surface, a landscape of competition and adaptation.

The electro-chemical properties of a neural membrane are, of course, put to two general kinds of use: making axons and synapses. An axon is just a tube of membrane with a fairly simple pore structure. There are pores which pump out sodium ions and pump in potassium ions to establish an initial state of electrical tension across the axon membrane, then another set of pores acts as a valve for the sudden release of this tension. The trick with the valves is that they are electrically sensitive. If depolarisation begins in one section of an axon, the change in potential will open the valves in the next. A spike or action potential is created as one bit of tubing after another depolarises in a chain reaction that flies all the way down to the end of the line. Because, physically, little moves – the ions simply step sideways across the axon wall – the process is highly efficient. Depending on the thickness of the axon, a spike can be sent a distance of several feet at a rate of several hundred miles an hour. The speed at which the axon can then be reset means that a cell can fire as many as a thousand spikes a second.

In keeping with its role as a bit of wire, the axon is the least plastic part of a neuron – although fatigue and growth changes can still change its operating characteristics. Where things get fancy is at the synaptic junction connecting two cells. The membranes on either side of this cleft are thick with a great many different kinds of pores and receptor sites, and how they react at any moment can be finely controlled by a whole range of chemical messengers and self-tuning feedback loops. The basic story of how a signal crosses a synapse is that when a spike of depolarisation reaches the tip of an axon, it causes a set of electrically sensitive calcium channels to open. The inflow of positive calcium ions triggers an enzyme reaction that eventually makes the axon tip eject stored bubbles of neurotransmitter into the junction. These messenger molecules simply float across – a journey of about a thousandth of a second – and bind to chemically sensitive sodium channels on the other side. The pores of the dendrite are forced wide open, so beginning the depolarisation of the next cell in the line.

But in practice, there is nothing certain about any of the steps in this chain. An axon may often not even release any neurotransmitter, despite being hit by a full-strength spike. The amount of neuro-transmitter spilled into the gap can also vary. Plus, there is a whole cocktail of other substances that may or may not be released at the same time. Then, what sort of reception the message gets on the opposite bank can alter from moment to moment. There might be magnesium ions physically blocking some of the sodium pores, or a longer-acting brain neurotransmitter may have subtly changed their response; often, chlorine channels may have been opened, letting in a negative charge that dampens the effect of any new input. So a spike might seem like a digital event – the all-or-nothing creation of a bit of information guaranteed to reach a known destination – but the same signal might one moment be met with an instant and enthusiastic response, the next only fizzle away into nothing, failing even to stir a cell's own axon tip.

Some of the variability in the behaviour of a neuron could be just noise – an unpredictability caused simply by the fact that a brain cell is an organic system depending on ions and molecules to bump about and hit the right spot. For example, on one occasion there might be 10,000 molecules of neurotransmitter secreted into a synaptic cleft; on another, it could be 9,000 or 11,000 – just enough sloppiness in the chain of transmission to create the odd glitch. If this was all that was happening, then a spot of clever design would always solve the problem. Brain cells might react only to the average of a train of spikes rather than any individual spike. In this way, a few stray signals could be ignored. But while there is undoubtedly a degree of noise in the brain, much of the variability looks deliberate. Neurons do not even seem to be trying to deliver a digital-like predictability in their response. Instead, they appear to thrive on being fluid. By using competition and feedback to fine-tune their workings, they can adapt their response to meet the needs of the moment. They can go with the flow.

This shows most clearly when scientists compare the behaviour of a synapse in an alert brain with that of a resting brain. When the brain is in a state of high vigilance, or if it is dealing with a stimulus that is interesting and new, the synapses along the way will respond with extra vigour. They will trigger easily and continue to buzz for some time after. But when a synapse is part of a pathway dealing with something dull, like the never-changing background drone of a fridge, then the transmission of spikes becomes much more haphazard and irregular. Experiments show that as few as one in ten of the spikes will even cause an axon to release its transmitter. It is as if the synapse knows whether the information it carries is important to the state of consciousness as a whole. When things do not matter much, its response is loose; the transmission of spikes can look rather erratic and noisy. But as soon as the message begins to count, the neural machinery tightens right up. Suddenly every signal starts leaping the gap.

This quite recently discovered fact is something very important. It had always been assumed that it was the action of billions of synapses that added up to make a state of consciousness. But consciousness – or at least, levels of attention and alertness – seems able to influence the response of individual synapses. In computer terms, the logic sounds alarmingly backwards. It is as if rather than the circuits creating the results, the results are creating the circuits. But there is really no great mystery if the brain is an evolving feedback and competition-driven system. The whole brain – both the settings of its circuits and its global state of organisation – would need to develop in concert during a bout of processing. One would impact the other, nudging everything along to some final balance of tension. Like tuning into a distant radio station, the responses of millions of brain cells would be twiddled a little bit this way, a little bit that way, until they began to produce a coherent signal.

This was the sort of logic that Hebb had been talking about. The problem for neuroscientists was that it was not until the 1990s that they started to get a clear sight of the actual feedback mechanisms by which a lowly synapse could be tuned. One crucial discovery was that a cell's output spike actually travels both ways: it runs down the axon, but also back over the cell itself and through its own dendrites. What this means is that the synapses are told whether or not they contributed to the last firing decision. So a synapse which played a part might be encouraged to react a little more strongly next time, while another which had not been active might be dampened to keep it quiet. Such tuning might last a fraction of a second, long enough to turn up the volume on a faint signal or a useful pattern, or it might lead to more permanent growth changes. The rebounding spikes could cause an individual dendrite to sprout extra connection sites, or to reabsorb redundant synapses. Tuning could become learning.

The backflow of spikes was just one of many feedback mechanisms that began to be recognised during the early 1990s. A still more surprising discovery was that a stimulated dendrite releases a small dose of a usually poisonous gas, nitric oxide (NO), to send a message back across the synaptic junction. Nitric oxide can be deadly because it binds to iron atoms and so can destroy the haemoglobin in blood cells. But it also appears to switch on certain enzymes in an axon tip, prompting them to ramp up the production of neurotransmitters. Nitric oxide also has the secondary effect of relaxing the walls of blood vessels, so its release probably increases the blood flow into an active area of the brain. Both directly and indirectly, a very simple messenger system could quickly improve the tone of a connection.

These two examples show the feedback mechanisms that exist even within a cell or across a junction. But then there are the feedback connections between cells that put them in touch with the wider picture. Here, Hebb turned out to be even more right than he imagined. As researchers found ways to record from cells and follow their signals, they discovered that feedback connections dominated the brain. They were everywhere, from short loops linking neighbouring cells to long chains in which a signal might bounce right around the brain before feeding back, much modified, to its source. The feedback connections were not even limited to which part of a neuron they made contact with. Some came back as part of the wash of input hitting a cell's dendrites, but others formed synapses on the cell body or close to the axon hillock where – being closer to the action – they could exert a much more powerful effect. It was found that some connections were even made right on the axon tip. If the connection was inhibitory, one neuron would be able to block another cell's spike even after it had been fired. So there was no shortage of pathways through which the activity of the wider network could feed back to influence the behaviour of its individual components. No part of the chain of transmission was immune from adjustment.

This meant there was a dilemma for anyone trying to draw parallels between information processing in the brain and in computers. Was the feedback-driven nature of brain activity merely a complication, or did it pose a more fundamental problem? It was all getting terribly confusing. Brain cells looked to have something digital about them. There was the all-or-nothing fact of a spike. And then each neuron made a precise set of connections. Signals were delivered to known locations. Yet how could a spike count as a bit of information if the next synapse might simply choose to ignore it? And where was the certainty in a connection pattern if synapses could be switched in and out of the action, depending on the needs of the moment?

But then again, the transmission of spikes did seem to tighten up when things began to count. And connection patterns were more stable than they were fluid. Even though it might be fine-tuning the activity at the connections, a cell remained joined to the same group of 10,000 or so other neurons. The growth changes needed to wrench itself away took hours or days. And even then, there was only limited scope for change. In a mature brain there was little room for movement by the cell body or the long filament of the axon, so all that could really happen was a slight shift in the balance of connections being made with the local group of cells clustered around either its dendrites or its axon tip.

As a biological organ, the brain could not help being a little noisy and unpredictable in its workings.. But presumably that did not matter as the brain would have the means to insulate its processing from mere noise. The brain then used feedback to adjust its circuits and competition to evolve its answers, which again introduced an element of unpredictability. But ultimately, all this feedback and competition appeared to be directed towards producing a well-organised response. And no one could say that spikes and connection patterns did not matter. To the computer-minded, the foundations might look soggy, but there did seem to be something concrete going on. The brain's circuits offered a processing landscape that might be plastic – it could adapt to its experiences – but which still had enough structural rigidity to make things happen.

The trouble with this charitable view was that there remained something fundamentally different about brains and computers. Any digitalism in the brain was a weak, blurred-edged, pseudo kind of digitalism. Spikes and connection patterns emerged out of a sea of metabolic and growth processes. Behind the scenes, everything had to be in some kind of dynamic balance to create a particular state of response. Computers, on the other hand, were digital by nature. They dealt only in defined bits of data and defined processing paths. There was no room for unpredictability. A transistor either worked to specification, or it was broken. So if a computer wanted to behave like a dynamic, feedback-tuned system, it had to fake it.

Being inherently predictable, a computer can only pretend to be basing its calculations on unpredictable or continuously varying processes. It is impossible to disguise the black and white nature of the computer, even with clever tricks. For example, to make the neurons in a backprop network seem more realistic, it is possible to program them so that instead of sending out a simple binary on or off message, they broadcast an actual value, some figure between the full-off of a 0 and the full-on of a 1. The nodes can be made to appear to be dealing in shades of grey rather than the unyielding blacks and whites of a conventional computer.

The problem is that a digital computer can only specify any given value to a limited number of decimal places. It does not have an infinite number of registers to represent a figure, so in practice every number has to be rounded off at some point. It is tempting to believe that this does not really matter. After all, even specifying the strength of its output signals to just a couple of decimal places would give a backprop network a hundred shades of grey with which to work. Plenty, it would seem. And computers can easily manage 32-bit or even 64-bit precision in their calculations, quickly pushing the available number of values into the millions. Surely, it would not take too many more decimal places to render the problem of rounding up completely irrelevant? A simulated neuron should be able to show all the rich variety of output of a real one.

However, in the late 1980s, just as the neural network bandwagon was itself beginning to pick up momentum, another new science was emerging that gave reason to think that the gap between the fake dynamism of computers and the fake digitalism of the brain might really count.

This new branch of mathematics – known first as chaos theory, then later as the study of complex non-linear systems – caught the whole of science by surprise, so totally unexpected was its message. Chaos theory originated in a general dissatisfaction within science with the same kind of overly reductionist, overly component-oriented – and, indeed, overly digital – way of looking at natural phenomena that was so dogging research in the mind sciences. Science had been founded on the belief that the proper route to understanding a complex system, such as the movement of the heavens, the mixing of chemicals, or the emergence of life, was to break it down into a collection of parts linked by simple mathematical formulae. You wanted a list of bits and the rules that put them back together again. And if the essence of a system could be reduced to an equation that fitted comfortably on the front of a T-shirt – something like Einstein's famous $e = mc^2$ – then that was perfect.

But reductionism depends on the assumption that the world is discontinuous, that it is made of discrete bits. However, real life does not have sharp boundaries. For instance, even our own bodies are not cleanly separated from their surroundings. The surface of our skin may appear to be a perimeter marking 'us' from 'not us' with digital clarity. It seems a binary distinction. Yet when viewed on a microscopic scale, when does an oxygen or water molecule stop belonging to the surrounding air and become part of ourselves? Or when does a skin flake or spot of grease become sufficiently detached from our body to count as just a passing speck of dust? From a distance, things can seem to have sharp boundaries, but get in close and those boundaries turn soft. The idea of the bounded object is really just a convenient fiction.

Of course, reductionism has served science well. The reason is that for most of the time scientists stick to situations, or scales of magnification, where the simplification does no real harm. When we talk about having a body, the fuzziness of its actual physical boundaries is normally quite irrelevant at our level of discussion. The odd skin flake or water molecule the wrong side makes little difference when our use of the concept captures at least 99.99 per cent of what we mean to say. In the same way, the normal laws of physics are as accurate as we need for most of the problems we face in life. When calculating the load forces on a new bridge design, the odd quantum blip affecting an atom in a steel girder will be lost among the statistical regularity of zillions of other atomic interactions.

There is a lot of science that can be done by concentrating on situations so close to being digital as not to make a difference. Yet there are clearly also a great many areas in life where the blurring of boundaries and the fluid nature of relationships cannot be ignored. The classic examples are the weather, economics, social systems, condensed matter physics, quantum mechanics, fluid dynamics, and anything to do with biology. Such systems are not just accumulations of components, bits of clockwork in which every gear is locked into a fixed relationship with its fellows. Instead, they are restless and evolving, driven by the pressures of their own internal competition. If such systems seem to have any stability, it is only because they have reached a momentary accommodation of tensions. Like soap bubbles, they have been stretched to some delicately trembling pitch of organisation. It should not be surprising, then, that attempts to break them into collections of labelled parts will destroy what seems most important about them. Reductionism is much too clumsy-fingered to perform such a task.

In the 1980s, chaos theory came along as a revolutionary way of dealing with systems built largely of dynamic interactions. Ironically, a lot of the maths was not actually that new. Some of the equations, such as the Navier–Stokes formula for calculating the behaviour of flowing liquids, had been around for a good century or more. But solving the equations was such a long-winded process that it took the invention of the computer – the ultimate expression of digital thinking – before mathematicians could really start exploring where they might lead.

Non-linear maths depends on the same calculation being repeated many times, with the results added back into the mix after each step. The formula remains the same, but the outcome becomes more and more complex as the numbers feed back into themselves. In this way, an equation modelling the collision of two jets of fluid would calculate the initial whorls of turbulence, then the smaller whorls produced as these whorls collided, and then the still smaller whorls that came after that. The first time round, such a calculation is fairly simple – perhaps a page or two of arithmetic. But as more and more feedback effects have to be included, the calculations become just too lengthy to bother with. It would have taken a Victorian mathematician a lifetime to pursue the outcome of a single example. Computers, of course, got rid of that problem. Suddenly, it was possible for researchers to plug the numbers into a machine and leave it to crunch its way through a thousand human lifetimes' calculating effort while they went off for a cup of coffee. Armed with this brute power, scientists began simply to play around with some of the old equations to see what might happen.

A lot of the time, the results were pretty boring. The echoes of feedback would soon fade away and the calculations would become locked into a rut. But just occasionally, when the balance was right, the equations started to generate stunningly intricate patterns. They would take off on a meandering, non-repetitive path which would gradually build into a structure with a surprising sense of order. Indeed, although the patterns were called chaotic because of their complexity, this was a dreadfully misleading term. The normal meaning of chaos is random, featureless behaviour – the very absence of pattern. But what researchers were demonstrating was how a certain class of simple feedback-driven equations could inflate over a few cycles to create a mass of swirling, turbulent, yet still perfectly lawful activity. Something that might look chaotic was actually the product of a deep-seated mathematical regularity.

Despite the misconception built into its name – which was only slightly helped by calling it deterministic chaos – chaos theory electrified science. It seemed capable of explaining all manner of order found in the natural world. The fronds of a fern, the ragged geography of a coastline, even the storms of a weather system could be generated by repeating the same loop calculation enough times. Chaos maths seemed like the secret law governing growth and natural form. It gave science a new tool with which to tackle a vast range of problems that could not be touched by ordinary linear maths.

The key to chaos theory was the idea of the attractor. The problem with a feedback-driven equation was that it could have infinitely many different outcomes depending on what values had been used to seed the initial calculation. So you could follow the fate of one set of starting conditions, but that would probably not tell you much about the general properties of the equation – the kind of outcomes it tended to produce no matter what figures were used at the beginning. Chaos theory took off when mathematicians began graphing plots of outcomes, using a computer to run an equation through hundreds or even thousands of different initial conditions and then stepping back to see if they fell into any sort of pattern. Often an attractor would emerge – a clustering of outcomes which showed the most common kind of fate for a system, regardless of where it might actually have started out.

The simplest type of attractor is a point attractor – a system within which no matter where you begin the calculation you will always end up at the same spot. A real-life example of a point attractor might be water funnelling down a plughole or a pendulum swinging to rest. It makes no difference what the water level or the position of the pendulum might be at the beginning of the system's evolution, there is only one possible outcome. The time it takes to arrive at the final balance might vary, but eventually gravity overpowers all other factors.

A slightly more interesting class of attractor is the limit cycle in which the set of allowed outcomes forms a line rather than a point. An example of this might be a marble rattling around inside the brim of a bowler hat. The marble might roll about from side to side a bit, but eventually it will have to settle somewhere along a two-dimensional path. Point- and limit-cycle attractors are all very well, but what put chaos theory on the map was the discovery of something much more exciting: the strange attractor. And the most famous example of a strange attractor was the first, the Lorenz butterfly.

In the 1960s, Edward Lorenz, a professor of meteorology at the Massachusetts Institute of Technology, was attempting to simulate weather patterns on a computer. It was still early days for computers and the only machine to which Lorenz had access was an ungainly vacuum-tube contraption that took a full second over each calculation. To get round the lack of power, Lorenz had greatly to simplify the Navier–Stokes fluid dynamics equations used for the simulation. And he did not have any fancy computer graphics to help him turn the results into actual weather charts. The machine simply churned out columns of figures representing the daily change in temperature and air pressure over a set of geographic co-ordinates.

As a simulation of a real weather system, it was pretty crude. But Lorenz soon noticed that his model had a startling property which made it seem like it had taken on a life all of its own. Lorenz began each run by seeding the program with a set of starting values for the temperature and air pressure – the weather conditions on day one. Then he would let the equations run and watch as a series of highs and lows bumped about his imaginary landscape. Not surprisingly, any time he used the same set of starting values, the outcome also remained the same. The maths would always produce an identical result. But one day, wanting to save time on reproducing an earlier simulation, Lorenz restarted a run halfway through. He plugged in the weather values taken from the middle of the old printout and then disappeared for an hour to escape the clattering of his ancient machine. Naturally, Lorenz expected the simulation simply to pick up where the original had left off, but when he got back he found that his model seemed to have jumped track and headed off down some entirely new path.

At first there looked to be an obvious explanation. The computer calculated values to an accuracy of six decimal places, but to save time Lorenz had restarted the simulation with figures rounded up to three decimal places. So his weather system had set off from a fractionally different set of values. However, Lorenz felt that such a minor change in the initial conditions should have led to an equally small change in the final outcome, not sent the system sailing off in a whole new direction. Furthermore, when he did add back a few of the decimal places, things did not settle down. Thinking about what was going on, Lorenz realised that the feedback built into the equations was the problem. It could take the most infinitesimal difference in starting values and amplify it within just a few cycles. Small differences did not stay small for long, but would fuel themselves to grow explosively.

This became known as the butterfly effect because it was like saying that the gentle fluttering of a butterfly's wings could be enough to tip the balance of a developing weather system and make the difference as to whether or not a hurricane eventually swept across a country on the far side of the planet. Rerun history without the butterfly and the gathering winds might have blown in another direction.

What startled Lorenz as a scientist was that the maths was both deterministic and utterly unpredictable. It was deterministic in that so long as the starting values were exactly the same, his program would always crank out precisely the same result. The system had no choice but to follow the equations. Yet the merest hint of a change in those values and immediately there was no telling where the simulation might go. The only way to find out was to let the program run – to wait and see.

This finding, which was fascinating in a mathematical model, became quite shattering once applied to the real world. Given that the real world is a continuous place, and so exact starting points can never be measured, this means that it is impossible – as a matter of principle – to predict the behaviour of a feedback-dependent system. Science had always known that the job was going to be tricky, but the assumption had been that more or better measurements would always solve the problem. By steadily increasing the number of samples, or the number of decimal places in each reading, the level of error would slowly, but surely, be squeezed out of the calculations. With a real-life weather forecast, for example, the readings from a thousand weather stations should be able to produce a much more accurate picture than the readings from a hundred. A million would be even better. But chaos theory said that the precision had to be infinite because feedback had the power to rise up and magnify even the smallest error in measurement, eventually wrecking the whole result.

This was dismal news for scientists wedded to a reductionist view of the world. It destroyed the belief that if you knew all the rules governing a system, you could then predict its future. Chaos theory said you could know the laws and still not predict. Of course, there were some systems so simple – such as those ruled by point- or limit-cycle attractors – where an outcome (if not the actual sequence of events) could be safely forecast. There might be a lot of turbulence in the water draining out of a bath, but in the end, gravity tolerates only one final ending. However, for a large class of natural phenomena, there would be an infinity of possible points of balance. The maths driving their progress could be completely lawful, but only observation would tell which of many possible fates lay in store. Excruciatingly, the future could be both preordained and unknowable.

Fortunately, chaos had an important saving grace. While the path of any particular system could not be predicted, outcomes had a tendency to group. Certain kinds of outcome would be far more likely than others. For example, in Lorenz's models of a weather system, there were patterns of pressure and temperature that were so exceptional they were unlikely ever to crop up. These would have been the equivalent of real-life climate extremes such as snow at the equator or thousand-miles-per-hour winds. On the other hand, there were conditions that the simulation would frequently pass through – the model's equivalent to monsoon rains in Asia or drought in the Sahara. A butterfly might be able to blow a storm off course, but some futures always remained more probable than others. Once a reasonable sample of possible outcomes for such a system were graphed to create a phase space portrait, this difference became easy to see. The graph would be densely packed with points in some areas, while others stayed almost empty. A truly random system would be equally likely to visit every point in the space of all possible outcomes. But a chaotic system would have some kind of attractor – a region it preferred to inhabit. And the more tightly packed the outcomes making up the attractor, the more it became possible to predict something about the future of the system.

So the concept of the attractor went some way to salvaging the loss of certainty that came with chaos theory. Even more encouragingly, there was the promise that science might discover that many quite different systems actually shared the same kinds of attractors. There could be a family resemblance linking natural phenomena as diverse as weather systems, the turbulence of a river, and the firing of a neuron. A study of attractor mechanics might end up uniting many areas of science.

The first of these chaotic attractors – called 'strange' because of their complex shapes – was in fact the product of a much simpler system than Lorenz's weather models. An attractor has to be built up from hundreds of outcomes and Lorenz's vacuum-tube computer was not up to doing that for anything complex. Instead, Lorenz used the same equations to simulate the most pared-down system he could imagine: the rolling convection currents that form in a pan of water as it is heated. And even then, he only had the computer power to model a two-dimensional slice through the system, a wheel of water rising in one direction as it was heated and tumbling back down the other as it cooled.

The wheels showed chaotic behaviour because every so often – and unpredictably – the direction of the flow would reverse itself. Small imbalances would build up to a level where they caused the wheel to start spinning in the opposite direction. Plotting the course taken by a single point in his simulation of such a system, Lorenz showed that it traced out a pair of connected whorls that looked not unlike a set of butterfly wings. In itself, the actual shape of the attractor was not that significant (although it chimed nicely with Lorenz's image of a butterfly causing a storm). One spiral followed the many possible locations of the spot as the convection current rolled it one way; the other tracked the spot as the direction of flow reversed. Any real-life chaotic system would produce a far more complexly shaped attractor than just a linked pair of loops. However, the picture was instantly memorable, because on closer inspection its simplicity quickly gave way to a sense of giddy infinity.

In places, the orbits were so clumped that they seemed to form a solid mass, but magnify a section and you would see it was made up of an immense number of strands. Magnify one of these strands and it, in turn, would prove to be made up of a vast number of smaller threads. You could zoom into a chaotic attractor as far as you liked and still you would keep finding orbits packed in alongside each other. This was visible proof of the point about initial conditions. Each orbit represented an individual history for a point – the path that the maths of the system would take it along. These orbits were both infinitely narrow and infinite in number. If you tried to stick a pin into the graph, you could never be sure which particular history you would land yourself on. You might think you were on one track, only to find yourself on a neighbouring history which suddenly veered off in a quite different direction. Each path might be mathematically preordained, but there was no practical way of clambering aboard any particular ride.

The implications of Lorenz's work were so shocking that it took quite a few years for the results to become widely accepted. It did not help that, as a meteorologist, he was publishing his papers in obscure titles such as the *Journal of the Atmospheric Sciences*. But by the early 1980s, a revolution was under way as scientists began to realise how much chaos theory might be able to explain. For example, fractal geometry, a derivative of chaos theory, suggested that much of the intricate structure of the natural world could be created using some very simple feedback rules. To the reductionist mind, the development of something like a snowflake or a coastline seemed too haphazard to be lawful. Where the next spur of ice would start to grow, or the next bit of rock fall off, was largely a matter of chance and so it seemed impossible to predict anything about the final shape. However, like raindrops splattering on a pavement, random events can end up creating statistically regular patterns.

When a storm first breaks, the rain hits the ground unevenly. Some bits of the pavement look almost as if they are attracting drops while other areas will stay dry for a surprisingly long time. This is because randomness does not mean that the rain will spread itself out, trying to avoid any spots where it has already fallen. The likelihood of hitting a dry or already wet area of pavement remains the same, so the pattern will build in a patchy way. Chance itself demands that some bits of pavement will get more than their fair share of drops while others get less. Fractal geometry described the properties of such patterns. For instance, it said that they were self-similar. In an echo of Lorenz's chaotic attractor, the ratio of wet spots to dry spots should look the same on every scale of magnification. Zoom in on a relatively dry bit of ground and it would turn out to host its own patchwork of wet and dry areas. The same was true of an eroding coastline. With the forces of weathering acting evenly upon it, a coastline should end up looking equally craggy on every scale of magnification. Each bump of rock would play host to its own smaller set of bumps. So again, a randomly developing system could show both unpredictability and inevitability. There was no way of telling how a particular puddle or coastline might grow, but you could be absolutely confident about some of its eventual geometric properties.

Once their eyes had been opened, scientists began to see the hand of chaos in all kinds of natural phenomena. Biologists used chaos theory to explain everything from the growth of patterns on snail shells to the branching of the body's blood vessels. Physicists saw chaotic patterning in the shape of clouds or the melting of ice. Earth scientists found chaos in the frequency of earthquakes and the tributary patterns of river systems.

Not surprisingly, computer scientists and brain researchers were also inspired by chaos theory. Computer designers wondered whether they could harness a chaotic attractor to drive a new kind of neural network. A network might be able to represent its memories or programs as an attractor state distributed across the strength of its connections. So rather than following a rigid step-by-step summation of weights to produce an answer, the system would be like a Hebbian feedback network in which input would wander about a bit before it eventually fell into some basin of attraction. An attractor could be seen as a mathematically more elegant way of describing a processing landscape – the clumping of orbits in certain regions of phase space giving the system a topography of bumps and hollows.

Neuroscientists saw the same link. One researcher, Walter Freeman of the University of California-Berkeley, even claimed to find attractor behaviour in the olfactory bulbs of rabbits. Using electrodes to record from the odour-processing centre of a rabbit's brain as the animal sniffed different chemicals through a mask, Freeman's results seemed to show that the olfactory bulb acted as a dynamic landscape. Each odour produced its own characteristic pattern of bumps and hollows in the electrical activity. So chaos theory appeared to fit very neatly with a Hebbian view, giving science a much more natural way of thinking about the business of information processing in the brain. What was even better was that chaos theory seemed capable of explaining both form and behaviour. For example, the branching of dendrites and axons was bound to be ruled by the same kind of fractal principles as those that governed the growth of blood vessels or river tributaries. So chaos theory could account for both the look and the activity of the brain.

Yet barely had the exciting potential of chaos theory begun to sink in when researchers were hit by something else. At the outset of the 1990s, chaos theory transmuted into the science of complexity, a school of thought that said chaos was just the first step on a road towards something much bigger.

The distinction between chaos and complexity can seem hazy at times, but, essentially, chaos theory describes how a simple, repetitive interaction, left alone to rub along, can produce something of rich structure. It is about the feedback-driven generation of complication.

Genuine complexity is something else, however. Shorelines, rain puddles and weather patterns have an intricate structure, but the really interesting things in life – systems like cells, economies, ecologies, and, of course, human minds – have extra properties such as an ability to adapt, to self-organise, to maintain some sort of coherence or internal integrity. These systems are not slaves to their maths, passively following a trajectory through phase space. Instead, they have developed some sort of memory or genetic mechanism which allows them to fine-tune the very feedback processes that drive them. They can change the attractor landscapes in which they dwell, and so reshape their own futures. A complex system is one that has harnessed chaos, rather than one that is merely produced by it.

In its most straightforward guise, complexity theory sounds no more than a restatement of classical Darwinian evolution, which is based on the simple statistical fact that what works has a tendency to outlast what doesn't. Given a range of systems or processes dependent on the same resource – whether it is plants competing for a patch of forest floor, or manufacturers chasing the same market niche – the best adapted will survive and the failures will fall by the wayside. And echoing chaos theory, over a number of generations even the smallest differences in fitness will tell. Apart from a competition, the only other ingredients needed by a Darwinian system are random changes – mutations – to seed the contest, and some sort of memory mechanism to preserve the changes that prove advantageous.

Evolution, then, boils down to a repeating cycle of random advance and non-random feedback. The parallels with the pattern-building equations of chaos theory are easy enough to see. A generation of a species is like one run through a non-linear calculation. The difference is that instead of the feedback always having the same value – an identical figure being plugged back into the calculation during every cycle – it is free to change. Genes give living systems a memory with which to save an endless variety of feedback values, so a slightly different, slightly better-adjusted balance of feedback can be applied during the development of each new generation. The equations are no longer steady-state, but adaptive. Memory makes the dynamism itself become something dynamic.

So far, complexity theory and Darwinian evolution seem to be saying much the same thing, just in different languages. However, complexity theory does have a twist. A key assumption of the traditional Darwinian story is that natural selection has to handcraft each little detail of a successful organism. Every freckle or dimple only exists thanks to the steady pressure of competition over many generations. But deterministic chaos brings with it a potential for a quite spontaneous and unselected eruption of order. The concept of the attractor means a lot of the organisation of life might require no crafting and come simply for free.

If a species of animal or plant were thought of as existing in a phase space – the space of all possible body forms and lifestyles – then Darwinian evolution argues that natural selection can push the organism to any corner of this space. Any outcome is equally plausible, so long as there is sufficient evolutionary pressure. A fish could grow feathers or an extra stomach if the circumstances were right. By the same token, natural selection is needed to make anything happen at all. A species cannot move to a particular spot in the landscape without being pushed every step of the way. Design is something that evolution imposes. However, a chaotic process has an inherent creative energy. The very fact that it can be described in terms of an underlying attractor means that a system has areas of phase space it is reluctant to traverse and, likewise, areas where it is more likely to be found. As with a weather map, there will be patterns which are almost inevitable, while others are near impossible. The result is that evolutionary change must be seen as a combination of the push of natural selection and the gravitational pull of chaos.

This is a big idea because it unburdens science of one of the most nagging problems about classical evolutionary theory. The odds against a great many important evolutionary events, from the development of a molecule as complex as DNA to the repeated development of eyes across many classes of animals, often appear rather astronomical. If life on Earth depends purely on chance to assemble the rich brew of organic molecules that were the precursors of DNA, then we would probably still be waiting for something to happen. On the other hand, if there is something inherently stable about DNA-like aggregations of molecules, then immediately the odds are slashed. They would need only the slightest nudge to evolve. So life would not have been a story of a competition between a vast range of equally improbable events, but of a competition within a much smaller pool of reasonably probable outcomes.

There is much more that could be said about chaos and complexity; they are huge subjects in their own right. But for mind science, the point is that their mathematics must shake the common conviction that computers and brains are fundamentally the same. Chaos theory says that being digital matters. There are consequences when decimal places get rounded off as small errors can soon develop into large differences. But much more importantly, there is a hidden energy in a feedback-driven, chaos-harnessing system. There is both the push of its competitions and the pull of its underlying dynamics – the places its attractors want it to go. So brains and computers might both process information, but as technology stands – even with the glamorous new field of artificial neural networks – they do it in a deeply different way. Quite simply, one has circuits that are alive, and the circuits of the other are dead.

The message of chaos hit home only slowly. It did not help that mind science had no direct involvement in the development of the new theories. All the ground work was done in other disciplines. Chaos theory grew out of fields like physics, chemistry and meteorology. With complexity, other fields like biology and economics provided the driving inspiration. However, gradually, mind scientists began to get the implications.

In the usual way of things, many of the early adopters of the dynamic view sounded rather idiosyncratic in their approach. They would invent unnecessary new jargon or come across as too fanatical about what they were doing. As a result, 1980s pioneers like Walter Freeman, Gerald Edelman of the Rockefeller Institute in New York, and Stephen Grossberg at Boston University were often met with a baffled, even openly hostile, reception. Their published work was widely read but also widely misunderstood. However, in the 1990s, a second wave of researchers – people like Karl Friston who were young enough to have grown up with chaos theory – could start bringing dynamism into the mainstream.

The problem for mind science was that it needed to develop its own special brand of dynamism as it was all too easy to apply the insights of chaos and complexity in an overly simplistic way. A few enthusiasts talked about complexity almost as a mystical force. They saw consciousness as a property that just popped out of a system once it had passed some critical threshold of information density. So the small nervous systems of starfish and worms could only support reflexes; the rather more complex nervous systems of snakes and rabbits could support sensation and learning. Then along came *Homo sapiens* with a big brain and billions of neural connections. It was the sheer scale of this connectedness that took us across an invisible threshold and into full consciousness. The evolution of the self-aware mind was not about the steady accumulation of many small processing tricks – a gradual, classically Darwinian construction job. Instead, consciousness arrived with a bang. The brain reached critical mass and was then thrust into awareness by some kind of spontaneous, attractor-driven reorganisation in the way it handled information.

But there is much more to the evolution of the brain and mind than happy accident. Complexity is about both push and pull – the tailoring hand of natural selection and the form-creating power of chaos. So consciousness cannot be understood unless some thought is given as to what purpose it serves in the economy of the brain. It must have been designed to do something, and this something must be reflected in the processing structure of the brain.

Even then, complexity cannot be the whole story. The brain still has its digital-like side. There is no escaping the all-or-nothing nature of cell firing, or the precision with which neurons make their connections. It could also be argued that some of the higher level properties of the brain, especially the ability to attend, have a kind of binary exclusivity. We can make the choice to focus our awareness on one object, activity, or event, and not another. Human brains can also – with perhaps a bit of a struggle – think logically. We can reason in a sequential, linear fashion which appears not unlike a computer program.

So, in the search for some sort of intellectual bedrock from which to launch the exploration of consciousness, psychology and neuroscience seem to need both computer science and dynamics. By themselves, neither is enough. Instead, a blend has to be found. Both ways of looking at information processing will have to go into the mix – and the way they eventually combine might surprise everyone.

CHAPTER FOUR

The Hunt for the Neural Code

In 1985, the journal *Science* carried an extremely disconcerting article. Robert Desimone, a hot young researcher recruited to set up a neurophysiology lab at the US National Institute of Mental Health near Washington DC, had been recording the responses of single neurons in the brain of a rhesus monkey. Desimone had inserted a hair-fine electrode into an area known as V4, a small patch of cortex believed to be the brain's colour centre – the place where raw wavelength information is turned into a conscious sensation of hue. Earlier experiments had shown that there were cells in V4 that responded to shape, surface texture and various other visual properties, but the great majority were tuned to seeing colours. These colour cells were also highly organised. Each could only be induced to fire by a particular shade appearing in a particular place in the visual field. In other words, they lined up to form a topographical map. A mix of cells covered each point of the 180-degree panorama taken in by the eyes and fired to tell the rest of the brain what proportion of yellow, red, or other coloured light was arriving at that spot.

Just knowing that brains had a centre for colour was a remarkable fact. Single cell recordings had already shown that the brain made a physical mapping of the world on V1, the primary visual cortex – a discovery later confirmed in more spectacular fashion by Tootell's autoradiography experiments. Now it seemed that there were higher level areas that fed off this first mapping to produce the actual qualitative experience of seeing. There was a hierarchy of processing in which V1 acted as the assembly point for incoming visual information and then ranks of specialist filters extracted a more concentrated representation of actual properties such as a sense of movement, coherent outline, stereoscopic depth, and, of course, colour.

A story seemed to be coming together. The firing of a brain cell was like a twinkling point of light, or a pixel on a computer screen. The all-or-nothing flare of a neuron created a data co-ordinate within a vast web of mapping. It stated unequivocally that a fragment of sensation existed at some locus of space. As one level of mapping fed the next, the pictures would become richer, brighter, sharper – conscious. It was a most attractive and widely believed theory. Then, in a paper entitled 'Selective attention gates visual processing in the extrastriate cortex', Desimone and a co-worker, Jeffrey Moran, had to go and spoil it.

Desimone was unusually placed for several reasons. First he had a background in psychology, whereas most of his fellow researchers were physiologists, more interested in the biological detail than the broad computational principles of the brain's neural machinery. Furthermore, he had chosen attention as his particular subject of expertise. At a time when to show a naked ambition to study consciousness was treated as flaky, the process of attention – dealing as it does in degrees of awareness – was just about the next best thing.

But above all, Desimone had the opportunity to do a different kind of experiment. The great majority of single-cell research was done on anaesthetised animals for the simple reason that an animal's eyes had to be kept still so a test stimulus could be projected to exactly the same spot on the retina throughout the many hours of recording. Otherwise, every time the animal shifted its eyes a different arrangement of neurons would be stimulated in the brain. But for Desimone to study the process of focused attention, he needed the animals to be wide awake during the recording session. This meant that on top of the months needed to teach a monkey an experimental task, he would also have to train them to fixate their eyes on a single point for a quite unnatural length of time. Nearly a year would be needed just to prepare the monkeys, and even then magnetic coils would have to be surgically implanted around their eyeballs so that any unwanted movement could be picked up, and recording aborted. Just in terms of effort and cost, working with a conscious subject was a huge undertaking and Desimone admits that only the National Institute of Mental Health could have given him the necessary backing: 'I wouldn't have liked to try to justify such a switch in techniques in a grant proposal to a university,' he says.

Having got the go-ahead, the question Desimone wanted to ask was very simple. He wanted to see what difference there was in the firing of a brain cell when the feature it represented fell outside the current focus of attention. Would a sensory neuron behave in a hardwired way, always sending out the same message regardless of the circumstances? Or would the strength of its signal change in some fashion?

The key to the experiment was to design a task in which the brain always saw the same thing. The actual sensory input did not alter during a trial; all that changed was how the monkey was supposed to view the data. Desimone set up a matching test in which a monkey had to watch a computer screen for coloured bars. A trial would start with the appearance of a target – say, a red bar – and then half a second later, a second bar would flash up in the same spot. If the two colours matched, the monkey had to signal this by releasing its grip on a paddle. To make the task a true test of concentration, there was always a second set of bars flashing alongside as a distraction. And the location of the target pair was switched back and forth every eight or sixteen trials so that the monkey really had to think about which spot it should be attending. With this arrangement, any V4 neuron would be sure of receiving the same pattern of stimulation – the constant sight of two sets of bars always flashing up side by side on the screen – but at regular intervals, the attentional demands would change dramatically.

Each recording session with a neuron had to begin with an outlining of its basic receptive field. The monkey would simply stare at the screen while Desimone presented bars in various combinations of colour and location to find out what kind of stimulus would make it fire in the first place. The results turned out to be much the same as recordings from an anaesthetised animal. The majority of V4 cells appeared tuned to reporting a single colour. They would fire flat out when exposed to a certain wavelength – say 640 nanometres, which happens to be a cheerful tomato red – and to either side of this peak, the response would fall away rapidly.

For example, while tomato red might provoke several hundred

spikes a second from a cell, a wavelength of 620 nanometres (an orangey, even yellowish hue) or 660 nanometres (a dark crimson) would be met with only a faint rustle of activity. The colour response took the shape of a narrow bell curve. Each V4 cell also coded for one small corner of the visual field. A tomato red cell would fire loudly whenever its chosen colour appeared within the two- or three-degree span of its receptive territory, but the response would die sharply as soon as a coloured bar moved outside this boundary.

Again, the pixel-like precision of the neural responses was tremendously impressive. Brain cells seemed to behave as feature detectors that lit up as soon as they saw their kind of thing. A tomato red cell would fire to tell the brain a spot of colour was occupying a certain place in the visual field. Conversely, when the cell was silent, this could be taken as an equally clear message that no stimulus was present. As the very term 'receptive field' suggested, there was a binary crispness to a neuron's response. A brain cell either witnessed a sensory event or it didn't. And the fact that a cell would respond even when an animal was comatose simply confirmed the idea that neurons fired with fixed meaning. The logic was wired-in, so nothing changed just because the machine was switched off.

Of course, this view of a receptive field was never actually true. For a start, a tomato red cell did fire weakly to colours on the fringes of its range – and what was it saying when it was willing to blast off the same few dozen spikes in response to two such distinct hues as orange and crimson?

Then there was the too easy assumption that the firing rate was what carried the message. The response curve thrown up by single-cell experiments suggested that the more spikes a cell put out each second, the more it was saying it was sure it was seeing its kind of thing. So all a scientist had to do was record a cell's average response for the duration of a stimulus. This was a reasonable interpretation only so long as – like the 0s and 1s flowing through the circuits of a computer – each neural spike carried equal weight. But if individual spikes varied in their impact, then just being a noisy cell might not carry such clear meaning. Coding for sensation would be a much murkier business than the concept of a receptive field implied.

On top of this, there were some troublesome biases built into the way neuroscientists approached single-cell recording experiments. Amplifying the faint pop of a depolarising neuron was never an easy task, so researchers tended to go in search of juicy responders – cells with a loud and stable response – passing by the many weaker or less stable cells along the way. Desimone himself says that when he first started single-cell recording, it always bothered him that cells could often be quite elusive in their behaviour. One minute they would show a textbook clean response, but the next, their receptive field might go a little fuzzy. In such cases, it was very tempting for a researcher to write off the cell and move on to one of the tens of thousands of other neurons populating the same square millimetre of brain tissue. If the responses of brain cells seemed fixed, it was at least partly because neuroscientists reported only the most well-behaved neurons.

However, in the end, researchers felt that what counted was aspiration. What was it that neurons, made of weak flesh and blood, were struggling to achieve? The receptive field concept held if a cell's peak firing rate did a reasonable job of signalling the presence of a sensory feature, and the evidence was that, generally speaking, it did. So Desimone went ahead and established the baseline receptive field of a V4 neuron in conventional fashion, then ran the matching task to discover if the directing of attention would have any effect on the responses of individual cells. The result was immediate. The two sets of bars – the target and the distractor – were placed close enough so that both would fall within the few degrees of a single cell's receptive field. Desimone found that if the target was red, and the cell coded for red, then there was no problem. It simply fired as normal to report the presence of a colour appearing at a place. But if the target bar was some other colour like green, then the cell shut up, or at least fired much more weakly than usual, even though it was seeing the red of the distractor bar.

The hardwired view of the brain said neurons were simple feature detectors; red light anywhere in a red-coding cell's receptive field should make it fire. There were enough green-coding cells in V4, so let them worry about the need to report the presence of a green target. But Desimone's neurons were acting as if they knew the brain wanted to focus on the green bar, and obligingly dropped their voices to a whisper. Somehow they became aware of both the green and red within their receptive fields and fluidly adapted their output to meet the demands of the moment.

The result was disconcerting, but only because in the mid-1980s, when Desimone carried out the experiment, the digital computer so utterly dominated people's thinking about how neurons should operate. Computers work by being discrete and modular – by breaking things down and packaging them up so that they run in splendid logical isolation. Even the most well-made computer is never physically perfect. A silicon chip cannot help but have some wires a few atoms fatter than others. The best power supply cannot completely flatten out the occasional current fluctuation. But there is a robustness in a computer's coding mechanisms – in the way it packages and transmits information – that allows it to rise above the shortcomings of the real world and conduct its processing in a realm of abstract purity. Physically, its os and 1s cannot be absolutely square-edged, but its logic gates can be engineered always to treat them that way.

The assumption was that brains must somehow be working the same trick. Concealed in the burble and fizz of cellular electro-chemistry, there must be a neural code – some stable, noise-defeating mechanism – that was responsible for carrying the message. The trouble was that everywhere researchers looked, there was noise and instability in the signalling activity of the brain. Synapses were not reliable. Individual spikes were not reliable. And now Desimone appeared to be saying that even receptive fields and firing rates were an unreliable measure of neural meaning. A cell's response was not fixed, but could be swayed to fit the current state of consciousness.

Not that there was any great mystery about what might be happening with Desimone's neurons. The results were easily explained by feedback between cells. Like all cortex cells, each V4 cell was tied to its neighbours by hundreds, if not thousands, of feedback connections. All it would take was for the green-coding cells in an area to tell any nearby red-coding cells to pipe down. In a sense, the red-coding cells would still want to say they were witnessing the presence of the red bar,

but the importance of getting out a message about the green target –
especially with the two bars appearing so dangerously close together –
meant it was probably better that the red output was suppressed.

Even a strictly computational model of a neuron as an integrate
and fire device would have predicted this kind of result. The story
was that cells fired to represent the sum of their inputs. But if each
cell had numerous feedback connections as well as straight sensory
input connections, then it should be obvious that the total pattern of
input would look very different when a brain was relaxed, with no
particular purpose in mind, to when it had been primed to achieve
some goal. Neuroscientists who had recorded from comatose animals
had by definition been measuring the response of neurons with little
internal guiding feedback. If receptive fields had seemed hardwired, it
was because the only thing driving the cells to fire was a raw stream of
sensory signals from the eyes. But as soon as a monkey was awake
and able to think about what it was doing, all sorts of competitive
cross-currents could begin to flow through its brain circuits. Each V4
cell would be hit by a much richer, more contingent pattern of input.
So the firing of a neuron would still represent the sum of its synaptic
activity, but it would be a view of sensory data seen through a filter of
the brain's wider attentional demands.

Equally importantly, the response of the brain to the cell's output
would probably also be changed. In a comatose monkey, a cell's
firing might seem reassuringly fixed, but was anyone really listening?
A cell could be driven to produce a signal, but plainly it was a signal
feeding into empty oblivion. With Desimone's V4 cells, however, the
signals might be fluid, but at least it was certain that the output
actually counted for something – that the signals were becoming
woven into a living tapestry of activity. Rigidity creates information
in a computer, but the opposite might be true in a system that relies
on dynamism and competition to power its processing.

From a different perspective, Desimone's result seems instantly
reasonable. Yet throughout the 1980s, there was such a strong
conviction that the brain needed a stable neural code to be able to do
its job that the inclination was to play down the result. There were
many ways of avoiding seeing a problem. After all, Desimone was

simply saying that the receptive field behaviour of a cell could be modified by an attentional state, not that it was changed out of all recognition. A V4 cell still coded for the existence of a colour at a place. And Desimone reported only that firing was diminished rather than abolished. In cells being 'gated', firing was typically reduced by about two-thirds, but the cells still fired.

Then there was the belief that attention would only affect firing at higher levels of the visual hierarchy. It was presumed that down in V1, the first proper stopping point for incoming information, neurons would be absolutely rock-solid in their response. They would report what the eyes saw – the raw physical reality – rather than the analysed, filtered, and focused view that became the psychological experience. Higher levels of firing might be fluid, but the foundation would be firm.

So nothing was derailed. Desimone's experiment showed that the receptive field of a cell could no longer be treated as hardwired. Attention could change a cell's output. But for most researchers, this just meant that the neural code – the rules by which neurons sent and received information – must exist at some more abstract level of mechanism. One suggestion was that perhaps each brain area used a special class of output cells to transmit its true message. Desimone might have been recording from rank and file V4 cells and missed a set of 'final layer' neurons that stood for what the colour filter thought it was seeing. Or it could be synchrony of firing that held the answer: only when cells were in beat did their spikes begin to carry weight. Theorists could think of many ways in which the brain might be extracting a fixed message from cells with plastic, context-dependent responses.

It was a confusing situation, but the very existence of stable conscious states seemed to say that there must be a reliable coding mechanism at the heart of it all. Subjectively, red always ended up looking red; a rose always smelt like a rose. So the brain must have a robust method for translating a pattern of raw sensory energy into a given mental response. There had to be something about an individual cell's firing that corresponded with the eventual conscious state. The business of neuroscience was to crack that code.

And the truth was that, for practical purposes, the computational approach worked. For a way of thinking about the brain which must ultimately be flawed, most of the time the computational view seemed to be getting things more right than wrong. From the 1950s, when single-cell recording first began, to the early 1990s, the idea of a digital-like code did manage to tell the story. And the irony was that it was an overly fluid, overly dynamic theory of how the brain might work which the computational view originally arrived to replace.

The modern era of neuroscience began in 1958 with a famous bit of luck. David Hubel and Torsten Wiesel of the Harvard Medical School were trying to get a response from the visual cortex of an anaesthetised cat by shining spots of light in its eyes. Like several rival research teams, Hubel and Wiesel had found that nothing seemed able to make the V1 neurons fire. Even the brightest light did not provoke an obvious reaction. Then one day, as another slide was being slipped into the projector after fruitless hours of tests with stimuli of all shapes and intensities, a cell suddenly went crazy. There was a loudspeaker hooked up to the electrode, and through it came a crackle of spikes like machine-gun fire. However, the cell was responding not to the black dot at the centre of the slide but to one of its edges. As Hubel and Wiesel tried to recreate their accidental triggering of the cell, they found it was so fussy that it would only fire when the slide was angled at about eleven o'clock and moving from lower left to upper right through one small corner of the visual field.

It was a finding that confounded everyone. Neuroscientists of the day were split between those who had expected the V1 neurons to behave like simple light detectors, firing to signal the presence or absence of illumination at a point in space, and a group of early dynamicists who believed that brain cells probably did not have any fixed coding responsibilities at all. Those in the dynamic camp felt that the wiring of the brain simply acted as a stage for the open interplay of information patterns. Neurons were general-purpose units with the necessary properties to support eddying firestorms of activity – Sherrington's enchanted loom of fleeting, never-abiding patterns – but lacking in individual meaning. The firing profile of a cell would be completely fluid, reflecting whatever pattern of activity happened to be sweeping through an area of circuitry at a moment. But what neither side was expecting was for a neuron to be so tied to such a carefully defined sensory feature – the sight of an edge, at an angle, at a place, and only a moving line at that.

The success of Hubel and Wiesel's experiment – a Nobel prize-winning achievement – opened the floodgates. Single-cell recording quickly became the principle tool of neuroscience. There were other ways of investigating the brain, of course. Neuroanatomists continued to stain and dissect; other researchers did lesioning experiments. They would sever nerves or cut out small chunks of brain to see what damage this did to an animal's intellectual abilities. But single-cell work seemed to offer the ultimate in precision. By measuring the input–output activity of individual neurons, researchers could discover exactly what a bit of the brain was doing.

As more V1 neurons were tested, it was found that almost all were edge-coding cells. They saw the world in snippets of line, but they did this in a variety of ways. Some responded only to bright lines on a dark background, some to dark lines on a bright background, and others only to the boundary between an area of light and dark. The cells also varied in their preferences for movement, some responding only to lines travelling in a particular direction while others would fire happily even when a line was not moving.

So the primary visual cortex had the machinery for a rich representation of line. The next big finding was that the cells were drawn up to form an orderly map. Recordings showed that each neuron shifted its preferred angle of orientation a few degrees from its nearest neighbours, so that an eleven o'clock cell would sit neatly between a ten o'clock and a twelve o'clock cell. With thousands of neurons to code for the same point in visual space, there would always be the right kind of cell, or combination of cells, to specify exactly what the eyes were seeing.

A worry was that V1 did not seem to have a way of handling colour. Only rarely did a line-coding cell also show some kind of preference for wavelength. The puzzle was eventually solved in 1981 when Hubel checked more closely and found that there were small clumps of specialist colour-coding cells studding the V1 at regular intervals. The cells did not have a line preference but responded to the wavelength intensities at each point in the visual field. These islands of colour information bobbing in an ocean of line processing had simply been missed by earlier electrode probes.

It was all very tidy and regular. An array of a billion or so neurons would encode the play of shape and shade that made up the visual scene. As researchers turned to the other senses, it was found that they each had their own topographically organised mapping. Touch was represented on S1, the primary somatosensory cortex, a strip of grey matter running down the side of the brain towards the ears. This map was organised in a crude outline of the body with the feet at the top of the strip and the head at the bottom. Like V1, the map was distorted so that parts of the body with more sensory receptors, like the lips and fingertips, took up proportionately more neural surface. There were also a variety of cell types so that some coded for sensations like tickles while others stood for jabs or deep pressure. The sense of hearing turned out to be mapped to a patch of brain, the primary auditory cortex or A1, behind the ears. It too was topographic – or tonotopic, to be more accurate. Frequencies were separated so that low-pitched tones lay at one end and high-pitched noises at the other – an organisation that directly reflected the way different wavelengths of sound were filtered out inside the tapering coil of the inner ear.

The arrangement for mapping olfaction – the senses of smell and taste – was a little more difficult to make out. Partly the problem was that the nose and tongue do not worry too much about the 'where' of a sensory event. The receptors are not organised to capture a spatial picture in the same way as the surface of the skin or the back of the eye-ball. Instead, the focus is on the 'what'. The nose alone uses about 10,000 types of receptor cell, each binding to a differently shaped molecule, to capture the many possible kinds of chemical in an odour. Each of these receptor families feeds into its own corner of the olfactory bulb to create a mapping that appears to be organised by dimensions of sensory quality. Scents are laid out along an axis from pungent to flowery, or fresh to rotten, rather than having the more obvious spatial co-ordinates of the mappings for sight, hearing, and touch.

Taste is a still more complicated sense because, as anyone with a blocked nose will know, much of the flavour of food actually comes from odours released during chewing. We smell our food as much as taste it. The experience of taste is also partly tactile. Taste is linked to qualities of mouth feel, such as the texture and temperature of food, whether it is crunchy or melting, greasy or acid. Given the way taste borrows from these two other senses, it is not surprising that it actually has two mapping areas in the cortex: one near the olfactory centres at the front of the brain, and the other alongside the somatosensory strip's representation of the mouth and tongue.

The cortex also turned out to have motor maps – maps of brain output – to match the many sensory input mappings. The most important seemed to be M_1, the primary motor map, which represented the parts of the body down a strip of cortex that ran parallel to the somatosensory cortex. A jolt from a stimulating electrode to an area of this map could cause the muscles in the relevant part of the body to twitch. But there were other output areas, including one near the front of the brain known as the frontal eye field (FEF), just for the control of the eyes. While the primary motor map was organised in the shape of the body – with the usual distortion to devote extra neurons to areas like the fingers or lips which needed to be more sensitively controlled – the eye field was like a visual mapping with the cells arranged so that each represented a co-ordinate in the visual scene. The firing of a cell would mark a position where a monkey intended to turn its eyes.

The detail of how all these sensory and motor areas represented information was a lot more complicated, of course. But it was the general picture that counted. Single-cell experiments seemed to show that neurons had sharply tuned receptive or projective fields; when a cell fired, it stood for something. These neurons then lined up to form topographically representative maps. The brain's circuitry was organised with a clear physical logic.

Again it was true that there was a greyness in the firing of every neuron. Even the chance V_1 cell that first started the whole single-cell recording industry was never exactly binary in its response. It would fire flat out to a line slanted at eleven o'clock, but it also fired weakly to angles of ten or twelve o'clock. It was most enthusiastic when an edge stretched the full width of its receptive field, but would still fire gently to a line that stopped a little short. And while it liked movement in a certain direction, it would fire at least a few spikes to lines moving along a reasonably similar trajectory. However, despite these quibbles, the message of Hubel and Wiesel's work was that every cell had its peak response point and these peaks lined up to make maps. As Tootell eventually showed with his dartboard experiment, and Kosslyn with the completely imaginary vision of a letter, you could even take snapshots of these firing patterns. The brain was not an unpatterned grid of connections as the dynamicists would have it. Instead, individual neurons stood for individual bits of information. Perhaps the slight greyness in a neuron's response meant that so strictly a computational approach was not perfect. But certainly it seemed more right than wrong.

Single-cell recordings continued to deliver revelations about the cortex's tidy organisation. The primary sensory and motor maps turned out to be merely the tip of the iceberg. In fact, the whole cortex was a hierarchy of mapping, with lower-level maps feeding into higher-level ones so that areas like V4 could represent the world in more complex and well-digested ways.

A few dynamicists clung to the idea that while the primary mapping areas might have a fixed, display-like organisation, the rest of the cortex was probably a free-form sheet of cells. Inputs and outputs would enter and leave through staging areas, but inside the cortex proper there would be a wild swirl of thoughts, memories, and associations. There was some support for this view because, under the microscope, the cortex certainly seemed a rather uniform structure. The human cortex is a thin rind of grey matter only a few millimetres thick, but once all its deep wrinkles have been unfolded, its size is gigantic: over a foot and a half square, or the spread of a large dishcloth. It holds over thirty billion neurons, and ten times that number of support cells to nourish and protect its structure.

Staining studies showed that the cortex had a highly regular six-layer design with each layer probably playing a different processing role. By tracing connections into and out of the cortex, neuroanatomists found that most of the raw input came into one of the middle layers, layer four. There was a special class of cells in this layer, the stellate neurons, which had broadly branched dendrites and stumpy axons that seemed tailored to the reception of new information. Layers two and three of the cortex were where most of the computation appeared to take place – where its maps actually developed. The top- and bottom-most layers, one and six, seemed to carry mostly feedback traffic, layer one dealing in cross-cortex messages and layer six bouncing signals off some of the brain's subcortical centres. And then output appeared to leave an area of cortex from layer five, heading either onwards to the next stage of processing or exiting the cortex as finished motor commands.

The cortex sheet seemed so homogeneous in its design that it was easy to believe it was just one enormous display surface. However, more careful measurement of the proportion of certain cell types and the thickness of the layers suggested that it might be broken into a mosaic of areas. Even in the 1890s, the German anatomist Korbinian Brodmann had identified more than fifty subdivisions of the cortex based on microscopic differences of structure.

Single-cell recordings soon answered the question about who was right. The cortex proved to be even more divided than Brodmann suggested, forming a patchwork of perhaps several hundred maps and maplets. Furthermore, these maps formed an ascending ladder of processing so that the raw sensory pictures at the bottom developed by stages into something much more abstract towards the top. Like a series of computational modules or program steps, they extracted meaning from each moment's wash of input.

The visual pathway was the best studied. Like the base of a pyramid, V1 provided a broad foundation of detail. It had the full picture, but did little analysis. However, stacked above V1 were a whole range of feature-extracting and meaning-creating filters. The first rung in this hierarchy was V2, an area that seemed to firm up the V1 picture in certain ways. The cells in V2 had a more powerful reaction to boundaries and wavelength changes, as if they were emphasising the principle shapes and blocks of colour present in the visual scene. Then both V1 and V2 fed into a set of maps that seemed to fine-focus the representation of various visual qualities such as motion, surface texture, a sense of depth, and, of course, with V4, colour.

The reason for calling V4 the brain's colour centre was that neurons lower down the visual chain responded to wavelength, not hue. The eye only has the equipment to see three colours in the first place. It has photoreceptors that trap light at the peak wavelengths of 420, 530, and 560 nanometres. The response of the retina is more like a sampling of the available spectrum of light through three slits. All the richness of subjective experience has to be put back in by the brain.

The process starts with the mapping of wavelength information in V1 and V2 using cells tuned to represent a red/green contrast or a blue/yellow contrast. So, for example, a green-coding cell will be switched on by predominantly green wavelength light hitting an area, and turned off if the mix of the three wavelengths is mainly red. But the response of the cell is too unrefined to say that it really sees a colour as such. It is only in V4 that the firing of cells begins to match the subjective experience of witnessing individual hues. Each V4 cell appears to combine the response of many 'red', 'green', 'yellow', and 'blue' neurons, and to mix it up with some general information about illumination levels – the amount of 'black' or 'white' in the picture – to produce the complex reaction of seeing an actual shade such as turquoise, brown, or pink. So what started out as a crude physical measurement – the wavelength response of a retinal pigment – could be turned by a chain of remapping and condensing into a highly specific psychological response.

The same principle applied to the other visual qualities. Down in V1, information would be represented in undigested but detailed form. Then, by combining the output of a selection of neurons, higher-level cells would begin to show responses that more closely matched the conscious experience of seeing. With motion, for example, cells in V1 simply fired whenever an edge with the right kind of slant and trajectory moved through their point in the visual field. At a slightly higher level of mapping, motion-sensitive cells became more picky, firing only if the movement reached a certain speed or if the edge was part of a larger shape. At still higher levels, there were cells that fired only to particular kinds of motion, such as the spiralling of a disk or the expanding image of a fast-approaching object.

After several decades of painstaking single-cell recordings, backed up by neuroanatomical tracing studies, eventually more than thirty visual mapping areas were identified. At the top of this hierarchy, cells appeared to code for actual objects – or at least components of objects. The workings of one of these areas, the inferotemporal (IT) cortex, was only unlocked by another famous experimental fluke. Like Hubel and Wiesel, a group of researchers had been trying every kind of stimulus to make an IT cell fire. Then, one of the scientists caught his hand in the projector beam and the cell went wild. It seemed that the monkey's brain had a neuron whose sole job was to signal the presence of hand shapes within the visual field.

Single-cell work also showed that these high-level neurons displayed interesting long-term behaviour. A neuron triggered by the sight of a particular event, like a hand, would often continue to fire for seconds, minutes, or even hours after the object had disappeared from view. Down in V1 or V2, turning your head away from looking at a red light or a wallpaper pattern would wipe the slate clean. The cells awakened by the sight would cease firing within a tenth of a second – something that would have to happen if they were to map the sights of wherever the eyes alighted next. But in high-level areas, cells kept going, as if standing for a memory trace of the event. These neurons were even found to stir in anticipation of an event. Researchers discovered that if they cued a monkey to expect a flash of light in one corner of its visual field, then the corresponding cells would begin to fire ahead of time. Or, in plain language, the monkey's brain was having thoughts about what it was about to see.

Moving across the cortex sheet, the other senses had their own processing hierarchies. In the vision-dominated primate brain, none of these employed quite so many levels of mapping. But the same principle of gradual feature extraction and meaning condensation applied. And even the motor areas had hierarchies – except, being output maps, they had the reverse logic. Activity started at the top of the pyramid in high-level maps which somehow represented an abstract urge for action. This then turned into a specific pattern of muscle commands as processing flowed down the chain.

So, by the mid-1980s single-cell recordings seemed to be telling a straightforward story. The brain represented the world through a pixel-like display of data points. At the lowest level, the mappings were heavy in detail but light on meaning. It seemed likely that the brain's activity at this level would be unconscious, or at least so fleeting and unanalysed that it left no lingering sense of having been experienced. The flickering of a V1 cell alone could not produce awareness. But as activity rippled up the processing hierarchy, each new mapping brought about an intensification of information. Specific sensory qualities would harden. Then, right at the top, neurons would begin to stand for whole chunks of experience. The sensory display became stable, tagged, and, presumably, consciously understood.

What was especially appealing was that the same basic topographical computing mechanism could rule the entire cortex sheet. There was no need to postulate some remarkable advance to explain the human mind. A low-level sensory map and a high-level thought-representing map would both use exactly the same kind of processing tricks. All that would change was the abstraction of the data being handled – the number of computational stages the information had already passed through. This meant that to evolve from animals to humans it seemed necessary to add only a few extra levels of processing to the stack. There need not be anything deeply different about *Homo sapiens*. As a species, we merely sprouted a little more cortex and colonised it with a few more rungs of meaning-extraction.

The findings kept coming. It was not just the cortex that was so richly ordered. The machinery of receptive fields, topographical organisation, and hierarchies of processing seemed to apply to other parts of the brain like the cerebellum, the basal ganglia, and the thalamus. The old dynamicists' idea that the brain was a featureless grid of connections was well and truly vanquished. Every neuron had a job.

Every brain area was a link in a chain of computation. Yet the discovery of hierarchies of processing raised some new difficulties. An obvious question was, how did the brain manage to tie together a buzzing stack of neural maps to create a unified state of consciousness? What trick of coding ensured that all the many fragments of a representation got bound together, so that the colour and shape of a passing car was tied to its sense of motion, the noise of its engine, and even the smell of its exhaust?

The second problem was known as the 'grandmother cell conundrum'. With its hierarchical organisation, the brain seemed to work by producing an ever more specific response to a stimulus. Down on V1, neurons coded for a very small point of space – as little as a quarter of a degree for cells representing the sharp centre of the visual field – but they fired to absolutely any stimulus crossing that area with the right line or wavelength properties. An edge could belong to the leg of a chair or a girlfriend's face, it was all the same. But by combining the responses of many earlier cells, higher-level neurons began to react to the sight of whole objects. They were no longer bothered about where in the visual field an object might occur. The researchers who stumbled upon the hand-coding cell in the inferotemporal cortex found that it fired whenever a hand was within the line of sight. But the cell only fired to the sight of hands. It had become semantically specific. By extension, this suggested there might be a high-level cell to stand for every kind of experience, and the firing of that cell would somehow fill our consciousness with its presence. So every time we saw or thought about our grandmothers, somewhere in our brain a grandmother-coding cell would have to burst into action.

A little thought soon makes it clear that this degree of specialisation is unlikely. After all, neurons die at a steady rate, and if the grandmother cell hypothesis were true, then each death of a high-level cell ought to knock rather a large hole in our ability to represent life. More tellingly, taken to an extreme, the hypothesis suggested we would need separate cells to represent every possible kind of grandmother experience – for example, our grandmother seen from behind, as well as face on, or with a new hair-do, or as recognised from a youthful photo. Even with the billions of neurons available in our heads, we would soon be running out of cells.

The idea of one cell coding for one sensory feature also faced quite a different kind of difficulty. For, astonishingly, the brain does much better at detecting sensations than the receptive field properties of its neurons would seem to allow. In standard psychophysical tests, the human eye can pick out very tiny details. We are sensitive to the faintest bend in a straight line, or the slightest gap between two almost touching balls. It would seem that to be able to code for such a gap, we would need to use at least three retinal cells: one to represent either edge of the gap, and a third to respond to the space in the middle. Yet experiments show we can see a gap five times thinner than the shadow it casts across a single receptor on the back of our eyeballs. Our vision far out-performs the physical dimensions of our light-detecting equipment. This is all the more extraordinary when it is considered that back in the brain, each V1 cell has to combine the information of many dozens of retinal cells to produce its edge-representing response. So even with a narrow receptive field of a quarter of a degree, a V1 cell seems much too coarse-grained in its resolution to be able to code for the existence of hair-fine visual features. So how can the brain be capturing and representing such information?

Luckily, the same computational trick was able to solve both the grandmother cell conundrum and the puzzle over acuity. Indeed, the explanation even turned a general coarseness in neural coding into an advantage.

The pixel model of how neurons worked – one cell coding for one feature – suggested that a sharp awareness of the world must depend on a high-resolution mapping of sensory data. The response of each neuron would have to be tuned as tightly as possible if the resulting conscious display were to be crisply defined. So it was a troubling fact that the firing of each cell usually followed a bell curve distribution of some kind. A cell would have its peak response point, but it would also react slightly to wavelengths, directions of movement, or presumably if it were a grandmother-coding neuron, to grand-mother-like experiences, somewhat to either side. For the pixel model, this off-centre response was noise. It only muddied the picture. But then it was realised that brain cells might actually be using a clever group coding scheme. The overlapping responses of a population of coarse-coding cells might actually produce a far more precise representation of a piece of data than a solitary neuron struggling to express the same information on its own.

To see how this works, take the problem of representing colour again. A tomato red cell in V4 will fire flat out to a wavelength of 640 nanometres, but it will also fire weakly to the sight of something orange or crimson appearing within its two or three degrees of the visual field. At a conscious level, the difference between orange and crimson is immense. There is no chance of confusing the two. And yet not only does a tomato red cell react to both, but its firing profile would be identical. A neuron firing at a third of its peak rate could equally well be signalling the presence of a colour twenty nanometres lower as one twenty nanometres higher. To any other cell depending on its output, the message of such a neuron would seem a little ambiguous.

However, what is a drawback for an individual neuron can become a bonus for a group of cells if their output is overlapped. For example, say a red spot of light fell across the receptive fields of three V4 cells, each with a slightly different frequency response. One neuron might be provoked to fire at 80 per cent of its capacity, while another fired at 40 per cent, and the third at 10 per cent. None of the cells would in fact be claiming to be 100 per cent sure about anything. None would be a tomato red cell, as such, but would merely be reporting how close the spot happened to come to hitting their own personal peak frequency. However, the particular combination of firing rates could only stand for a single wavelength. Only a hue of 640 nanometres could provoke the mix of an 80, 40 and 10 per cent response within this small group of cells. From a population reaction, a pattern of voting by a spread of cells, would emerge an exceptionally precise trajectory of firing.

A population code would work just as well with representations of grandmothers, or the existence of a hair-fine gap between two balls. In a high-level area of the brain, a group of cells tuned to vaguely grandmotherly experiences would each vote on how much the current pattern of light falling on the retina resembled an image of your grandmother. Some cells might fire wildly at 100 per cent, while others, perhaps coding more for other kinds of experiences such as the sight of your grandfather, or just elderly-looking people generally, would show a more lukewarm reaction. The response would be variable, but overall, the combination of firing ratios should point unmistakably to one answer.

As well as offering great precision, a population voting mechanism would have the apparently contradictory benefits of being both robust and plastic. Robust because the death or even misfiring of a few neurons would not make much difference with performance depending on the combined response of thousands, perhaps millions, of cells. A population vote would also be inherently flexible because it could evolve to fit the situation. For instance, if we happened first to catch sight of our grandmother with her face turned away, there might be an initial uncertainty in our brain's response. Cells coding for grey hair and other grandmotherly attributes might fire enthusiastically, but cells associated with a more specific memory for her face would remain quiet. However, as she started to turn, the balance of firing would shift to a more generally agreed response. What might have started as a thought about the likelihood of a grandmother sighting would turn into a vote for certainty. So unlike the pixel model, population coding naturally allowed for shades of certainty and meaning. The same topographically organised surface could represent both vague inklings and definite impressions.

Like all good ideas, population coding had actually been bumping around the back pages of science journals for many years. Versions of the theory had been advanced under various names such as tensor networks and ensemble coding. But it was only in the 1980s that the desperate need for a way to rescue digital-like clarity from basically noisy-seeming components became apparent. The pixel model based on 100 per cent firing rates and ignoring off-centre responses was not working. So when neuroscientists realised that there was a mechanism that actually made a virtue of coarse-grain reactions, they seized upon it with gratitude. All that was needed was some proof that the brain might actually use the mechanism.

This proof came quickly from researchers studying motor mappings. Several teams of scientists looked at parts of the brain that were known to guide the movement of the hands or eyes to a particular spot in space. It was already known that the neurons in these maps fired to signal a trajectory. Each cell represented a command to move in a certain direction with a certain force. The question was whether single cells took sole responsibility for an action or whether the targeting was more flexibly and accurately achieved by a vote spread across a larger number of cells.

One problem was that neuroscientists were not even geared up to answer such a question. The very conviction that the responses of individual neurons were what counted had led to a self-reinforcing situation where neuroscientists only ever recorded from a single cell at a time. They simply had not developed the equipment to record from many neurons simultaneously. So the first population studies had to cheat a bit and record from a series of cells as an animal kept repeating the same action. A monkey would be trained to keep looking towards the same spot of the visual field, or reaching to touch the same target on a computer screen. It seemed safe to assume that the cells would behave much the same way on each trial, so a sequence of readings could be combined to create a population picture. Despite such fudging, a convincing example of population coding was reported in 1986. Recording from hand-controlling cells, it was shown that, individually, the neurons were rather approximate in their firing activity. But when the overall weight of firing was calculated, there was a tight match with the direction of the hand movement.

Unlike Desimone's discovery about receptive field plasticity in V4, population coding so clearly did something that people wanted – salvaging precision from noise – that it was swiftly embraced. And yet thought about carefully, it still undermined the computational view in quite a subtle way. A large part of the appeal of a topographical model of brain processing was that the business of representation seemed so tightly tied to the firing of an array of neurons. If a tomato red cell or a hand-flagging cell burst into life, then the chatter of spikes stood for the positive existence of that feature. But population coding said single cells could no longer be sure of anything. A neuron only voted on how closely the current spread of input matched its peak response point.

And then there was the even bigger problem of how these votes were counted. Unless you believed that neural firing alone was enough to produce a state of consciousness – the physical act of display taking on its own magical, self-appreciating glow of awareness – then the meaning of any particular cortex map could only become apparent in the impact it had on the next level of mapping. There had to be a reaction to the reaction. But the whole point about population coding was that this reaction would itself have to take the form of a vote smeared across many cells. There could be no single cell in V4 acting like a polling officer to decide what a group of several thousand V1 and V2 neurons were trying to say. The V4 response had to lie in a further pattern of firing by individually uncertain components. So representation might still depend upon what cells were doing, but with the brain only ever having votes about votes, it would somehow have to be tied to the way that the responses all hung together. It would be how a pattern of firing eventually settled across the entire hierarchy of mapping.

This made the idea of population coding a double-edged sword for computationists. It now meant that even the 'off-centre' responses of neurons could be treated as carrying sharp meaning. But this meaning was no longer directly represented in the response of other neurons. A reassuring sense of precision gained in one way was lost in another. But barely had neuroscience begun to absorb the implications of population votes than there was another big idea to digest

The discovery that the cortex was a stack of mapping had immediately raised the new problem of how all the fragments of representation got tied together. What tied the colours to the shapes to the movements, sounds, and smells? It was possible that no particular mechanism was needed. If firing happened at the same time in various corners of the brain, it would automatically be experienced as a cohesive whole. But there was good reason to think that binding must be an active process. For example, there were some well-known illusions where, with the rapid presentation of two objects, the colour of one could bleed into the other. The wrong hue would get bound to the wrong shape.

And then there seemed to be a need for some binding-type mechanism to distinguish events in the bright focus of consciousness from those on the dim periphery. Obviously, our minds show varying degrees of awareness for the many sensations and thoughts filling each moment. We have a centre of attention where mental events appear represented in some well-lit, high-contrast, maximum-comprehension sort of way. But there is also a shadowy fringe to awareness filled with vague inklings, stirrings, and presentiments. This background of activity might be dim, yet it would still have to be represented in a neural response.

The difference between the weakly conscious and the strongly conscious could lie just in the volume of firing. Cells speaking loudly might make a deeper impression. But it was becoming clear that raw firing rates could never be the whole story. Population coding now said that even partial responses should count for something. And Desimone's V4 results showed that attention could control firing rates; attention might make the rate as much as the rate made the state of attention. Whatever the story, some extra mechanism to help define 'in focus' activity seemed to be required.

In 1989, an experiment carried out by Wolf Singer and Charles Gray at the Max Planck Institute in Frankfurt suggested that synchrony of firing might hold the key; only cells spiking together at precisely the same moment would hang together in a bound and focused way. The pair were recording from neurons in a motion-mapping area of the brain of an anaesthetised cat as a bar was moved back and forth across a screen. The bar was designed so that it could be either moved as one piece, or split into two and the halves moved in opposite directions. They then recorded from pairs of neurons to see how their reactions compared when they were both coding for the same bar, and once the bar was separated. In terms of their raw firing rates, nothing changed. Each motion-sensing cell simply reported what was happening in its small corner of the visual field.

But whenever the cells happened to be coding for the same object – a bar long enough to stretch across both their receptive fields – they began to fire in time. Singer and Gray reported that the synchrony was not perfect – in the way of all neural responses, it was a little approximate with the occasional missed beat – but statistical analysis proved a rhythm to be there.

Other experiments followed. The synchrony was often hard to spot because it erupted so fleetingly and was always slightly irregular. It was also much more difficult to find any evidence of synchrony in awake animals or in the more advanced brains of monkeys. But gradually the findings mounted. Synchrony was reported in other brain areas such as the olfactory bulb, between cells on either side of the cerebral hemispheres, and even spanning levels of cortex mapping. The way that neurons were believed to react to input also supported the idea of a synchrony. The electro-chemistry of a cell's membrane made it likely that a cell would respond much more powerfully to two spikes hitting it at the same moment than the same two spikes coming a fraction apart.

The evidence was suggestive rather than conclusive. But, as with population coding, synchrony seemed just too good an idea for the brain not to be doing something of the sort. It meant that a cell would be free to fire as fast or as slow as it liked. A cell would simply respond to its inputs and produce a certain number of spikes as output. But whether the spikes had any impact would depend on their timing. A cell might have to be part of a network of cells all firing on the beat for its message to win through to the next stage of processing. So synchrony would both bind a pattern and ensure its transmission. At any moment, the brain has to deal with a huge flood of information. A person looking at Singer and Gray's display of a moving bar would also be seeing – and hearing, and feeling – many other things. Across the cortex, sensory cells might be offering up a vote about the pressure of a chair pressing against the legs, or the sight of the wall behind the computer screen. This information would be represented. It would be in the system. But only neurons whose votes came sufficiently on the beat might have the drive to send a wave of activity rippling all the way up the hierarchy of processing, especially to the highest levels where sensations appeared to become most focused and recognised.

Of course, there were many questions to be answered. Was synchrony an addition or an alternative to the kind of rate-controlling effects Desimone was reporting? Potentially, both seemed to be ways of deciding which cells counted most in forming a state of consciousness. And then, was synchrony imposed on cells or did it emerge through competition? It was possible that the brain had special cells or a central clock mechanism to drum out an entraining beat. Like the conductor of a neural orchestra, one bit of the brain might set the rhythm. But it was equally possible that synchrony was a side-effect of a sudden flurry of feedback messages passing between an active group of cells. If cells coding for one end of a bar started sending out signals to other cells doing the same thing, then their spikes might start to bunch up simply because any tendency to co-incident firing would be self-reinforcing. There would be a rhythm which might have the happy effect of giving the firing pattern more weight, but it would be an emergent phenomenon.

So synchrony created its own puzzles. However, neuroscientists had the feeling of steady progress. They were getting nearer to an understanding of the neural code – of what distinguished signal from noise in the brain's computations. Receptive fields, topographic maps, and processing hierarchies told the basic story; population voting and synchrony refined it. It was now even not too difficult to fit the existence of feedback connections or Desimone's attention effects into the picture.

By the 1990s, neuroscience was coming to realise just how dominated the brain was by feedback connections. Even the stellate cells in layer four of the primary visual cortex – the input cells to an input area – were connected more to each other than the outside world. Only a fifth of a stellate cell's 2,000 or so inputs actually came from the retina. And this was certainly not unusual. It had become a rule of thumb that almost everywhere in the brain, for every line projected forward, five to ten times as many lines would come winging back. However, this dense mesh of connectivity was not a problem for computationists if it was assumed that the purpose of all this feedback was to stabilise the information-processing machinery of the brain. Being a biological organ, the brain was inevitably going to be a bit sloppy, so it could do with intricate feedback mechanisms to tighten up the performance of its circuits. It was reasoned that feedback was there to sharpen the delivery of the neural message.

Desimone's attention effects were also becoming much easier to accommodate. All Desimone was showing was that the brain could impose a global control on its processing elements. Originally, many had been uncomfortable with this finding because it went against the old pixel model idea that the firing of a neuron meant something concrete – that it lit up its own dancing mote of consciousness. But population voting now made it clear that even the partial response of a cell could carry crisp meaning, while synchrony said even a brightly firing cell might fail to make an impact unless it was part of a coherent network of activity. So there was no longer such a worry that a top-down influence on the behaviour of a cell undermined its ability to represent. The business of coding for information was a little more sophisticated than that.

And yet all the time the computationists were steamrollering forward, buoyed by their success, the evidence kept growing that there might be more to the brain. It was plain that the brain could not be the kind of free-form grid envisaged by the dynamicists of the 1940s and 1950s. It did have a processing structure. Yet still, remarkable examples of the brain's fundamental plasticity kept on turning up.

CHAPTER FIVE

A Dynamical Computation

Desimone published his paper on attention effects in 1985, but such was the difficulty of working with awake, thinking animal subjects that it was another three years before either he or anyone else could get organised to follow up the finding. Yet eventually many labs were reporting examples of ways in which the brain's expectations or intentions could quickly change the firing of a cell, either amplifying it or subduing it. For example, John Maunsell of the Baylor College of Medicine in Houston, Texas showed that movement-coding cells in a monkey's visual pathway would double their usual firing rate whenever they were involved in tracking a target that was being attended.

Neuroscientists continued to try to reconcile these discoveries with a computational view. The next line of defence was to argue that attentional states probably only affected the behaviour of higher-level neurons, and down at the bottom of the pyramid of processing, in places like V1 where information was first mapped, the responses of cells were likely to be solidly hardwired.

The case for this was argued especially strongly by Francis Crick, the DNA discoverer who had become neuroscience's most high-profile theorist on the issue of consciousness. After sharing the honour of a Nobel prize in 1962 with James Watson, Crick had taken up a chair at the Salk Institute in San Diego, a privileged job that left him more or less free to pursue whatever took his fancy. Crick had already made one big career switch, abandoning physics – having studied 'the dullest problem imaginable': the viscosity of water at different pressures and temperatures – to join the biologists at Cambridge University in the hunt for the secret of life. Finding himself comfortably ensconced at the Salk Institute, Crick felt he could do worse than play out his retirement years giving the secrets of consciousness a go as well. And unlike many other outsiders to neuroscience, Crick was made welcome because he used his clout to flush interesting ideas out into the open. He was not peddling some private crackpot scheme, only saying the kinds of things other neuroscientists would say if the overly puritan culture of the field would let them.

Throughout the early 1990s, in articles in journals like *Nature* and *Scientific American*, Crick railed against the idea that low-level mappings like V1, or even V2, could play any direct part in a state of consciousness. He believed that awareness was the output of some limited part of the brain – perhaps the synchronised or 'in-focus' firing of a small selection of high-level cells. As an input reception area, almost by definition V1 neurons should not show the kind of plastic adjustment that was the hallmark of attentional focusing. The firing in V1 was supposed to represent the raw visual picture, the state of play before consciousness-producing analysis of the information got under way.

But by the mid-1990s, even V1 no longer looked sacrosanct. Nikos Logothetis, another Baylor College neuroscientist, had been investigating the mechanisms behind stereoscopic vision – how we fuse the views of our two eyes into a single conscious experience imbued with a feeling of depth. Experiments had already shown that the information from each eyeball was kept separate when first projected to V1. Signals from the left and right eye were channelled to the same point of the mapping, but they formed parallel bands of activity rather than being merged. This made sense, as the slightly different view seen by each eye would have to be represented on V1 before the higher levels of the visual hierarchy could begin to make their calculations about distance and depth.

Proof that the activity of V1 must lie outside the eventual conscious picture seemingly came from the phenomenon of binocular rivalry. The brain normally has little trouble fusing the images from our two eyes because they show only slightly different versions of the same scene. However, if a stereoscope is used to project a quite different stimulus to each eye, then the clashing information cannot be fused and our attention has to switch back and forth between the two views. For example, if the stimuli are a pair of zebra-striped grating patterns, identical except for the fact that they slope in opposite directions, then we will see one pattern for a few moments before our brain appears to tire and we switch to the other. The presumption was that the two gratings were being represented on V1 but the later stages of processing had to make a choice about which set of signals would be presented to consciousness.

To discover what was going on, Logothetis and his team recorded from the brain of a monkey as it reported which of two grating patterns it was currently seeing. Would the firing of its V1 cells reflect the raw sensory input from each eye, or somehow fall into line with its conscious state? The answer, as expected, was that most of the V1 cells did keep firing as normal. But on the other hand, a significant proportion – about a fifth – did not. As the monkey's awareness shifted from one side to the other, the response from these neurons was suppressed.

Crick was hardly about to admit defeat. In commenting on this result in *Nature* in 1996, Crick took refuge in the fact that the majority of V1 cells did behave in an 'unconscious' way, whereas at higher levels of the visual hierarchy almost every cell reflected the conscious picture. Crick added there was good reason to believe that it was only neurons in the 'output' layers of the V1 mapping, those in layers five and six, which were being suppressed. So even if he had to concede a degree of plasticity in V1, he was sure that its computations began with a rigid coding of visual information. Yet even as Crick was drawing this new line in the sand, within neuroscience the tide was finally beginning to turn against the very idea that any part of the brain needed to be so hardwired to give processing a solid foundation.

Indeed, measuring the activity of the cortex across broader timescales, researchers already knew V1 and other low-level mappings to be extremely plastic in their representation of information. In a series of famous experiments during the 1980s, Michael Merzenich at the University of California in San Francisco studied the way the brain of a monkey mapped for hand and finger sensations. Merzenich and his team wanted to see how neurons in the primary somatosensory cortex would react when the input to the brain was changed in some dramatic fashion, such as after the nerve supply from a finger was cut, after two fingers had been sewn together, or after a monkey had been trained to trail a finger on a rotating disc for a couple of hours each day for a few months.

If the cortex mapping was hardwired, then the experiments should have little effect on the physical layout of the monkey's brain. Perhaps the cells belonging to a particular finger representation might shrink or die once their stimulation was cut off, but not much else would be expected to happen. However, Merzenich found that given a few weeks to adapt, the brain's circuits took on an entirely new topographical arrangement. For example, if the sensory nerves from a finger were cut, then not only did the representation of that finger disappear, but the abandoned area of cortex was taken over by the mappings of neighbouring fingers. Cells that might originally have been part of an index finger representation would switch allegiance to swell the representation of the remaining digits. Or when two fingers were sewn together to force the monkey to use them as one, the brain adapted by losing the previously sharp division between their mappings. The cortex representations fused to form a single superfinger – and, presumably, the monkeys also began to experience them that way.

But the most remarkable result came from monkeys trained to trail a finger on the edge of a spinning disc. Merzenich found that the mapping of the finger not only grew, as if the constant stimulation was making the digit hyper-sensitive, but it stole cells from a neighbouring area responsible for mapping the monkey's face. The cortex neurons were so plastic that, given sufficient demand, they could switch the part of the body they were representing!

As with everything else, when Merzenich's monkey work was first reported, the tendency was to downplay the results because they seemed to contradict accepted ideas of how the brain had to be if it was going to do information processing. Neuroscientists consoled themselves with the fact that it took weeks, if not months, to bring about a radical change in the receptive fields of primary-level somatosensory neurons. But by 1992, other researchers had demon-strated that the cortex could redraw its boundaries in minutes. In one experiment, a laser was used to burn a small hole in the back of a cat's eyeball. Recording from V1 to discover how cells would respond to this sudden loss of input, it was found that almost immediately they began to expand their receptive fields so as to 'see around' the destroyed area. The neurons managed to fill in the gap by switching to a wider angle of view.

Such results showed that far from being wired into place, representing a fixed co-ordinate on the great map of consciousness, every neuron was part of a dynamic web. It had to jostle for its position. As with the attention effects reported by Desimone, the receptive field of a neuron was under constant pressure from the shaping feedback of other cells. A neuron said what the greater context allowed it to say. This plasticity of coding might be more obvious at higher levels of the brain, where neurons responded to the cross-currents of feedback and competition almost instantly, but given sufficient time, like a sticky gel, even V1 or the primary somatosensory cortex could adapt. Change the balance of forces in some way and the whole network of representation would slowly shift until it found a new point of equilibrium.

Neuroscience had a problem, because the computer model worked so well up to a point, then broke down rather catastrophically. On the one hand, the brain was a lot like a computer in having a highly structured way of dealing with information; on the other, its circuitry was basically fluid. As Hebb had made clear all those years ago, if there was a neural code, it was having to be scrawled across a living surface that responded to the very act of writing. These two ways of looking at the brain, the computational and the dynamic, seemed so contrasting that they did not appear any more fusible than Logothetis's stereoscope images. However, in the 1990s this started to change rapidly as people began to understand the idea of complexity and saw how it could embrace exactly such a hybrid system.

As said, complexity is not so much a new concept as a realisation that the principles behind Darwinian evolution could be expanded to form an intellectual framework for tackling a great many of the things in life that fit so poorly into the traditional reductionist mould of scientific thought. A complex system is one that is fundamentally dynamic, even chaotic, but which has found ways of harnessing this dynamism. It has developed a machinery that channels activity down certain predictable and constructive paths to inflate a state of organisation.

So complexity is already more than just naked dynamism. A purely chaotic system, like the weather, simply bounces about the infinite orbits of its attractor. It does not evolve, just unfolds mathematically. But a complex system has some memory-type mechanism for tuning its feedback flows – for plugging a new and slightly better adapted value back into the equations during each cycle. It is a dynamism that comes under tight evolutionary control.

A well-adapted system becomes meta-stable. It is built of dynamic material, yet a network of internal checks and balances holds its structure firm, or at least firm-ish. Like the brain, it looks solid enough – it hangs together for sufficient time to do the job for which it is designed – but it is not solid really. And more importantly for an explanation of the brain, a complex system can begin to evolve hierarchies of structure. Having evolved a first level of organisation, rather than remaining stuck where it started, it can go on to add layer upon layer of fresh structure to develop into something that is super-complex.

Life is the most familiar example of this. The DNA molecule is a device for milking structure from the randomness of organic chemistry. By themselves, a vat of proteins would bump about and react, but they would not develop any higher level of organisation – at least not regularly and reliably. Each collision would be as likely to erode structure as to create it. A strand of DNA is no more than a recipe for churning out a particular brew of proteins at a given time. But this list has been so carefully tuned over millennia of evolution that when it is transcribed, the resulting mix cannot help but self-assemble to form a cell. The proteins coming off the DNA manufacturing line will jostle about on individually chaotic trajectories, their fate still apparently ruled by chance, yet the blend is so precisely specified that the proteins will find themselves falling into the same old conjunctions, time and time again.

While it is not usual to describe it as such, DNA acts as a kind of digital bottleneck. It does information processing on a generational timescale, trapping knowledge about what kinds of protein mixtures work best and applying that to each new cycle of birth and growth. DNA is computer-like in that it has the necessary rigidity to make particular kinds of structure as soon as the winds of chemical chance begin to blow through it. But it also has just enough plasticity to act as a memory, to be changed by events, and to milk chemistry for a slightly different brew of proteins the next time round. And once DNA proved that it could work at one level – that it could build cells – this became the foundation for yet greater elaboration. The disorder of chemistry was forced to begin constructing organs, bodies, and whole ecosystems.

The brain is similar in that it has a machinery – a processing structure tuned by millennia of evolution – which can milk organisation from the disorder of the moment. But whereas with life everything seems to come back to the one molecule, the brain appears to have hierarchies of digital-like bottlenecks for distilling order from a flow of events. The channelling starts right down at the synapse level, perhaps even at the membrane pore level. A synaptic junction has the right mix of rigidity and plasticity to turn a soupy mix of electro-chemistry into the transmission of a signal. There may be shades of grey in what takes place at a synapse, but overall, something reasonably definite happens. An axon spike either is or is not converted to some level of dendritic activity. The synapse also learns in some way from the experience, becoming either fatigued or sensitised and, over a longer timescale, generally strengthened or weakened.

Stacked up above the synapse are layer upon layer of further bottlenecks. Neurons turn a complex mix of synaptic activity into a crisp output decision. A spike is triggered and delivered to a fixed set of locations. The neuron also learns something from the moment, both about what kinds of input patterns to expect and what kind of feedback reactions are likely from the rest of the brain. After neurons come topographical mappings and hierarchies of such mappings, each doing something to focus the flow of information passing through them, and also remembering something from the flow. Above that, the whole brain seems organised around yet one further bottlenecking step. Synapses, neurons, and cortex mappings are sufficient to raise the level of activity from raw electro-chemistry to orderly neural representations, but for these maps to be woven into a graded field of consciousness – an awareness with a bright, focused foreground and a dimly perceived yet still organised periphery – they must feed through some form of system-wide, whole-brain transition. And as will be seen, this last bit of sorting out is probably achieved by the cortex sheet – the higher brain – using a clutter of sub-cortical organs as a bottleneck to connect back to itself and so focus aspects of its own activity.

Regardless of this last step, the point is that the brain has real structure. It has many levels at which it wrings information from basically dynamic processes. But like DNA, this structure has to have a built-in plasticity to allow it to learn from what it has just done, and thus to do a better job the next time round. That is why synapses, the receptive fields of neurons, and the organisation of brain maps are such slippery customers to study. Their neural code, the way they represent information, can never be completely divorced from their own recent history – from what they have learnt about how they should behave to make a positive contribution to the brain's overall state of activity. So change the demands placed on any brain 'component' and it should show signs of adapting. A state of attention can silence a cell or suddenly guarantee the transmission of a spike at a synapse. A physical change like a hole burnt in the retina or a severed nerve will provoke an adaptation in the receptive field of a neuron or the topography of a cortex map. Computation is about stability; dynamism is about instability; but complexity is about getting just the right pitch of meta-stability out of a basically dynamic system so that it keeps moving forward, never falling back.

In such a system, the act of processing – what the structure does during any particular moment – has to be thought about in quite a different way. Computers process information by setting up careful chains of input and output. There is a sequence of digitally isolated events in which each step must be completed, its meaning stabilised, before the next step can begin. The brain, by contrast, evolves its way to a state of output. Everything has to happen in concert. An individual synapse, neuron, or mapping area can only arrive at a meaningful level of activity after the entire stack of representation across the brain has had time to settle down. All the eddies of competition and feedback have to have the chance to play themselves out and reach some kind of fleeting balance.

So when talking about the issue of coding, it was important to realise that the first few moments of firing from a cell, or mapping by a cortex map, might not actually count for much. A dynamic view would say that it was just the opening gambit in a negotiation of forces. As Friston remarked to Kosslyn while standing by the pond, the first spreading ripples of activity would be just the start. The echoes of feedback would need time to bounce back off the pond walls. What was happening at a point would have to come to feel some impact from the overall shape of the vessel. In a computer, information was something compartmentalised and stable. But in the brain, information had a trajectory – it had to develop. And by the time a neuron or map had settled into some kind of output state – although by now, even the word 'output' had dubious connotations – this activity would as much represent what the rest of the brain thought about the response as what the cell or mapping area felt about the stimulus that provoked the firing in the first place. Information only became information when each fragment of brain activation came also to reflect something about the whole.

On the open day to launch the Institute of Neurology's scanner lab in London in the spring of 1996, Karl Friston sat on the fire escape outside his smart new corner office and contemplatively tamped a wad of tobacco into his pipe. Colleagues talking about Friston cupped their hands to their heads and made motion a throbbing brain to indicate their respect for his intellect. But perhaps a more tangible sign of his status was the postbox-size metal ashtray which had been screwed into the outside wall for his personal use. Puffing to get the pipe lit, Friston gathered his thoughts about what it might mean to make measurements of an organ where bright activity and meaningful activity were not always necessarily the same thing.

As one of the original team of PET pioneers at the Hammersmith Hospital unit during the early 1980s, Friston had been in on the ground floor of the scanning revolution. He had arrived from Cambridge University, a maths whiz who went on to qualify in medicine and psychiatry. To round out his education while the machines were being readied for operation, the Hammersmith sent him to spend a couple of years in California with Gerald Edelman, the visionary, if controversial, consciousness theorist and Nobel laureate. Now, after a decade of proving itself as a technique, scanning was ready to go big time.

At the Hammersmith – as with their great rivals over at the Washington University in St Louis – Friston and his colleagues had been restricted largely to doing 'proper' medical research, studying mental diseases and recovery from brain damage rather than consciousness. Indeed, the neuroscientists had had to share time on the PET system with heart specialists and other kinds of researchers. But with £20 million in backing from the Wellcome Trust, the world's biggest medical charity, and other sources, the Hammersmith group had been able eventually to set up its own laboratory – the first scanning facility to have the stated purpose of tackling the mysteries of consciousness. And not only did the new Wellcome Department of Cognitive Neurology have the latest generation of PET camera, it also had a functional-MRI machine, the equipment to do EEG record-ings, and the money to buy an MEG system if need be – the one that used liquid helium-cooled superconducting sensors to pick up the brain's faint magnetic fields.

Not surprisingly, envy hung heavy in the air on the open day as visiting neuroscientists wandered about the lush furnishings and glittering marble of the lobby, or ran a hand over the machinery in the sunlit atrium housing the scanners. It was a different world in every way. Traditional neuroscience was either years of being locked away in a basement room dissecting pickled brains or else doing experiments on animals – and no one really liked what was involved in electrode recordings or brain lesioning work. The use of animals could be justified by the fact that a better understanding of the brain should lead to new treatments for illnesses like schizophrenia and Alzheimer's. And certainly, taking place at universities and medical schools, any work was subject to the closest professional scrutiny and much moral debate. There were strict controls on suffering and the prevention of unnecessary experiments. But no matter how solid the arguments, animal experimentation tainted the field.

The stigma was undoubtedly one of the reasons why neuroscience appeared so reticent about the subject of consciousness all through the single-cell recording years. Physicists were carving twenty-mile-long rings in the Swiss Alps to smash particles together. Astronomers were building huge telescopes on the tops of remote mountains to peer at the origins of the universe. Engineers were sending rockets to the moon. But experiments on the animal brain had to take place in urine-soaked fortresses, deliberately anonymous buildings with few windows and entry-phone-controlled doors to safeguard against the threat of animal rights protestors.

While NASA and CERN had enormous publicity machines and courted the press at every opportunity, neuroscientists felt forced to skulk in the shadows, hoping not to attract too much notice. They could hardly invite the TV cameras into a recording session where an anaesthetised cat dangled in a steel frame, its lungs punctured to prevent the gentle motion of breathing from disturbing the position of the electrodes in its head; or to film the bumbling attempts of a monkey to insert its hand through a narrow slit several months after having had some small but crucial part of its motor centres cut away. In such an atmosphere, talking brightly about what the results might mean for an understanding of consciousness would seem danger-ously frivolous.

Scanners changed the study of the mind completely. They were a swell, big-ticket way of doing science that hurt nobody. The machinery even looked the part: clinical masses of metal which clanked and whirred in one room as people in white coats huddled round a flickering bank of monitors in another. With scanners, neuroscience chiefs could finally begin to bang the drums for consciousness research. The technology seemed worthy of the quest.

For those accustomed to animal work, there was an even bigger culture shock. It was just so easy to do a scanner experiment. A green student could have a good idea, scan half a dozen volunteers the next week, and have a paper submitted within a couple of months. Such speed was unheard of in an animal lab. It would take that long just to wrangle authorisation for an experiment out of a university's ethics committee. With all the time it took to train animals, do the recordings, and collate the results, a major project could take three to five years to complete. The grinding slowness of animal work was yet another reason why neuroscience had seemed such a cautious field. Researchers could not afford to take too much risk with an experiment because it would take so long to recover from any false start. A wrong move could sideline a career. And certainly, no one was about to let enthusiastic Ph.D.s loose on the equipment just to see what they might find by playing around.

So the opening of the Wellcome Laboratory seemed the beginning of a new era. The proud brass nameplate on the door, the soft lighting and magazines in the foyer, the army of fresh-faced youngsters waiting to flood the journals with new findings as soon as they had been pointed in the right direction, everything looked set for the big push. It was time to find out how brains made minds. However there was a problem. As Kosslyn's experience with mental imagery had already shown, scanning researchers might not yet know how to record the activity of the brain in a way that was truly meaningful.

Perched on his fire escape, looking out across the back roofs of the hotels and restaurants crowded around the Institute of Neurology, Friston admitted that scanning had fallen rather too heavily for the cognitive science view of the mind as an amalgam of faculties or processing modules. During the first half of the 1990s, labs had bulldozed ahead with little critical thought about what they were doing.

'It got like the human genome project,' said Friston. 'We had developed a standard method, using subtractive techniques, for mapping mental abilities to parts of the brain. You just cranked the handle and turned out the results. When functional-MRI came in, everybody got in on the act. It became a production-line technique, a new form of phrenology where we were trying to pin a label to every bump of the brain.

'The belief was that eventually every part of the brain would get done. Just as the human genome project will reach closure by identifying every last gene, so we would end up mapping the whole brain. Last year it was the cingulate and prefrontal cortex that everyone was chasing, this year it might be the insula and precuneus cortex. But people have begun to realise how little all this actually tells us. What we need now is a theoretical framework that lets us talk about the dynamics of the brain – how each area connects and interacts.'

The key to a new approach, Friston explained, was to start looking for correlations rather than raw strength of activity. The cognitive model had suggested that each bit of the brain would 'do' a processing step. Kosslyn had been expecting to find perhaps just three areas lighting up in his mental imagery experiment: one to store the memory trace of a letter, one to retrieve it, and one to display the results. But it had become clear that the brain acted as a dynamic whole. All the storing, fetching, and displaying did produce some clear centres of activity. Most significantly, there was the topographical activation of V1, proving the point that mental imagery must piggy-back on the machinery of sensory processing. Yet Kosslyn also saw that vast areas of the brain lit up at least dimly in the course of the task. It seemed the brain could not create a disturbance of activity in one spot without provoking ripples of adjustment everywhere else.

The subtraction technique could remove a lot of this collateral activity, but what if it stood for 'the shape of the vessel'? What if the activity being washed out of the scanning images was critical in giving the brighter bits of activation their form? Alternatively, just because activity was diffuse or weak, this did not necessarily mean it was not central in some way. Sparse firing by a population of cells might actually be a sign of efficiency rather than disinterest. A fiery rash of firing might show that an area of brain was struggling to cope with the given task. Once you began to think about it, there were a lot of reasons why the equation between metabolic heat and meaningful activity seemed a trifle simple-minded.

A better route would be to use scanners to highlight the parts of the brain that rose and fell in concert during the performance of a mental task, regardless of the scale of activity. If the brain always had to pull together a hierarchy of mapping areas to do anything, then what mattered was which bits were most strongly connected when that action took place. To this end, Friston had already developed a few statistical tricks to extract correlations from existing ways of collecting scanning data. But happily, just in time to impress the guests at the open day, he was ready to reveal his first experiment based on a ground-up rethink of how neuroscience should be pursuing the increasingly elusive concept of a neural code.

As Friston had come to realise, the main difficulty was that any unit of measurement carries with it assumptions about what kind of brain activity counts as meaningful. If an experimenter tallies spikes or measures percentage changes in blood flow, it is because a busy cell or a busy brain area seems to be saying something that a silent cell or brain area is not. And even with more sophisticated concepts like population voting and synchrony coding, a neuroscientist is still presuming that only certain forms of activity are information-bearing – the cells which fire in time or which are at least partially activated by a stimulus. It was digital thinking, because it cast the activity of the brain in a black and white, either/or situation.

A dynamic view said nothing was ever really so clear-cut. Yes, some aspects of the brain's reaction might bear more weight. Synchrony or the sheer bulk of firing might create the most obvious peaks in a landscape of processing. But dynamics said that underneath, everything remained connected, everything counted at least a little towards the final state. So unless neuroscientists were already sure about what kind of activity really mattered in the brain, then their methods of measurement ought to make the fewest possible assumptions about what counted. That was a discovery that should be left to emerge from the data itself.

In practice, this meant a researcher had to take some kind of blanket reading, capturing every kind of change over a certain spread of space and time. Allowing for the right time window was particularly important if it took time for any brain activity to develop into a state of organised significance. A lot of spikes in the first split second might not tell the full story. Likewise, if meaning lay in an overall population response rather than the twinkle of individual cells, then any unit of measurement would have to be spread over enough neural terrain to capture even the fringes of this voting activity. It was a big mistake just to insert an electrode and go looking for the single juiciest responder. And the same went for synchrony. It seemed almost certain that synchrony mattered to the brain, but it was unlikely to be the kind of digitally precise mechanism that most neuroscientists were assuming. Instead, it would be a looser, statistical phenomenon where cells hitting the beat would probably count the most, but even neurons which only showed a slight tendency towards synchrony would contribute something mean- ingful to the final state of play.

So any measurement had to have a wide base. Friston's solution was to look for something he called a neural transient, a sort of evolving super-spike produced by a population of cells within the same small area of brain. Rather than chasing individual spike patterns or signs of synchrony, the brain would be recorded in some broad spectrum way and then complex mathematics could be used to identify any fleeting change in behaviour, regardless of the precise mix of firing characteristics needed to produce the activity blip. Crucially, the way he would know that the transient pattern was actually meaningful was that it would be precisely correlated with something else happening in some other part of the brain. Dynamics said an act of processing evolved as areas of the brain fell into some kind of competitive balance, so if Friston's measurements showed two areas reaching a crescendo of activity in tandem, then it was likely there was real communication going on between them.

Struggling to put his ideas into words, Friston held out two fluttering hands to suggest a pair of brain areas resonating in dynamic harmony: 'Say you've got two areas of the brain. Each will have its own dynamic. The way the neurons are connected locally will create a rest state attractor. And because cells are always firing, there will be a level of tonic activity going through the network to keep things cycling. Before anything starts to happen, each area will already be in a certain state of balance which can be described mathematically in terms of the surface of an attractor manifold.

'Then some input hits this area' [he wiggles one hand] 'and changes the balance so that the activity starts to move into a new state. This change disturbs the dynamic in this area over here,' [he wiggles the other hand] 'and produces a change too. You get coupled wavelets of activity in both areas as the two surfaces modulate each other's firing in a characteristic way.

'But the big thing is that these two transients can look very different from each other. At the moment, most people are looking for evidence of neural coupling either in the fact that both areas will show a matching rise in firing rates, or the emergence of some shared firing rhythm. However, because each area inhabits its own attractor space, the appearance of the wavelets could be quite unalike. For instance, a high-frequency pattern of activity in one area might produce a low-frequency pattern in the other. How an area reacted would depend on its own local dynamics. If you were looking for significant brain activity using an ordinary measure, such as a coherent oscillation or correlated rises in firing, this would make the connection easy to miss.'

So there would be matching change, but the appearance of these transient peaks might be considerably different. To prove the case, Friston had just tested this idea against real data. The PET and MRI scanners in his own laboratory were not actually very good for the job because they lacked the speed to capture the rapid evolution of a transient. They could see the surges in metabolism that suggested two areas were connecting every time a particular task was being executed, but they did not have the fine-grain, sub-second resolution to show how the connection developed over time – when it began and when it reached a peak of meaningfulness (or mutual information, to use the technical term). So instead, Friston used data from an experiment carried out on a new MEG machine in

Germany, MEG offering the combination of a reasonable spatial resolution with the unbeatable millisecond level temporal resolution of EEG recordings.

Seated inside a metal Faraday cage to shield against stray magnetic fields and with a helmet of supercooled sensors strapped to his head, a subject was asked to perform the simple act of moving a joy-stick in one of four directions at regular two-second intervals. The nature of the task was not so important, apart from the fact that Friston knew from PET experiments that making the choice should activate two critical planning areas in the brain: the posterior parietal cortex (PPC), where the brain appears to crystallise its sense of space and direction, and a motor decision centre in the frontal cortex. So his expectation was that each time the subject thought about which way to push the joy-stick, a transient change in activity would have to link the two areas. And after applying some heavy-duty maths to the MEG readings, this is what Friston found.

The first important discovery was that it took the brain at least forty milliseconds – about one-twenty-fifth of a second – to develop a significant level of connection between the two areas, and the linkage then persisted for about half a second. This put an upper and lower boundary on the period during which any disturbances in cell activity could be treated as meaningful. Still more tellingly, the change in activity in the two areas did indeed look quite different. Both areas became markedly more synchronised in their firing behaviour during the fleeting moment of connection, but the beat had a high frequency in the frontal decision area and a low frequency in the area representing the direction of the planned joy-stick movement.

A simple input–output model of the way cortex maps connected had led many neuroscientists to believe that sharing the same beat would be what actually bound two levels of activity together. But Friston was showing that while rhythm might be a mark of significant activity, it was probably a strictly local phenomenon that reflected the dynamics of a particular area of circuitry. Linkage resulted in the emergence of rhythm, but it did not dictate its pace. And the point was that something which Friston took as proof of a transient connection would have been either completely missed by conventional methods of measurement or else treated as evidence of the exact opposite. Faced with such a clash of beats – the eruption of high frequency synchrony on one map and low frequency on the other – most researchers would probably have concluded that the two cortex areas were not talking to each other at all.

In itself, Friston's result did not reveal anything startling about the brain. He already knew the two areas would play a part in the joy-stick moving task, and the actual parameters of the transient – how long it lasted and what form it took – did not tell him much about what a connection between two brain areas performing a different task would look like. The self-organising nature of such interactions meant that every transient could have a quite different profile. If there were any general principles ruling the way transients formed, it would take a great many more experiments to discover them.

But what mattered was that Friston was approaching the brain from a very different direction. He was assuming it to be a dynamic entity and then letting everything – from the style of theorising to his methods of measurement – follow suit. The computational model of the brain had taken neuroscience a long way. It seemed to get more right than wrong. But now a real shift in thinking was required if science was going to get to grips with the brain and the phenomenon of consciousness.

The good news is that once the brain's activity is viewed as an evolutionary competition carried out by structured, yet basically fluid circuits, a lot of existing neuroscience begins to make much more sense. Results like Desimone's attention effects in V4 or Merzenich's evidence of rewiring in a somatosensory map suddenly slot into place. And in particular, the dynamic approach helps explain one long-standing puzzle in neuroscience, an experiment carried out back in the 1960s by the Californian physiologist Benjamin Libet, which led him to make the astonishing claim that consciousness lags half a second behind the unfolding of life.

CHAPTER SIX

Benjamin Libet's Half-Second

Well, if there is one thing that seems certain about consciousness, it is that it is immediate. We are aware of life's passing parade of sensations – and of our own thoughts, feelings, and impulses – at the instant they happen. Yet as soon as it is accepted that the mind is the product of processes taking place within the brain, we introduce the possibility of delay. It must take time for nerve traffic to travel from the sense organs to the mapping areas of the brain. It must then take more time for thoughts and feelings about these messages to propagate through the brain's maze of circuitry. If the processing is complex – as it certainly must be in humans – then these delays ought to be measurable, and even noticeable with careful introspection.

As it happens, the conduction speed of nerves was the very problem that got the new science of psychology off the ground in Germany in the mid-1800s. It had been thought that nerves would act more or less instantaneously. But in the 1840s, the brilliant young German physiologist Hermann Helmholtz showed that nerve impulses actually travel surprisingly slowly. Helmholtz applied electric shocks at a series of points down the spinal nerve leading to a frog's hind leg. From the slightly faster reactions nearer to the muscles, he worked out that its nerves must conduct impulses at about sixty miles per hour. Turning to humans, Helmholtz carried out much the same experiment by asking subjects to push a buzzer as soon as they felt him touching their legs at different points. From the slight changes in reaction times, Helmholtz calculated that human leg nerves must carry signals at about 200 miles per hour – fast, but nowhere near instant.

Modern research has since shown that human nerves actually conduct at a whole range of speeds, the rate depending on the size of

120

the axon and also the thickness of a fatty insulation material, known as myelin, wrapped around it. The nervous system is like a road network with a few fast motorways and many winding country lanes. Large, heavily myelinated nerves, such as the muscle and sensory nerves which must run the length of the body, transmit their impulses at up to 240 miles per hour. But the congested network of small unmyelinated nerves which make up the bulk of our brain work much more slowly. Once inside our heads, impulses tend to crawl along at between two and twenty miles per hour. What such conduction speeds mean is that while consciousness might be fast, it cannot be instant. It takes a minimum of ten to twenty milliseconds for any sensory message to reach the brain. After that, the brain must spend yet more time in evolving a response.

The question of exactly how long was tackled by Wilhelm Wundt, Helmholtz's assistant at the University of Heidelberg. Wundt carried out mental chronometry experiments in which he tried to measure the speed of thought processes using reaction times. By noting how fast subjects responded to a buzzer or flash of light, Wundt could get an idea of how long it took to form an impression of various kinds of sensation. Reaction-time tests could also be used to measure higher-level processes. Differences in the speed with which a subject could find an object among a clutter of distractions, or name the capital city of a country, were used to gauge how rapidly people could shift their attention or recall a memory.

From this work, Wundt developed a theory of the mind based on what he called perceptions and apperceptions. Perception was an early-forming, pre-aware response to the world that allowed us to perform reflex actions like hitting tennis balls and driving cars. In this phase, we can react quickly and unthinkingly – or if there is thought, it is of an impulsive or creative nature. Then, after perception, comes the fuller, more reflective consciousness of apperception. This is awareness with the mind sharply focused on the meaning of a moment and our response properly supervised and perhaps even a little ponderous.

Excited that science seemed able to get inside the fine-grain structure of an instant of consciousness, the field of psychology blossomed. One of the most important early discoveries was that even the process of forming a sensory picture was smeared out over about a tenth of a second. Experiments showed that the brain tended to fuse together events like two noises or two flashes of light if they followed in quick succession. This could cause powerful illusions. For example, when a pair of bulbs are lit in alternation, an observer sees a single spot of light bouncing back and forth through the air. This impression of motion, christened the 'phi effect', is familiar from theatre-front billboards which use rows of blinking lights to create moving displays. It also makes films and TV pictures possible. When a movie is projected at the rate of twenty-four frames a second, the information in each frame blurs to form a smoothly flowing experience. Televisions and computer displays work on the same principle, using an electron gun to paint the screen with a rapid succession of stills.

The amount of detail in a film picture means that the illusion begins to break down and the motion becomes jerky if there is much more than about fifty milliseconds between each frame. But simple shapes or flashes of light can seem like one moving object even when they are separated by as much as 200 milliseconds. And the phi effect is not just confined to the sense of vision. Fusing also happens with our sense of touch. If our arm is tapped at three or four points in quick succession – each tap following within fifty to a hundred milliseconds – it feels as if a single object is being trailed down towards our hand.

The brain can fill in with other properties besides motion. In one famous experiment, subjects were asked to watch a pair of light bulbs, one red, the other green, placed an inch apart. The two lights flashed within fifty milliseconds of each other. Not only did it appear to subjects that there was a single spot of light bouncing back and forth between the two bulbs, but as near as they could judge, the spot also seemed to switch colour halfway across. This meant that the imaginary bouncing light appeared to turn from green to red, or red to green, before the bulb on the other side had even come on! The brain was filling in with the appropriate colour in advance of the event.

Such illusions are revealing because they catch the brain in the act of covering up for its own lagging processing. If consciousness takes a certain time to build, it is inevitable that sensations will arrive almost on top of each other while a moment of awareness is still under construction. The brain's job is to fit the jumble of input into a coherent story. This means choices have to be made. If two separate waves of sensation appear similar enough – like two light bulb flashes, or frames of film – then it will make sense for the brain to read them as fragmentary glimpses of the same rapidly moving object or fast-changing scene. The motion-mapping areas of the visual cortex would come to tag the stationary stimuli with a velocity and direction. And in everyday life, of course, the brain would be almost bound to be doing the right thing. It is only in the laboratory that the brain's habit of assuming a connection between two very close events is made to look foolish.

So with just a one or two quick experiments, Helmholtz, Wundt, and their followers revealed more about consciousness in a few years than armchair philosophy had in centuries. Buoyed by this success, psychology took off. By 1900, more than a hundred laboratories had sprung up around the world. But almost as soon, the emphasis on the timing of mental events disappeared. Psychology came under the sway of behaviourism and the belief that people's claims about when or whether some event entered their awareness was much too subjective a form of evidence for the purposes of science. The study of reaction times and perceptual responses never actually died out, yet as a field, psychophysics became a backwater. Research had a solidly practical feel. Certainly psychophysicists did not try to use the detail they gleaned about sensory integration times or other aspects of mental processing as a launch pad for a more general assault on the problem of consciousness. Wundt's theories about awareness developing in distinct stages were left to gather dust.

Given this state of affairs, it is not hard to see why, when Benjamin Libet popped up in the 1960s to suggest consciousness might take as long as half a second to dawn, few people knew quite what to make of his claim.

*

Peering owlishly through thick glasses, a small, frail figure in an ancient pastel blue safari suit, Libet weighs his words with extreme care. An old-school neuroscientist, he wants to talk only about established fact. Attempts to draw him into a speculative discussion about consciousness are met with the driest of chuckles. Yet for decades, Libet's own work has been the focus for puzzlement and even frank disbelief.

In the 1950s, Libet was a physiologist at the University of California School of Medicine in San Francisco doing thoroughly routine research on nerve transmission and neurotransmitters. Then, in 1958, a surgeon friend happened to set up a neurosurgery unit at a nearby hospital. This doctor, Bertram Feinstein, was experimenting with a new operation to control the tremors, tics, and spasms caused by degenerative brain conditions such as Parkinson's disease. Feinstein's plan was to use electrodes to destroy small parts of the brain's motor system and so stop these tics at their source. Such lesioning was an extreme measure, but in an age before drug treatments became available, surgery was about the only option.

Feinstein's operation was a major procedure, but because the brain feels no pain, the surgery could be done under local anaesthetic with the patients fully conscious throughout. Feinstein and Libet realised that this gave them an almost unique opportunity to do experiments on a living human brain. With the consent of the patients, Libet could take a few minutes to stick in an extra electrode, give their brains a small tweak of current, and see what kind of experience the artificial stimulus would produce.

Libet was not actually the first to try this. In the 1930s and 1940s, Wilder Penfield, a Canadian neurosurgeon specialising in operations to treat epilepsy, had made a name for himself by probing the brains of his patients with an electrode. Penfield's original aim had been simply to locate the diseased regions acting as a focus for the seizures. Before a fit, many epileptics experience an aura – a characteristic sensation such as a clanging noise, flashing lights, or a burning smell – that warns them an electrical storm is brewing. Penfield felt that if he could artificially induce such an aura, he would be able to pinpoint the area of brain needing removal.

But once he got started, Penfield became fascinated by the vivid responses he could provoke and began to probe the brains of his patients all over. When their primary sensory areas were being stimulated, the patients would report only something like a brief flash of colour or a chirruping noise. When the needle was pushed into higher-level areas of the cortex, however, they often had sudden dream-like snatches of experience. One patient said he saw robbers coming at him with guns; another heard a mother calling to a child. Over the course of more than a thousand operations, Penfield gathered evidence that the cortex was much more organised than many people had believed. Indeed, several years before Hubel and Wiesel began their single-cell recordings with cats, Penfield was already hinting that it had a hierarchical and topographical logic.

Although such experiments seemed an enviably direct way of tackling the connection between brains and conscious states, few other neurosurgeons felt inclined to follow Penfield's lead. The ethical issues aside, most simply did not have the background to become consciousness researchers. They were doctors, not theorists. Nor were they willing to let some third party into their operating rooms, risking the health of their patients with a lot of fiddling about with electrodes. But, fortunately for Libet, Feinstein felt the chance would be too good to miss.

The scope for experimentation was actually quite limited. Whereas Penfield had lifted the entire top of the skull off his patients, Feinstein's operation only needed a coin-sized hole over the motor centres. This meant there was little chance of exploring the higher levels of cortex function. But Libet still had access to the primary somatosensory cortex's mapping of touch sensations, and also, by drilling a few inches down into the brain, the parts of the thalamus and brain stem which carried the nerves leading into the mapping. Making a virtue of his restricted access, Libet planned a modest series of tests. A problem with Penfield's work had been that while it was wide-ranging, it had not asked some basic questions about exactly what form of electrical stimulus it took to produce a conscious reaction. Libet decided to examine this carefully.

Over the course of five years, Libet sat in on nearly a hundred operations. He quickly found that to get any sort of reaction from the patients, he had to use a pulsed current. A simple blast of electricity did nothing. But apart from that, the exact nature of the stimulation – the frequency of the pulses, or the polarity of the current – appeared to matter surprisingly little. So long as the pulse train was persistent and gentle, the patients reported some remarkably realistic sensations, such as a furrowing of the skin, taps, jabs, brushes, and occasionally a flush of warmth or coldness on some part of the body. Occasionally, the sensations were very specific. Patients would exclaim that they could feel a drop of water trickling down the back of their hand, or that it felt as though they had a ball of cotton wool pinched between their fingers. But turning the current up above a few micro-amps did not make the sensations stronger; it just turned them into a harsh tingle, a feeling much like pins and needles.

As he continued his testing, Libet noticed something else. It normally took about half a second of pulses before the patients began to report anything. If the pulse train was stopped short at 460 or 470 milliseconds, then the patients did not seem to notice that the current had ever been turned on. But the instant Libet set the dial past the 500–millisecond mark, suddenly there was something to feel.

Of course, the result was not quite so clear cut. The half-second figure was only an average; for some patients the time was characteristically either a little shorter or longer. Also, if a person happened not to be concentrating, the delay could stretch right out to a second or more. And the half-second rule counted only for gentle current strengths. If the current was bumped up to the level where it would produce tingling rather than realistic sensations, the patients could notice a pulse train as short as a tenth of a second in duration. However, Libet felt that he had stumbled upon something of fundamental significance. It was as if processing had to be kept going for a certain length of time in order to manufacture an organised conscious experience.

To prove that the result was not just some sort of freak effect of stimulating the cortex directly, Libet went on to stimulate the other brain centres along the input path for touch sensations. By pushing the electrode a little deeper into the brain, Libet could reach the thalamus, a nut-shaped organ through which all sensory lines to the cortex must pass. Then, deeper still, the electrode would penetrate the brain stem, the first thickening of the spinal cord as it enters the brain. When he turned on the current, again Libet found that it needed half a second for the patients to begin to feel a realistic sensation. And this time, the cortex was reacting to messages arriving over its normal nerve pathways. So if there was a half-second delay, it was likely to be due to the time it took the cortex to evolve its response rather than because a naked current was an unnatural form of excitation.

To discover more about what might be going on at the magic half-second mark, Libet decided to do some recording on top of his stimulation experiments. Using the newly fashionable technique of evoked response potentials (ERP) recording, he began to check if perhaps there was something tell-tale about the activity of neurons at the moment everything got settled and came together.

As mentioned, the evoked potential method was an advance on plain EEG recording. The problem with EEG electrodes was that they were unselective, picking up every last crackle of activity. They heard the roar of the brain rather than tracking individual neural events. But by recording the brain responding to exactly the same stimulus several hundred times in a row, then averaging the signal, the background activity would eventually cancel itself out. Libet had the bonus that he would be laying his electrodes directly on the brain, so the result would be exceptionally sharp, with none of the usual distortion or damping caused by the bone of the scalp.

As a stimulus, Libet used a quick electrical shock to the back of a patient's hand. He then recorded the response from the somatosensory cortex over the course of some 500 trials and averaged the readings. The result was a rollercoaster set of squiggles marking the rises and falls in electrical potential as the cortex went about its job of representing the faint buzz on the skin. And, as Libet had hoped, the squiggles kept on going for a surprisingly long time given the fact that the patients claimed to be conscious of the shock at almost the instant it was delivered.

The trace actually showed the brains of the patients making a first surge of response very early – within ten to twenty milliseconds. This was precisely the time needed for the sensory messages to travel the distance from the hand to the brain, so it seemed there was no particular delay in the arrival of the necessary information in the somatosensory mapping area. Then, over the next tenth of a second or so, the mapping area saw its most dramatic convulsion. About fifty milliseconds into the processing of the moment, the recording needle showed a sudden and massive reversal in polarity from positive to negative. There was a plateau, and then at about 150 milliseconds the electrical activity slumped again. Next came a second, more gentle but persistent, phase of activity. At about 200 milliseconds, the trace rose slowly in a shallow arc that extended until at least half a second. And while the first big bump at a tenth of a second seemed a strictly local response, the second wave of activity represented a much more diffuse reaction. Part of the reason for its shallowness was that Libet was picking up the weak and variable echoes of activity taking place in far-flung corners of the brain.

This seemed striking evidence of an evolving mental state in which an initially fierce bout of mapping in the somatosensory cortex eventually gave way to a feedback-tuned, whole-brain state of focused attention. The trouble was that plenty of other interpretations could also fit the data. Perhaps the patients really were conscious of a buzzing on the back of their hands as soon as the first blip of nerve signals hit the cortex. The long-winded response might be just the result of the patients having thoughts about that awareness. Or it might even be some kind of neural housekeeping activity, such as a rebalancing of the brain's circuitry after a burst of firing.

The truth was that Libet had no way of telling. Indeed, he did not even have reasonable grounds for speculation given that virtually nothing was known about the hierarchical and feedback-driven nature of brain processing when he was doing his experiments in the 1960s. However, the one fact of which he could be sure was that it took quite some time for the brains of Feinstein's patients physically to show a global level of response to a hand stimulus, no matter how quickly they might have felt they were aware of the jolt.

When Libet published these results in the *Journal of Neurophysiology* in 1964, they attracted great attention, particularly from John

Eccles, a maverick Australian who had just won the Nobel prize for his own work on nerve synapses. Eccles was a controversial figure because not only was he one of the very few neuroscientists keen to talk about consciousness, he was also a passionate Cartesian dualist, pushing the scientific heresy that the brain was merely an inanimate vehicle for the conscious soul. Eccles liked Libet's results because they appeared to drive a wedge between events in the brain and the dawning of awareness. For most scientists, Libet was saying consciousness arrives late and the appearance of simultaneity with events must be the result of some kind of psychological illusion. But Eccles turned the logic round to suggest consciousness really was instant and that brain processing followed in its wake. The soul saw, and then instructed the circuitry of the brain to reflect the same sensory state.

This championing by Eccles both helped and hindered Libet. Almost immediately it got him invited to speak at a rare Vatican-sponsored conference on consciousness, where he found himself rubbing shoulders with luminaries of the field such as Wilder Penfield and Vernon Mountcastle. But it also made him look too much like Eccles's protégé, so mainstream neuroscientists who could not stomach Cartesian mysticism also tended to deride any suggestion of a half-second processing delay. The general opinion was that Libet would have to come up with a rather more convincing demonstration of his findings before anyone took them too seriously.

Returning to the lab, Libet struck upon the idea of exploiting backward masking, a strange sensory phenomenon that had been troubling psychophysicists for many years. Backward masking is an illusion in which a strong stimulus can block awareness of a weaker stimulus occurring just beforehand. For example, if a dim flash of light is followed about a tenth of a second later by a brighter flash in the same place, then the first flash will go unseen. It is as if the mental processing of the first flash is interrupted before it can run to completion. The effect is strongest if the two events are separated by about 100 milliseconds, but can work for a gap as long as 200 milli-seconds. Complex visual sensations, like pictures and printed words, can be blotted out, and even other kinds of sensations like sounds or touches. A tap on the arm can be masked by a second, harder tap.

Making the phenomenon still more intriguing, psychophysicists had found that a third stimulus could be used to mask the masker. If a further flash followed the second, then the very first flash now entered consciousness and it was the middle flash that became masked!

While a little unnerving, it is easy to see that backward masking is merely the reverse of the phi effect. If the brain needs time to sort out an impression of the world, then its interpretations of a jumble of sensory events can always go two ways. With the phi effect, the brain is deciding that two successive flashes of light are best read as a single dancing dot – the most probable explanation given the chances of two very similar events occurring in such close proximity. By the same token, when a weak flash of light is quickly followed by a second more positive flash coming from the same place, the brain would be safest to read this as just one event. Indeed, the best demonstrations of backward masking come when the second stimulus is not only more powerful, but actively clashes with the first – for instance, if a black disc is followed by a black ring encircling the space where it has just been. As polar opposites, the disc and the ring cannot be merged into a single sensory event, so it is not surprising that one should block the representation of the other.

Of course, such explanations only made sense if it was already agreed that brain processing was smeared out in time at least a little. It said that the brain tarried, taking a good tenth of a second or so to gather the sensory input needed to produce a settled moment of awareness. But Libet saw that he could probably use backward masking to prove that the real processing time was nearer half a second.

His experiment was straightforward. He would shock a subject on the hand and then, a split second later, use his electrode to zap the same place on the somatosensory mapping. Libet's reasoning was that if it actually took half a second inside the brain for the original hand sensation to be fully processed, then the late blast of stimulation might be able to interfere with the still budding experience even quite deep into its processing cycle. And this was what he found. The cortex stimulus was able to mask a skin stimulus as much as 300 milliseconds later – three times longer than normal.

This seemed powerful evidence that consciousness takes a relatively long time to build, and that any experience of it being instantaneous must be a backdated illusion. We only feel that we are there as events happen. But the critics were still not swayed. Their reply was that a much simpler explanation for Libet's results was that the late cortex blast merely wiped away the memory of the earlier hand stimulus. His subjects would have experienced the buzz almost as it occurred, but then this fleeting awareness would have been lost amid the noisy clamour of the electrode-generated activity.

Libet had an answer to that. In a follow-up experiment, he showed he could not only produce a late masking, but also a late enhancement of a skin sensation. This time he would exploit the phi effect to 'top up' a hand twinge with a little extra cortex stimulation. For enhancement to work, the electrode jolt now had to be weak enough so as not to seem like a competing event. The experiment also had to be set up in a way that psychologically encouraged subjects to read any extra cortex excitation as part of the initial skin stimulus. Libet told the subjects that their task was to judge the relative strength of two hand sensations coming about five seconds apart when, unknown to them, both skin stimuli would actually be exactly the same, the only difference being that on some trials Libet would be throwing in a secret burst of cortex stimulation.

As expected, even when the cortex jolt trailed a buzz on the hand by some 300 or 400 milliseconds, there was a blurring of the input. The hand sensation suddenly felt a bit stronger. And this time there could be no question of a memory of an earlier experience being erased. Indeed, like the phi experiment in which a leaping dot apparently changed colour before the bulb on the other side had time to switch on, Libet's subjects seemed to be experiencing the effect of something before it physically occurred! Awareness of the strengthened stimulus was dated to a moment before Libet flicked the electrode switch, so either there were spooky goings-on, or the brain really did take about half a second to reach a settled interpretation of the world and then backdated the results to foster the illusion of simultaneity.

*

The masking and enhancement experiments were reported in the mid-1970s, and even though the enhancement results were not circulated so widely, Libet felt he had made his case. However, he found that a half-second processing time was still a result that did not fit. Too many researchers took the view that one day someone would spot the flaw in Libet's approach and the whole problem would simply evaporate like so many other scientific oddities. Then sadly, in 1977, Libet's work with Feinstein was brought to an abrupt halt when the latter fell ill and eventually died. Libet was faced with a choice of either admitting defeat and letting the issue drop, or discovering some new way of continuing his experiments that did not involve brain surgery. Casting around, Libet found a route. But it was to lead him to become embroiled in a confused, often bitter, philosophical row over the nature of human free will.

Hearing that Libet was seeking a fresh direction, Eccles drew his attention to some studies carried out by Hans Kornhuber, a German pioneer of ERP recordings. Kornhuber had discovered that before people make a voluntary movement, such as tapping a toe or reaching out to grab something, there is a surprisingly long build-up of neural potential in the parts of the cortex responsible for motor control. The brain begins to stir as much as a second ahead of time. For complex actions, such as when a pianist gets ready to run through his scales, the build-up is even longer. Kornhuber called this evidence of mental preparation the 'readiness potential' (RP).

Clearly Kornhuber's results raised questions for those who believed it was consciousness that commanded the body, and that human beings made decisions, such as choosing the moment to flex a wrist or bend a finger, through an instantaneous act of will. Typically, psychophysicists did not make a big fuss about Kornhuber's results and what they might imply for an understanding of consciousness. But Libet saw that the readiness potential offered him a new line of attack. Already well versed in the art of EEG recordings, all he needed was some way of timing the exact moment when subjects felt they had willed some movement. This could be done with almost millisecond precision by asking the subjects to note the position of a rapidly circling dot of light on an oscilloscope as their impulse struck.

Six college students were recruited and sat in a chair. A trial would start with a 'get ready' bleep from a buzzer. The students would then wait with arms outstretched for the sudden urge to lift a finger or a wrist to overtake them. Because Libet wanted to catch the very first moment of entry into consciousness of the idea to act, the students were asked to report any trials in which they detected prior deliberation, inner speech thoughts along the lines of 'Shall I go now? OK, come on then', so they would not get included in the final results. Only quick and spontaneous impulses were to count.

After several hundred trials, the EEG data were averaged. Timing themselves, the students reported they first became aware of an intention to act about 200 milliseconds before they moved their hands. Yet the ERP trace showed that even with such a simple and impulsive decision, their motor-planning areas had begun to stir even earlier, showing flickers of activity around 500 milliseconds before any movement was made. In other words, the decision to flex a finger or wrist must have started out subconsciously and only emerged as a conscious wish about a third of a second into the processing cycle.

This was Libet's simplest experiment, involving no brain surgery and just a month or two of work. Yet because the experiment questioned some basic philosophical assumptions about human free will, it created easily the most controversy when published during the early 1980s. Libet's results seemed to imply we are not in charge of our own minds. By the time we know we want to do something even as trivial as lifting a finger, a decision will already have been made by low-level brain processes outside our control. It was as if awareness was just there for the ride.

In an attempt to head off the worst of the confusion, Libet pointed out that although awareness for the gathering impulse may have kicked in with only 200 milliseconds to go, this still seemed to give the mind time to block the movement. The conscious self might not have initiated the decision to start flexing the hand – which surely, after all, was the definition of an impulsive action – but it was there in time to exercise a veto, so preserving the cherished notion of free will. But for many commentators, the feeling that consciousness was an instant phenomenon was just too insistent. It seemed impossible to accept that each moment of awareness might come with a hidden inner structure, that even states of thought and intention might have to evolve, and that this evolution could then be concealed from our experiencing selves.

Ironically, the problem was quite the opposite for some of Libet's more informed critics. They could see that even a straight computational model of brain processing said that consciousness had to emerge by stages. But the stumbling block for them was the suddenness with which he seemed to be saying that consciousness clicked on. If a state of awareness had to evolve over half a second – although most would bet that a tenth of a second was a more reasonable estimate of the processing time – then the dawning of subjective experience should be a gradual thing. The students doing the experiment should have progressed from being a little bit aware of a rising impulse to being very aware. So Libet's insistence that they look at a clock and mark 'a time of awareness' was probably creating an artefact. He was not measuring when consciousness emerged, just when consciousness became sufficiently certain to be reported with some confidence.

The contrasting reactions showed that some people were happy with the idea of a sudden clicking on of awareness, some with a slow dawning of awareness, but no one liked Libet's combination of a slow yet sudden crystallisation of a state of subjective experience. Or, to put it in terms of the battle between the digital and the dynamic, consciousness could be a binary state or it could be an evolving state, but how could you have an evolving system that suddenly went through some kind of late binary transition? In the 1980s, before complexity theory, mind scientists and philosophers just did not have an intellectual framework for thinking about such a state of affairs, and their confusion showed.

The truth was that the results of Libet's free will experiment were never as unsettling as they sounded. The very first psychologists – mental chronometers like Wundt – had been quite at home with the idea that consciousness might come gradually, and also by discernible stages. Indeed, Wundt's theory of perception followed by apperception seemed to fit Libet's work pretty neatly. If sensing and acting were brain processes – states of information that had to evolve – then there would always be a time course, an arc of development. And it also seemed logical that the early stages in the history of a moment would lack in strength, organisation, selectivity of focus, or some other critical quality. Therefore they would be subconscious. Or, rather, pre-conscious – a wave of activity on its way to becoming a settled state of consciousness.

But being pre-conscious does not mean that the early stages can do no profitable work in the economy of the brain. In fact there is every reason to believe that much of our mental lives are lived in a twilight world of not properly conscious impulses, inklings, automatisms, and reflexive action. The standard example of an intelligent, yet heavily automated, mental process is driving a car. When people are first learning to drive, each gear change or cornering manoeuvre demands their full attention. Their awareness can be so absorbed in the sheer mechanics of the task that they hardly have time to take in the road ahead, let alone notice the passing scenery. Yet after some months or years of driving, the skills become ingrained enough to be second nature. People manage to drive home from work while chatting to a friend, listening to the news on the radio, day-dreaming, or planning a shopping list. Not only are they no longer conscious of the details of gear changes and steering, but they may become so caught up in non-driving thoughts that they are barely aware of having turned at the lights, or having made a tricky lane change.

This automation of quite dangerous decision-making and skilled action is not the exception but the norm. Our brains seem designed to handle as much as possible at a subconscious level of awareness and execution, leaving focal consciousness to deal with tasks which are either particularly difficult or novel. Not surprisingly, when cognitive scientists thought about the distinction between these two levels of brain activity, they found it natural to view them as two separate modes of processing, and thus likely to be inhabiting different pathways in the brain. But if the brain evolves its states of processing, then automatic behaviour and peripheral levels of awareness would fit into the story in quite a different way. There would be no specialised pathways, but instead a whole-brain reaction to the moment that began with many pockets of localised response and escalated to some more globally focused reaction. The brain would begin with a preliminary shake-down of all arriving information to try to deal with the routine business of the moment – stuff that is easy or habitual, like changing gears. This would then clear the way for attention to be paid to whatever was proving especially difficult or interesting. So there might be the appearance of two modes of processing, but really it would be two different depths of processing carried out across the same brain structure.

Already at the time of Libet's free will experiment, a number of psychologists were beginning to have thoughts in this direction. In particular, Bernard Baars of the Wright Institute at the University of California-Berkeley had developed what he called his global workspace model of consciousness. A cuddly, bearded bear of a man with honey eyes, Baars felt that the idea of an escalation from some local and pre-conscious level of processing to a broad-scale, whole-brain attack on a problem could explain an awful lot about the mind. So Baars was one of the few who always believed that Libet's results made perfect sense. However, Baars could also see that there was a deceptive simplicity to Libet's finger-lifting task which was to blame for much of the controversy. People were taking the experiment to mean that lower-level brain processes generate our thoughts and the conscious-level mind arrives too late in the day to do more than supervise the results. Yet that was not the case at all.

Two points had to be remembered. Firstly, the students went into the experiment knowing what kind of action they were expected to produce. If they were at all uncertain, they would check with Libet, flexing a finger and asking, 'You mean, go like that?' So they were primed with a conscious-level image of what was meant to happen, and had even warmed up their motor centres with rehearsals of the movement they might make. Much of the thinking needed to deal with what was in truth a fairly novel situation – taking part in an experiment with electrodes stuck to their heads and making an otherwise pointless motion – was done well in advance. Then, even as they were doing the experiment, they had to be alert to how well they were doing what was asked of them. Were they keeping their minds sufficiently blank to allow the required impulse just to happen? As Baars put it, there was always a conscious-level context in place, framing whatever occurred.

The second point was that Libet was asking the students consciously to notice something that in ordinary circumstances their brains would ignore. The very essence of automatic behaviour is that we are confident enough about its execution to let it go unremarked. We do not have to note the fine detail of a gear change; the context will tell us when to change, and habit will make the action smooth. But Libet wanted to make clear at exactly what point we can become globally aware of a bubbling-up impulse. He was asking for the escalation of a minor brain event precisely because he wanted to expose the time course of such a processing step.

With the free will experiment, Libet was starting to get right inside the moment, seeing the cycle of development that leads up to a state of settled consciousness. But by now fast approaching retirement age, he only had time for one more experiment. What he ought to do was obvious. The general reaction to his work had made it clear that most people still took his half-second finding to mean that no worthwhile brain activity could be completed in under half a second, not that there might be successive stages of pre-conscious then conscious processing. So the next logical step would be to demonstrate that the brain could react in some subconscious way to a lesser period of electrode stimulation – that while it might take half a second's worth of pulses to sustain the activity needed to produce the phantom sensation of a furrowing in the skin or water dripping down the back of the hand, perhaps less than half a second might still have some quiet effect on the brain.

This final test was difficult to set up, as it meant going back to brain stimulation and Feinstein was not around to provide a ready supply of subjects. However, a University of California neurosurgeon, Yoshio Hosobuchi, happened to be experimenting with implanted electrodes as a novel method of pain relief. Certain kinds of severe pain, usually caused by damage to the spinal nerves, do not respond to drug treatment but can be blocked by electrical stimulation of the part of the thalamus bringing in the pain messages. Patients have the electrodes permanently fixed in place, then strap a battery pack to their belts so they can give themselves a quick jolt whenever the pain gets too bad. The fact that the electrodes were implanted allowed Libet to do an experiment without even going into the operating theatre.

Libet planned a variation on a standard subliminal perception test. In this experiment, a sub-threshold stimulus – one too weak to be consciously seen, such as the fleeting projection of some word or picture on a screen – is presented during one of two time intervals. Because the image flashes past much too rapidly to be noticed, the subjects feel that they saw nothing during either of the test periods. But if told not to worry and just to have a guess in which interval something might have occurred, more often than not people will guess the right answer. The success rate is certainly not high: hits usually range between 60 and 70 per cent, which is not much more than the 50 per cent they would score by chance anyway. Yet still, something must be happening at a subconscious level to tip the balance by even this much.

For the purposes of Libet's experiment, the stimulus became a jolt delivered through the pain-control electrodes buried deep in each patient's thalamus. The subjects were sat down with a pair of lights to mark the two time intervals. Then, during one of the intervals, Libet would switch on the electrode for either longer or shorter than half a second. When the pulse train lasted for more than half a second, all the subjects were quite definite they had experienced something, and when it was less, they were equally sure nothing had happened. But as Libet predicted, if asked simply to have a wild guess, they turned out to be right about 65 per cent of the time. In some cases, as little as 150 milliseconds' worth of thalamic stimulation could make a difference.

Plainly, the idea of a subliminal effect itself raises some thorny issues. There is the problem of how a slight eddy of sensory input – one too faint to be positively detected – can still cause enough disturbance in the brain occasionally to tilt a verbal response in the right direction. However, Libet's experiment at least confirmed that lesser periods of electrode stimulation had some slight impact. His half-second result could not be taken to say that the brain stood idle and incapacitated while it waited for a state of consciousness to snap belatedly into place.

Libet's last paper, published in *Brain* in 1991, received a much more sympathetic hearing because the neuroscience world was changing rapidly. If nothing else, it helped simply that researchers were talking about consciousness again. For as soon as consciousness came out of the closet, a lot more attention started being paid to shades of consciousness. The difference between focal-level processing and more automatic or pre-conscious levels of processing became a fashionable topic. But the emergence of a dynamic approach to the mind also made the general tenor of Libet's results seem increasingly acceptable. As with Desimone's evidence of attention effects, a finding that was heresy one year became almost scientific orthodoxy the next. All it took was a subtle shift in the prevailing point of view.

Yet still, there was a grave sticking point with Libet's half-second work. The trouble was that awareness simply did not feel delayed, and Libet's talk about the final picture including some backdating illusion sounded a little weak. There had to be a better explanation of how the brain papered over its processing gaps. The answer, it turned out, lay in anticipation.

CHAPTER SEVEN

A Moment of Anticipation

Concentrate. Keep your eye on the ball. Watch it right on to the face of the racket. How many novice tennis players have vainly struggled to follow such advice? A beginner hitting another air shot would find it all too easy to believe Benjamin Libet when he says our awareness of the world comes half a second late. Yet if Libet is right, even professional tennis players must experience the same processing gap. Given that on the men's circuit, serves are regularly banged down at 120 miles per hour and so take less than half a second to travel the length of the court, this means that the last time many players might have conscious-level information about the location of the ball would be while it was still in their opponent's hand!

Sports psychology is one of the few areas in the mind sciences where researchers are forced to confront the issue of mental processing times. The field has only really existed since the 1970s, paid for by the growth of professional sports. Yet the practical problem of helping athletes hit balls or quicken their reflexes has made sports psychologists ask the kinds of questions that the rest of psychology has side-stepped. Researchers have had to get inside the conscious moment to discover why some people are slow and uncoordinated, while others – the gifted few – seem able to conjure with time.

If asked what makes someone a sporting star, the natural assumption is that the person must benefit from some basic speed advantage. They must have quicksilver reflexes or a faster-reacting eye, or perhaps the motor control centres of their brains turn round decisions much sooner. So it has been a huge surprise for sports psychologists to find that top athletes score virtually the same as the average person in reaction-time tests, tests of visual acuity, or any other raw measure of mental processing ability. When sat at a lab bench and asked to hit a button as soon as they see a light flash, gifted baseball or tennis players might have fractionally faster reactions. They might average 200 milliseconds compared to an average of 220 milliseconds for a control group of ordinary people. But such differences are too small to explain a huge gulf in athletic ability. And besides, sports psychologists believe the twenty-millisecond advantage is probably due to other factors such as the athletes having the muscles to move their hands a bit faster, or their competitive natures might make them concentrate harder, preventing the occasional lapses that would bring their average down over fifty or so trials. In short, the evidence from reaction-time tests is that the period needed to form an awareness of a sensory stimulus – or rather, as the work of Libet suggests, an early subconscious-level detection – seems fairly standard for the human brain. If there is variation, it is surprisingly slight.

A few sports psychologists speculated that the apparent quickness of an athlete's reflexes might be something which only showed on the playing field. Perhaps with training, people would hone the pathways that dealt with seeing balls or dodging a lunging opponent, allowing them to cut normal sensory development times on these particular skills. However, careful experimentation ruled out even this possibility.

One of the best-known studies was done in 1987 by Peter McLeod, then a researcher at the Applied Psychology Unit in Cambridge, England. This famous laboratory, formed during World War Two to help with pilot training and instrumentation design, has long been a bastion of psychophysics research. Following in the lab's tradition of practical experimentation, McLeod got together a group of cricket players, all internationals from the England team, and filmed them in slow motion to discover how they coped with various kinds of deliveries from a bowling machine.

Like other ball games, the dimensions of a cricket pitch are not accidental but have evolved to test a player's reactions. The distance between wickets has been dictated by the speed with which a bowler can hurl a ball, so that hitting the ball becomes difficult but not impossible. In top-class cricket, a fast bowler can send down deliveries at ninety miles per hour, meaning the ball will reach the batsman in 440 milliseconds. And even a medium-paced delivery of sixty miles per hour will still cover the distance between the bowler's hand and batter's crease in just 660 milliseconds. However, the time constraints imposed on a batsman are actually much tighter than these figures suggest. The raised seam of a cricket ball means it can kick sideways as it bounces off the ground in front of a player. Cricket balls may also develop a late swing just before impact. By roughing one face of the ball to increase drag and then angling the seam against the angle of the flight, a bowler can make the ball bend just as it begins to slow in the air. The result is that batsmen will often find themselves having to make hurried adjustments to a ball doing strange things just a few feet away from them.

The batsman's choice of shot also brings its own time constraints. The safest shot to hit is straight down the line of an incoming ball. As long as the delivery does not move too far off-pitch, it should eventually run smack into the face of the bat. But more attacking shots must be made with a hook or a cut across the ball's flight. In these cases, even a very slow delivery of forty-five miles per hour will give a batsman just a four-millisecond margin of error for getting his bat into the right place at the right moment. A few thousandths of a second early or late with the swing and he will find himself swiping at thin air.

To discover how cricketers manage to fend off hostile bowling, often for hours at a time, McLeod scrutinised the slow-motion footage of his group for clues. First he found that the batsmen began their strokes in a highly stereotyped way. Each kind of shot, of course, required a somewhat different preparation in terms of placement of the feet or backlift of the bat. But once a player had decided on a particular stroke, such as a hook, the body would turn and the bat would be taken back to exactly the same position as if the action ran along a fixed groove. There was a metronomic precision to the preparation that contrasted greatly with that of even a group of club-standard players.

Next, McLeod looked at how the top batsmen dealt with the unpredictability of the actual delivery. Using a ball machine, McLeod fixed the speed and angle of each delivery. Then, to make the bounce testing, he laid a bumpy bit of matting just in front of the batsmen, ensuring the ball would take the occasional wicked kick. To hit such a ball, a player would have to make a hasty mid-course adjustment in the trajectory of his swing.

Poring over the record of hundreds of strokes, McLeod found that the batsmen never reacted immediately to the changing flight of the ball. Instead, their swings would continue going straight down the original line for a full 200 milliseconds before suddenly veering sideways to make the correction necessary to intercept the ball on its new path. What was surprising was not just how late, but also how sharply this adjustment was made. McLeod had thought there might be a smoother change with the players slowly bringing their bats round as they watched the ball start to kick away, but instead the batsmen leapt straight from one path to another, as if they were jumping tracks having just completed a lengthy set of recalculations.

McLeod also found it remarkable how the 200–millisecond lag in reaction time was such a constant figure for all the batsmen involved. McLeod says: 'It didn't matter whether the correction was big or small, all the times bunched at around 200 milliseconds. I never saw a hint of any change in less than 190 milliseconds. I videoed some real cricket matches off the TV and checked the pictures with a ruler to check it wasn't just something to do with the set-up I was using in the laboratory. The times were just the same. If the ball did something nasty less than 200 milliseconds away from a batsman, he would act as if he never saw it.'

McLeod's results are graphic evidence that brain-processing lags are real. Libet had suggested that full consciousness needs about half a second to develop, but McLeod's experiment – and the many hundreds of other sports studies like it – demonstrate that even rapid pre-conscious processing takes up an unexpected length of time. The very quickest reaction of which a human is capable is to the crack of a starter's pistol at the beginning of a race. The bang of a gun is a simple event to process. It is not like judging the changing flight of a cricket ball in which there must be at least some time taken up in seeing the deviation begin to happen. And crouched in the blocks, a sprinter's response is already planned; there is no need to spend time calculating trajectories. Yet despite such a low processing overhead, it still takes more than a tenth of a second for a sprinter to respond. Indeed, this time is so in-built that in international competition pressure sensors in the foot blocks are used to rule any movement in under 120 milliseconds an automatic false start. Brain processing has an absolute limit, and this has become a fact enshrined in the rules of modern sport.

So sports psychologists were faced with the problem that the emergence of awareness was an incompressible factor. You could not be better by being faster mentally. Worse still, the brain did not react to anything in less than a tenth of a second, so regardless of what you believed about Libet's half-second claim for full consciousness, there was a puzzle over how the brain managed to deal with the last few instants of a ball's flight or a late lunge by an opponent, in any game. Where could the advantage of a top-class athlete lie?

Part of the answer is that the physically gifted must have special motor skills. When an international-standard cricket player sees a delivery turn, or a tennis professional is faced with a ball skidding low off the Wimbledon grass, something about their balance and co-ordination allows them to organise a tidier, more fluid response than the average person. There is an economy and a precision that buys them time. Electrical recording of the muscles of top athletes has shown this to be literally true. Electrodes were taped to the arms of both novice and expert players in several sports to record the EEG crackle of the messages being sent to their muscle fibres. It was found that whereas the novices produced a barrage of nerve signals, straining every muscle to bring about a movement and often causing opposing muscle groups to wrestle against each other, the limbs of the experts moved almost in silence. Their brains appeared to know exactly which muscles to pull to get the job done, making their movement silkily sure. To an extent, such economy always comes with training – the brain can learn efficiency – but sports psychologists believe that some people are lucky and have brains that are better at honing down the motor template needed to execute an action. During the moment, their brains might not work any faster, but the gifted are more receptive to the practice that will eventually allow them to work smarter.

Yet there had to be more to the story of how people manage to execute skilled acts when their brains lag behind the moment. And as sports psychologists pored over their slow-motion replays, checking to see when good players first started preparing for a stroke or moment of action, they soon realised that anticipation must hold the key. The brains of the best were making earlier and more accurate predictions about what was about to happen, and it was this that was carrying them through the moment, allowing them to behave as if they could feel the ball right on their rackets when making a feathery drop shot, or to take a delicate, last-second decision to glance away a turning cricket delivery.

An easy example to study was the return of serve in tennis. Facing a fast serve, players have barely 400 milliseconds in which to see whether the ball is headed for their forehand or backhand, and then to make any late adjustments for unexpected skids or jumps of the ball off the court surface. Given that simply turning the shoulders and lifting the racket back occupies a third of a second, and that it takes about half a second to reach wide for a ball, anticipation has to have a role. Even if awareness were actually instant, it still would not be fast enough to get a player across the court in time.

Tests were carried out in which novice and professional players were shown film clips of a person serving. The film was stopped at different stages of the server's action and the subjects were then asked to guess whether the ball was going to land on their forehand, backhand, or smack down the middle. Neither the novices or experts had any trouble predicting where the ball would go after seeing just 120 milliseconds of flight. This showed that they could all anticipate. They did not have to see the ball land. But the significant finding was that the professionals were able to guess the direction of a serve with fair accuracy if the film was halted forty milliseconds before the ball was struck. The seasoned players were gleaning hints from the way the server was shaping up during the ball toss, and not having to wait to sample the actual flight of the ball.

Dozens of other such experiments have since confirmed that

sports players buy time by learning to read the body language of their opponents. Bruce Abernethy, a sports psychologist at the University of Queensland in Australia, has shown that top badminton players can tell a lot from seeing an opponent's chest and shoulders begin to move a full 170 milliseconds before the shuttlecock is struck. Likewise, Abernethy filmed cricket batsmen and found that they were stepping forward in anticipation of a short-pitched delivery some 100 milliseconds before the bowler released the ball.

Another key point about this habit of prediction was that it was never an all-or-nothing affair. It was not tied to one particular moment in an opponent's ball toss or wind-up, but instead took the form of a dynamically narrowing cone of probability. Each player began with broad expectations, usually dictated by their knowledge of the capabilities of their opponents or thoughts about what their opponents might need to achieve due to the state of the game. Then, watching their opponents shape up would start to give them general hints about how to prepare – perhaps enough for a cricketer to decide whether to step on to the front or back foot, or a tennis player to begin swivelling left or right. But the guessing games never stopped. Tests showed that seeing the first 100 milliseconds, then the second 100 milliseconds of the ball's flight would lead to a steadily more accurate idea of what to expect. The skilled players were refining their state of expectancy right until about 200 milliseconds before contact, by which time, as McLeod's experiment showed, the brain could no longer physically react. If something happened to a ball that late, even the most accomplished player would swing and miss.

Frustratingly for the sports psychologists, who obviously wanted to be able to teach the secrets of good anticipation, none of the top players could explain what it was they were actually looking at to get their clues. When questioned, they said they did not feel they were watching anything in particular. Indeed, most said they had not even been aware they were making guesses ahead of time. They believed they had simply been concentrating hard and making sure they watched the ball right on to their bat or racket, so were conscious of the shots pretty much as they happened.

*

Libet's half-second results say that awareness is smeared out after the event. First there is a pre-conscious phase of processing, and then some kind of conscious-level resolution. But anticipation stretches our ideas about brain processing in the opposite direction. It says that predictions ease our passage into the moment. In some sense, we are conscious ahead of time. We do not notice the large gap in our awareness because our brains move seamlessly from a state of intelligent forecast to a state of confirmed sensory expectation.

Plainly, the habit of prediction is not something reserved just for special-case situations like hitting a tennis ball. Every moment is processed within some prior context – a framework of hopes and fears, intentions and expectations, memories and goals. These form the backdrop against which the events of the moment will be judged. And the more we get right about the coming moment, the less work there will be to do during it. If the thinking has already been done, most of our actions can be carried out on automatic pilot, leaving the brain to focus its attention on whatever turns out to be truly surprising, novel, or significant about an instant.

A little introspection makes it clear that even our most trivial activities come freighted with a dynamically tapering cone of anticipations – some which may be consciously explicit, but many which are either dimly conscious or apparently even unconscious and implicit. For example, when we reach out for the gleaming brass handle of a door, our brain will not only be predicting the instant of contact and the correct angle at which to hold our hand, but it will also be second-guessing how the handle should actually feel as we touch it. It will be predicting the sensory parts of the experience. The fact that we are riding such a wave of predictions would soon be brought home to us if we were to reach out and discover that the handle was made of something sticky or mushy. At some subconscious level, we would already have formed the expectation of touching cold, unyielding metal. Indeed, if someone the other side happened to snatch open the door at the moment our fingers were about to close on the handle, we might even catch a ghostly impression of what we were just about to feel with our hands. We would experience the fleeting edge of our own sensory forecast.

It is almost impossible to imagine a moment without a context. There is always something about what has just happened that predicts what is likely to happen – or not to happen – next. Even sitting in our homes, loafing in a comfy chair and apparently not thinking about or doing anything in particular, we would still be deeply embedded in a set of expectations. There would be a mental backdrop that told us what kind of events were most probable. So we might be half-expecting our partner or cat to stroll through the door, but not our boss from work, or an armadillo. Likewise, we might half-expect the phone to ring or a breeze to flutter the curtains, but not the walls to change colour or the carpet suddenly to start making snide personal remarks about us. We would feel oriented to our surroundings by a carefully graded sense of possibility.

Of course, life can catch us out. The unpredictable does sometimes occur. As we lounge in our chair, an armadillo might wander through the door, or more plausibly, a burglar. Or perhaps something which has become familiar might stop. The neighbours may turn down a droning radio, or the dull hum of our fridge could cut out. At such moments, the mismatch between our state of expectation and the turn of events will often cause a baffled double-take. We will find ourselves floundering for an instant, struggling to reorient ourselves in the new situation – needing perhaps as long as Libet's half a second to take in events and get back on track. Yet the very fact that we can feel caught out simply confirms we must have had a set of expectations in the first place. Surprises have to have something to contradict.

The question then is, how does the brain generate a state of anticipation? From a computational point of view, anticipation looks a very difficult ability to explain. A computer works by fetching data and then executing instructions. It is a step-by-step style of processing in which something happens first – the data arrive – and then the system sets about trying to make sense of it. Of course, a clever designer could always program a computer to make predictions and mobilise data in advance of the next cycle of activity, but any expectations would have to be spelt out in concrete fashion; each element would have to be mobilised individually, so it would take a lot of effort to generate predictions of much detail. The almost instant whipping-up of a flexible, open-ended yet constantly hardening state of readiness is not something that would seem to come naturally to a computer. The clunkiness of their processing style suggests that no matter what the number-crunching power of their circuitry, life would always remain a succession of surprises to them.

The strict fetch/execute logic of computers meant that cognitive science never quite got to grips with the idea of anticipation. Of course, there were honourable exceptions such as Bernard Baars, who made context the cornerstone of his global workspace theory. And more particularly, Ulric Neisser of Cornell University in New York, who was actually one of the founding fathers of cognitive psychology, was always very clear about the fact that the brain goes into each moment fully primed. In his 1976 classic *Cognition and Reality*, Neisser argued at great length that perception was the result of a cycle of processing in which anticipations blur into confirmed sensation, so bridging any processing gap and also making the whole business of representation more efficient.

But it was hard for such insights to shape a generation of researchers brought up to think about the mind as a collection of modules and functions. Most of the cognitive scientists touching on anticipation tended to treat it as either some tacked-on feature – an ability wheeled out to deal with special cases like hitting tennis balls – or else described it using rather abstract terms, such as 'perceptual schema' or 'mental disposition', which disguised the fact that anticipation was a phenomenon that existed in time. A schema or a disposition was something that sounded as if it could be fitted to the data at any point following their arrival. There was no implication that it was a state of information that the brain must rouse ahead of each coming moment.

With the move to a more dynamic, evolving model of the brain, however, anticipation immediately became much easier to understand. Rather than being an extra feature that must somehow be laboriously welded on to the processing of a moment, the generation of expectations began to look inevitable. It was something a dynamically constructed brain would do for free.

The best evidence of what might be going on to produce a state of expectation at a neural coding level again came out of the laboratory of Robert Desimone at the US National Institute of Mental Health. Desimone's recordings from colour-coding cells in V4 showed how states of attention – or, perhaps more accurately, states of intention – could tailor the firing response of an individual neuron. Desimone and his team then went on to explore the mechanisms of these effects in more detail. And one experiment in particular, reported in *Nature* in 1993, seemed to offer a lot of clues about the production of anticipations.

In this experiment, Desimone's team recorded from cells in the inferotemporal (IT) cortex of a monkey while it waited for a target image to appear on a screen. Desimone already knew that the IT area had maps coding for the sight of complex visual objects. It was the place where researchers in the 1970s found that waving a hand would produce a response from an anaesthetised monkey. More careful experiments had since proved that while some IT neurons were tuned to react to highly specific phenomena like hands and faces, most coded in a more general way. They did not code for particular experiences like grandmothers, but represented the perceptual elements – the assortment of shapes and textures – that might be needed to paint a population vote of a grandmother's face.

The most painstaking research had been carried out by Keiji Tanaka at the RIKEN Institute in Japan. Tanaka's method was to show an anaesthetised monkey a photograph of some natural object, such as a tiger's head, which would get a lot of cells firing, and then progressively simplify the picture until some chosen IT cell ceased to respond. So with one cell, for example, the tiger's head was reduced to just a white square with two small black rectangles roughly where its ears would be. The neuron appeared tuned to representing this precise conjunction of features because when the square or rectangles were shown alone, there was no response. With enormous patience, Tanaka followed this procedure for a great many cells and found they combined to represent a whole spectrum of visual primitives. There were cells that fired to T-shapes, stars, pairs of touching balls, and, of course, other common fragments of experience like hand shapes and face shapes. Other cells seemed to specialise in coding for surface textures such as hairiness or smoothness. The IT neurons were also topographically arranged so that neighbouring cells had the same basic object preference, but with a slight shift in orientation, size, or proportion. For instance, within a group of cells responsive to star-shaped patterns, some would fire at a peak rate to fat or many-armed stars, while others might prefer skinny or sparsely armed stars.

So the IT area seemed perfectly set up for population voting. Any kind of visual conjunction could be represented by a blend of firing. And Tanaka even showed that this grid of representation was adaptive; experience could produce long-term changes in the tuning of a cell. In an experiment reminiscent of Merzenich's finger stimulation studies, Tanaka spent a year training a monkey to pay special attention to twenty-eight target shapes. When he tested the monkey at the end of this time, he found that many more of its IT neurons now reacted to the shapes. The cells had shifted their tuning curves so as better to fulfil the demands of what had become a frequent sensory task.

Tanaka's work revealed a lot about the representational logic of the IT cortex – the principles behind its organisation. But Desimone wanted to discover what happened to such cells when they were called into action and were helping to form part of a real state of consciousness. So an awake monkey was given the task of seeking and finding a series of pictures. In the experiment, a trial would begin with the display of a target, which might be a drawing of a small sailing boat. This would disappear and then, just three seconds later, reappear along with a second picture of perhaps something like a man's face. The monkey was supposed to make a choice and signal recognition of the original image by flicking its eyes to look straight at it, the usual apparatus of eye-position coils picking up the direction of its gaze.

The design of the task meant that at a global level the monkey had to do a number of things. It had to note and remember a target picture. Then it had to find it again and focus on it to the exclusion of all other stimuli a few seconds later. The resulting activity of the IT neurons displayed several interesting features. The first significant finding was what happened to cells that were not involved in coding for the target, but instead coded for the ignored picture. A cell tuned to the sight of a face would burst into life every time a face appeared alongside whatever happened to be the target picture for the trial. The cell would fire at the rate of some twenty spikes a second to tell the brain what it was seeing. But then suddenly, within the space of 200 milliseconds, the firing of the neuron would be suppressed. Its firing would fall back to only about six or seven spikes a second.

This was, of course, a repeat of Desimone's original V4 finding. Any cell coding for a potentially distracting sight would be hushed up and physically pushed into the background of awareness. The brain appeared to create a consciously focused state of representation by turning up the volume on what it wanted to hear and turning down the volume of any surrounding activity that might interfere. But the new point was that it took a little time for this attention effect to show itself. The face-coding cell fired brightly enough for nearly 200 milliseconds, signalling the presence of the distraction, and only later became damped by the more global needs of the brain.

In fact, the same slight delay had been present in Desimone's V4 results, he just had not mentioned it at the time. The colour-coding cells had begun by firing brightly and then switched to a tuned response only after 100 to 200 milliseconds. However, it was a crucial discovery because it said that the brain began with a raw response to the moment. Everything started with the chance of being represented. A tracery of mapping would course its way up the sensory hierarchy, so laying the foundation for at least a peripheral or pre-conscious level of awareness. It was only after a phase of basic sensory integration that the more global cross-currents of feedback and competition began to flow and have their effect on a cell. States of focus had to evolve.

For an explanation of anticipation, however, what was more interesting was the behaviour of cells actually coding for the target of a trial. As might be expected, on first seeing the target presented alone, and again when the monkey had to distinguish it from a distractor, these cells fired at maximum strength. And tellingly – given Libet's half-second claims – the presumably consciousness-producing firing always persisted for at least 500 milliseconds, even when it meant that a neuron was still going well after the picture had already been switched off! But what was more important from the point of view of anticipation was how the target-coding cells behaved during the short wait between the two exposures. Desimone found that their firing rate dropped, but they never actually went quiet. The cells kept up a chatter of six or seven spikes a second, as if coding for a state of memory or expectation. There was a template of activity, a gentle warming of the pathways in the IT area, which would match the coming experience.

By itself, Desimone's experiment did not actually prove anything. Recordings from solitary cells could only hint at how the IT neurons were interacting with each other – or, indeed, with the rest of the brain's processing hierarchy – in representing a state of information. And simple firing rates were not everything anyway; the relative timing of each spike was likely to play a role as well. Yet the preservation of a slightly raised state of firing did make sense if the brain was seen as a dynamic system.

The computer model suggests that the brain is an inert lump of circuits awaiting input. Sensation is fed in one end and a hierarchy of mapping cranks out a state of consciousness at the other. But the dynamic view paints a very different picture. It says that the processing structure of the brain only exists in the first place because it has achieved a prevailing balance of tensions. Continual feedback pressure is needed to shore up everything from the transmission properties of an individual synapse to the mapping properties of a patch of cortex surface. So unlike a computer, things are going on even when the brain appears to be doing nothing. A quiet brain is still having to produce a state of tone. It has to give new input a surface to disturb.

This brings up yet another uncomfortable feature of neurons which neuroscientists usually try to skirt around. They are in fact always firing. Every one of the billions of cells in our head is popping off at least one or two stray spikes each second. When developing theories about neural coding mechanisms, the temptation has been to dismiss this constant background rustle of activity as meaningless noise. The message was believed to lie in the bright or synchronised firing, and the odd pop of a neuron was seen as just the inevitable consequence of trying to do computing with sloppy biological components. As watery bags of ions, cells could not help but leak a little current when not in use. This was no great problem because it was easy to imagine that the brain's coding mechanisms would include some sort of threshold setting to make sure that this idle tick-over firing was screened out when the time came to count an area's final population vote.

Yet, as dynamicists like Karl Friston were beginning to realise, this background firing might not be so random after all. If brain cells are woven into a network of feedback relationships – connections that give their own firing meaning – then the popping off of a neuron would not be noise but an expression of an underlying state of organisation. Cells would be triggering each other with skitters of activity in a way that reflected their connections. Or, as Friston put it, an area of mapping would be cycling in an attractor state, some general balance of tensions.

This simple fact causes a 180–degree switch in perspective. A computer represents a state of nothing doing by doing nothing – by having silent circuits. The arrival of input then forces it to go from a nothing to a something. But the brain works the other way round, starting with a state of firing that in some vague fashion represents everything ever experienced by an area of circuitry, then tilting towards some specific state of firing. It goes from a defocused representation of all it knows to a focused response to new input. A cloud of everything condenses to become a something.

What this means is that even in a state of rest, at its most defocused, the brain is in some way prepared. Its circuits stand poised to be tipped into a more definite reaction. At a subjective level, we might experience this state of tone as a sense of readiness or potential. It is notable that when we shut our eyes, we see not blackness – an absence of information – but instead a shimmer of shape and colour. The rustling of our visual areas appears visible. They are already halfway to going somewhere. And because this state of tone is an active construction – a feedback alliance – it would be easy to begin tilting the balance of firing in a certain direction. By lifting the background activity of a group of cells slightly – say, neurons coding for the experience of seeing a picture of a boat – a bias could be set in place. Then, when the time came to run the competition to discover what was actually in the moment, this slight edge of priming would nudge activity in the chosen direction. Through the power of feedback to amplify small differences, a slightly higher tick-over firing rate in a group of boat-coding cells would be enough to ensure that the claims of rival object-coding cells were drowned out as soon as an area like IT was driven into a mapping response. A monkey would find the sought-for target being thrust into view.

With a computer, an active decision would have to be made about what kind of anticipatory bias to load into its circuits. But the brain already has all its information loaded; it is merely a question of how much to push the focus towards some specific experience. And there would be huge flexibility in creating a state of priming. The brain could rouse a wide area of circuitry to create a general readiness – a gentle pre-warming of all the pathways most likely to be involved in the coming bout of processing. Or it could raise the firing profile of a select group of cells to catch some more particular event.

But the real beauty of the system is that a state of anticipation would not prevent the brain fixing on something else. If someone had slammed the lab door while the monkeys were doing the experiment, the stimulus would be powerful enough to override the sight of the target. The dynamic model says that an anticipation merely produces a fleeting tightening of the brain's processing landscape, perhaps deepening the basin of attraction for things like pictures of boats while raising the threshold for other experiences, such as pictures of faces. So a pattern of activity will fall more easily into a certain groove, but only if it was passing that way already. The brain always remains free to head in other directions if the moment does not work out quite as planned.

If Desimone's work gave some clues about how the brain represents a state of expectation or intention, there was still the bigger question of how the brain generates such a state. Where does the brain get its ideas about exactly what to anticipate? A dynamic view of brain processing gives a very simple answer: predictions would flow quite automatically from whatever has just been the brain's last point of focus.

Anticipations are there to get us into the moment, to allow us to deal with a flood of sensation with great efficiency. Every instant comes packed with a vast amount of detail. Desimone's monkeys faced not just the sight of two pictures; their senses were being assaulted by all the sights, sounds, smells, and feels that go with being in a laboratory cage, watching a computer display. But if much of the thinking and experiencing has in some sense been done in advance, then most new input will slot straight into place. A ripple of adjustment may still have to be evolved, but it can take place locally and pre-consciously. There will be no need to call on the global resources of the brain for a deeper, more considered reaction. As Baars argued, attention would be reserved for whatever part of the moment could not be dealt with quickly and cheaply at a local level – or, alternatively, because the brain had set out to catch the event when it eventually happened. Escalation into focal consciousness would occur either because something was particularly expected or unexpected, significant or surprising.

So the brain would arrive at a focus. Anticipation would act as a filter to screen events, and would leave some aspect of the moment standing proud. The brain would be left with a certain area of its memory and sensory pathways feeling sharply stimulated by what had just happened. Then, being sharply stimulated, these areas would begin to rouse further thoughts and associations – surrounding areas of memory – that would quite naturally warn the brain about what might happen next. So a growing state of anticipation would be also what we got out of the moment.

Take an example like the simple act of walking into our house one day after work and hearing our grandmother's voice coming from the living room. The cycle of processing would begin with an act of recognition. The brain's mapping hierarchy would pick up the pattern of noise trapped by the ear and draw it up through a stack of filtering to produce an organised state of representation. At the bottom of the stack, on A1, the primary auditory cortex, there would be a tonographic map of a set of frequency densities. Then, as this information was pushed through further layers of mapping and voting, it would begin to hit high-level areas where it would become identified as the sound of a voice – and a particular, known voice at that. In the auditory equivalent of IT, there would be a population vote suggesting that there was a very high probability we were hearing our grandmother speak.

Of course, the dynamic model says there is rather more to establishing a meaning-imbued pyramid of mapping. Signals do not just flow up through the mapping hierarchy – a one-way, bottom-up traffic in information with the peak level areas somehow ending up doing all the experiencing. Instead, the many levels of mapping grow into a focused state of representation in concert. They evolve together through the reinforcing effects of feedback.

Raw sensation may arrive in the lower mapping areas to start the ball rolling, but their first response would be a little ragged and untuned. It would need the high-level areas to begin voting for grandmothers for their activity to become confirmed. The dawning recognition at the top would then feed down the chain to sharpen and strengthen the pattern of activity at the bottom. From memories of our grandmother's voice, we would be better able to separate the sound of her words from any blurring background noise, or bring out certain characteristics, such as a slight croakiness in her speech. Hand in hand, over the course of a tenth of a second or so, the whole hierarchy will move from a rough preliminary network of voting to a crisp, stable, and highly interpreted state of representation. Our mental experience would become an inseparable mix of what our ears heard and what we felt they ought to have heard.

At the same moment our brain would be mapping many other sensations, but something about the unpredicted nature of hearing our grandmother's voice would be enough to ensure its escalation into the spotlight of attention. Her representation would be kept burning as other aspects of the moment faded or, as Desimone's work suggested, became actively suppressed. Then, with this lingering firing and a cleared deck would start to come a spreading stain of associations. There would be time for the flickers of feedback to rouse the various areas of circuitry to which our grandmother thoughts were connected, so bringing the right kind of new thoughts to the surface.

With population voting, the seeds of these ideas would already be present in the original response. The vote would stir a range of high-level cells, some of which might be considered 100 per cent grandmother neurons, firing flat out to the sound of her voice and little else. But many others might have fired at only 60 or 30 per cent, being tuned more to experiences such as the sound of our grandfather's voice, the voices of other close family members, or even just the sound of an elderly voice generally. So in rousing enough cells to get a decent bearing on our grandmother, we could not help but fish up the faint corner of thousands of connected experiences.

For an instant, at least, nothing much would come of these links. The state of representation would be drawn up too tightly around the sharp stab of the experience. But as the brain began to relax again, the excitement of the still-firing cells could seep out to create an inflamed halo of associations around the original act of mapping. Not only would there be a broader arousal of our sound-representing pathways, but the activity should cross over to stir areas in the other sensory modalities. The sound of our grandmother might rouse neurons that coded for grandmother-related sights, touches, and even smells and feelings – anything that we strongly connected with her. We might only have heard her voice while walking down a corridor, yet already some of the key cells needed to process the sight of her face or catch the waft of her usual perfume will have been alerted. Given that we now would also be beginning to think about the room in which she must be sitting, the stage would be set for generating a whole range of predictions.

If the spreading ripples of activity remained confined to the uppermost levels of sensory mapping, then it is hard to say what kind of feelings might be engendered – perhaps not much more than a sense of preparedness to have certain kinds of experience, a vague and contentless foreboding. But the fact that the cortex is a feedback-based system, with more paths returning down its hierarchy than heading up, means that the grand stack of mapping can be turned on its head. Sensations might work their way in from the bottom. But there is no reason why a jangling of neurons at the top should not cause a cascade of mapping in the opposite direction. The logic would go into reverse with sensory detail being added, rather than extracted, as the wave of activity ran back through the various mid-level filters and low-level maps. If pushed all the way down to the primary sensory surfaces, a high-level inkling ought eventually to become fleshed out as a fully fledged sensory experience – a vivid mind's-eye feeling of almost witnessing the real thing.

So, on hearing our grandmother's voice, almost immediately – within half a second, anyway – we might find a specific image flashing before us. An outward and downward rush of association might produce the fleeting impression of seeing her bent forward in an armchair, her face turned towards us with the usual wry smile as we open the door. Drawing on all that was most characteristic of our grandmother – and also on our knowledge about the look of our living room – our brains might whip up a complete synthetic experience that would slot fairly seamlessly into our actual experience a moment later.

Clearly, the more unvarying an experience has proved to be in the past, the more accurate our predictions will be. For example, the sensations that are part of everyday actions such as reaching for a door handle, or changing gears in a car, can be anticipated in total detail. Indeed, our own movements will cause many of the feelings we experience, and research has shown that the motor parts of the brain transmit their intention to move to the sensory areas a fraction ahead of time to give them actual warning. But after changing gears or opening doors many thousands of times in our lives, the likely sensations will be very familiar anyway. In such cases, the cascade of priming activity back down through the processing hierarchy would form a cone that ran narrow and deep. Our memory banks would cast a sharp shadow across the primary sensory areas a split second before our hands came into contact with the door knob or gear stick.

But often the situation will be more open-ended. We will not really know what to expect, so any cascade of pathway-rousing activity would have to be shallower, more diffuse. For instance, if instead of our grandmother we had heard the voice of some unknown elderly person on entering the house, then this would have generated a much more general state of anticipation in our minds. We would have recognised the elderly nature of the voice and so become primed for the sight of wrinkles and grey hair, but we would be much less likely to experience one particular anticipatory image. Yet the point is that sharp or general, the generation of a state of anticipation would be automatic. The brain would isolate the most important event of one moment, then simply by lingering on its representation a fraction longer, it would begin to create a glowing halo of priming that prepared it for the next.

So dynamics brings the hierarchical organisation of the brain alive. The same neural machinery can be as quick to generate states of information as it is to extract them. Yet there are still some puzzles to be answered. So far we seem to have been talking about a conscious-level development of an anticipatory state. We become focally aware of some significant fact – such as the sound of our grandmother's voice – and start then to experience conscious-level expectations. Yet sports psychology studies seem to suggest that a lot of anticipation is done at a pre-conscious level. When players are asked what they look for in an opponent's ball toss or run-up, they cannot reply. They are certainly aware of a few things at a conscious level, such as the fact they are playing a game of tennis or cricket and that they need to concentrate, but even this self-urging to concentrate amounts to no more than an attempt to keep their mind clear, to rid it of the kind of specific, consciously experienced thoughts that only seem to get in the way of a quick reaction.

It appears that automatic or reflex actions also have their own unthinking level of anticipation. And once more, this is not just a feature of playing sports. Even getting down a corridor to open a door involves a lot of subconscious skill, and therefore subconscious predictions. At some level, our brains would have to be churning out a stream of anticipations to prepare our feet for accurate contact with the carpet, or our hand for gripping the door handle.

However, as with Libet's free will experiment, it is important to remember that none of this spontaneous activity – either the motor planning that picks up our feet or the sensory anticipations that guide their fall – can occur without some kind of prevailing context in place. Taking a footstep or reaching for a door handle may seem like acts unconnected with any thoughts we might be having about the conscious experience of hearing our grandmother, yet they are fragments of processing that only exist because of our greater goal of getting ourselves into the living room.

In other words, implicit in the fact that our minds are prepared to jump straight to an expectation of seeing our grandmother is the belief that shortly we will manage to find our way into her presence. Our minds could detect no reason to expect any intervening obstacles, so the predicted impression of our grandmother becomes our prevailing context – the guiding image which will shape our actions for at least the next few seconds – and walking and opening doors then become activities that organise themselves to fit. As well-practised and easy-to-anticipate skills, there would be little need for us to bother with the details. The brain would deal with them at a quick, pre-conscious level without requiring the kind of escalation, focal sharpening, and prolonged exploration of possibilities that would make for a conscious state. It would only be when something went wrong – if the door turned out to be locked, or a ruck in the carpet tripped us up – that we would be forced to retarget our attention.

By now it should be coming clear that anticipation is as much about the control of motor output as it is a preparation to deal with sensations. Plans and intentions are really just another way of looking at the generation of an expectation – an expectation about what we will do rather than what the world is going to do. And there is even one further riddle that is solved by an understanding of anticipation, that of mental imagery. A mental image is simply a state of expectation that does not get matched to an actual sensation. We go through the first half of the perceptual cycle, getting ourselves mentally ready to see or feel something, but that something then never turns up, leaving us with the ghostly glow of our own sensory priming.

This explanation makes sense of the often tantalising nature of our mental images. As has been seen, psychologists have had great trouble getting to grips with imagery. It took the evidence of a PET study even to persuade cognitive psychologists that images probably use the same topographical pathways as ordinary perceptions. Many thought that as a high-level mental process, imagery should have its own brain areas and possibly even its own abstract neural code. Kosslyn was able to settle this argument by demonstrating that the act of imagining a letter generated a network of activity that ran all the way down to V1. But this still left most with the rather clunky, computational notion that imagery was a form of memory trace replay. If, for instance, we wanted to imagine a grey rhinoceros, then our brains would dredge up a rhino outline from one memory file, take a splash of dusty grey from another, load both memory traces into a high-level buffer, and finally project the resulting picture across the display tube of the lower visual areas. However, a mental image is a far less concrete state than such a cut-and-paste model would suggest.

For a start, most people find it impossible to keep a particular picture fixed in their heads for more than an instant. Almost as soon as an image appears, it begins to slip out of sight or transmute into something different. We might get a glimpse of a close-up on a rhino's dust-caked face, even seeing its ear flicking away a fly. But for most people, the image will be bright for just a split second before it starts to fade and, quickly, some other rhino image swells to take its place. Our minds might flit to a long-shot scene of a pair of rhinos stamping around a mud-hole, or a particular memory of a rhino seen at the zoo. Like a slide show, a succession of images will run through our heads, never giving us time to dwell on any one impression.

Anticipation explains this in-built restiveness. The brain was never really designed for contemplating images. Our ability to imagine and fantasise is something that has had to piggyback on a processing hierarchy designed first and foremost for the business of perception. And to do perception well, the brain needs a machinery that comes up with a fresh wave of prediction at least a couple of times a second, or about as fast as we can make a substantial shift in our conscious point of view. So while we can drive the brain briefly into an artificial state of anticipation – a state of sensory expectancy that we know is not going to be answered – it would be unnatural for the brain to linger and not move on.

Anticipation also accounts for the often nebulous character of mental imagery. Some images are undoubtedly sharp and vivid. In these cases, we would expect to find the priming activity reaching all the way down the sensory hierarchy so as to pick up the maximum amount of detail. And this is, of course, exactly what Kosslyn demonstrated with his PET experiments, where topographical patterns were found on V1 itself. However, Kosslyn's tests were designed to produce well-fleshed-out states of imagery. His subjects were asked to visualise copies of letters they had only just viewed. In everyday life, the same kind of vivid state of priming would be stirred if we were searching the hallway for a missing set of car keys, or about to make thudding contact with a cricket ball – situations where the details are highly predictable. Yet an expectation can equally well be broad and shallow. A gentle and diffuse spread of activity might leave us with just the feeling of being generally oriented towards the idea of seeing rhinoceroses. We might not have an actual rhinoceros image in mind, but we would have a strong sense of potential for moving towards such an image as soon as the need arose.

So anticipation and imagination are fundamentally the same. The difference is that an anticipation is a prediction tied directly to what is happening around us at the moment, but with a mental image we are putting ourselves in some other place and asking our brain what life might look like from there. From years of visiting zoos, watching TV wildlife documentaries, and reading *National Geographic* magazines, we will have well-stocked memory banks. All it takes is to activate the right spot and then let the spreading flow of activation do the rest. The brain would need no special circuitry or cortex areas. It has one hierarchy, but it is a hierarchy that can be exploited in many different ways.

This is not a new idea. In the 1970s, long before the Kosslyn–Pylyshyn debate ever took place, Ulric Neisser was able to write that 'images are indeed derivatives of perceptual activity. In particular, they are the anticipatory phases of that activity, schemata that the perceiver has detached from the perceptual cycle for other purposes.' But it was only with the emergence of a more dynamic understanding of the brain in the 1990s – with a general shift in context – that mind scientists began to feel that such explanations appeared rather obvious.

CHAPTER EIGHT

The Needs that Shape the Brain

Expectations rule even the science of consciousness. It is clear enough in the gossip at conferences or in the tone of most theoretical writings that there are some strongly held beliefs about the sort of advance which would have to happen to crack this perhaps deepest of all problems. Most commentators start from the position that despite a century of intensive psychological and neurological investigation, science is not even close to a glimmer of an answer. And if there is to be an explanation, it is going to have to come from an Einstein-like breakthrough. Through a flash of insight or a lucky experiment, some lone genius will stumble upon the one key processing trick or one key brain area that reveals everything. Suddenly we would all be able to see what turned inanimate matter into thinking, experiencing flesh.

This dream of one person with one fact succeeding where tens of thousands of researchers and millions of facts have failed is a fantasy born of reductionism. If the workings of a system can be reduced to a set of components, then science's job is to discover the vital bit – the gimmick – that makes consciousness possible. Dynamism, however – or, more correctly, the science of complex, adaptive systems – creates quite a different set of expectations both about how the problem of the mind should be approached and also about what may eventually count as an answer. Dynamism says it is not the bits that count but the way they hang together. In a complex system, everything is intimately connected. So, like a whorl in a stream, a brain state cannot exist separate of the pressures creating it. In a sense, there are no bits, just a plastic medium being moulded by evolutionary demands. Take away the pressure to do something – cut off the surging current of events – and the whorls will die. The structure will collapse.

This can be seen happening with consciousness. Put a person in a sensory deprivation tank, where there is nothing to see, hear, or feel, and awareness soon begins to crumble. After half an hour bobbing in the salty water of a darkened, muffled chamber, volunteers lose their feel for time and space. The sense that they inhabit a body disappears. Their thinking becomes desultory and disorganised. No longer buoyed by the pressure to map an exterior world – to internalise its structure – people begin to lose the capacity to represent even its basic dimensions. Being in the world, and being forced to respond, focuses us. It stretches us into a state of mental being.

As has been seen, the brain does have a processing architecture. Neurons have receptive fields and the cortex has its hierarchy of maps. So a computational view of consciousness is not totally wrong, but even a complete list of all the brain's components or a complete wiring diagram could never explain it. It would be too static a picture to give a feeling of understanding for what is a living system. If the brain is being shaped by what it is having to do, then from a scientific point of view, it is best to consider the brain as a fluid bag of circuits that gets sucked into a particular arrangement by the demands being placed upon it. Theories about brain processing should not be about bits of mechanism but about the actual problems that a brain has to solve during each moment of awareness.

Good timing is the secret to a lot of things. Late in the 1980s, a Nobel prize-winning immunologist trying to revolutionise the mind sciences with a more dynamic approach to consciousness managed to attract an awful lot of bad press. In a gathering of neuroscientists, the mere mention of Gerald Edelman's name was enough to send eyes rolling heavenward.

A New Yorker, famous for his intense manner, patrician bearing, and highflown tastes (he once trained to be a concert violinist), Edelman won a Nobel prize for medical research in 1972. Then, like Francis Crick, Edelman decided to branch out and tackle the bigger question of human consciousness. But whereas Crick spent most of his time visiting other people's labs, learning about what they were doing and helping to tease out the implications of their research,

Edelman wanted to have his own set-up and pursue his own vision of the brain.

A charismatic figure, Edelman proved adept at drumming up sponsorship. Eventually, enough money was raised to found the Neurosciences Institute, a $16 million 'monastery of science' built into a hillside at the Scripps Research Institute in southern California. The cash paid for wet labs to study neurology and dry labs for computer simulations of brain circuits. Edelman could also afford thirty full-time staff. For other neuroscientists, it was bad enough that a complete outsider was getting his own lab, free of any of the usual teaching responsibilities or funding constraints. But what really raised hackles was that Edelman was talking as if he had already discovered the secret of consciousness. Interviewed in the *New Yorker* magazine, Edelman said that the mainstream of mind science – dominated as it was by the computational model – was so far off the mark as to be 'not even wrong'. His own big idea was that brain processing was based on neural competition. States of consciousness evolved through exactly the same kind of classical Darwinian mechanisms that drove the evolution of life.

In a trilogy of densely argued books published between 1987 and 1989, Edelman laid out his manifesto for Neural Darwinism, or TNGS, the theory of neuronal group selection. Edelman wrote that just as animals compete for food and living space in the struggle for life, so sensations had to compete for space on the mapping surfaces of the brain. Every moment would begin with a battle in which some networks of activity would blossom, gaining the neural territory needed to become conscious-level percepts, while other, weaker, nerve ensembles withered away. Over the course of about a tenth of a second or so, there would be a struggle in which only the 'fittest' patterns survived. Crucially, one of the factors determining the success or failure of a new sensation was the support it received from higher levels of the brain. If a sensation was anticipated, or deemed important in some other way, positive feedback from higher areas would help swell the mapping activity, elevating it above the general clamour. Edelman called this feedback between levels of processing a 're-entrant circuit'.

Another essential point that Edelman felt the rest of neuroscience was missing was that it was quite wrong to think of the brain as a lump of fixed hardware. He saw that it was incredibly plastic – plastic on every timescale, from the fleeting instant of a neural firing pattern, through the minute-by-minute and day-by-day rewiring changes that create memory traces, right out to the developmental competition needed to wire the brain's pathways during childhood, and even the genetic timescale changes of ordinary evolution. The brain might look to have a stable design, but – as Merzenich's work with finger maps in monkeys had showed – it is in a constant state of flux, forever changing its connection patterns as a result of internal selection and competition. In short, brain circuitry flowed to meet the processing demands being placed upon it.

This struck Edelman as a beautiful idea. It meant that there was a continuum tying consciousness back to biology. Exactly the same selectionist principles ruled at every level of the brain's operation and only the pace of the adaptation varied. Genetic-level adjustments in the organisation of the brain happened with glacial slowness, while the neural competition to map the contents of a moment would flare and collapse in the blink of an eye. Yet the identical mechanism of 'processing through adaptation' applied.

This hierarchy in adaptive levels made consciousness seem a lot easier to explain. The story of the animal brain would be one of a gradual increase in the speed of its competition-based adjustments. Simple animals, like worms and jellyfish, could only adapt their nervous systems over generations. Slightly smarter animals, like slugs and snails, could make the nerve junction growth changes that gave them a rudimentary ability to remember and learn – a sort of reflex-level awareness. Then further up the evolutionary scale, animals become capable of creating fleeting mental maps of the world. In under a second they could evolve a change in brain state, a disturbance to a memory landscape in response to a fresh wave of sensory input. So consciousness turned out to be just a further twist in an old evolutionary tale. It was the result of speeding up the adaptive abilities of the brain until its circuitry became able to adjust to the events of a single moment.

This evolutionary angle on consciousness came naturally to Edelman because of his own work in immunology. Edelman's Nobel prize had come from helping prove that immune cells are produced through a selectionist competition. The problem for the body is that it cannot predict what kinds of bugs or viruses it might face in life, so its solution is to manufacture a huge and rather random variety of immune cell types. All these cells float in the bloodstream. Then, when there is an invader, whichever immune cell happens to have the right response characteristics will be stimulated to reproduce itself in great numbers, rapidly swamping the disease.

The principle of selection seemed so elegant and powerful that when Edelman decided to give the problem of consciousness a go, it appeared obvious that as a biological system, the same sort of logic must be at work in the brain. So Edelman was rather shocked when he found that the Darwinian model did not seem to feature at all in the computer-obsessed thinking of psychology and neuroscience. With typical vigour, Edelman tried hard to point this out. But he got little thanks. In a snub that typified the general response, Crick coolly remarked that Edelman was an enthusiast notable more for the exuberance of his ideas than for their clarity. Backs were turned on him, and researchers resumed their search for the neural code or topographical mapping principles by which the brain must process information.

It was true, of course, that Edelman made some presentational mistakes. He invented a lot of his own jargon for concepts that already seemed to exist, using words like 're-entrant' which did not appear to add a lot to the idea of ordinary feedback. Still more of a problem was that his use of the Darwinian metaphor made him sound curiously outdated. By the end of the 1980s, anyone thinking seriously about neural competition or evolution was already hurrying to jump aboard the new intellectual vehicle provided by the sciences of chaos and complexity. So while for some Edelman was a step too far, for others he already seemed a step behind. A much broader revolution was in the offing. For Edelman, the timing was all wrong.

The tale of Gerald Edelman is not important apart from what it says about the state of the mind sciences at the outset of the 1990s. The history of science is littered with individuals who seemed to have had a considerable personal understanding of the dynamic nature of the brain and the way it works. Brilliant theorists like Wundt, Hebb, and Neisser – and many others not so far mentioned, such as the Gestalt psychologist Wolfgang Köhler, or the Russian physiologist Evgeny Sokolov – have briefly opened the door to a way of thinking that might have led the study of the mind in quite a different direction. Yet the time was never ripe. To most people, the standard reductionist view of information processing always seemed to be working well enough – it was getting more things right than wrong – so there was no recognition of a need for change.

However, just a few years into the 1990s, the mind sciences did begin to see that groundswell shift in opinion. Everywhere they looked, researchers were discovering troublesome facts. Scanning showed too much activity in the brain; single-cell recording showed too much plasticity; the simple models of cognitive science were just not fitting. And suddenly, the sciences of chaos and complexity were offering a genuine alternative to reductionism, a way of thinking about an evolving system that was backed up by solid maths.

The change was certainly not universal. As is the way with all revolutions, it was the young researchers with nothing to lose who made the more fervent converts. And the change did not show through with great clarity. It was easy enough for academics like Friston or Kosslyn to get across what they meant in private conversation, where they could flutter their hands a bit or point at a nearby pond, but much harder to say the same things in the formal language of peer-reviewed articles. However, gradually a more dynamic view filtered into the mainstream of thought so that by the mid-1990s it had become difficult to find anyone who would disagree with at least the essence of what Edelman had been trying to say.

One of the points Edelman probably did make more clearly than anyone before him was that the plasticity of the brain's circuits meant that the tasks it was trying to achieve were likely to be stamped into its design. So, understand what the brain needed to do during a moment of processing and you would be a long way to under-standing the organisation of its pathways.

One such grand principle ruling the brain's layout had already become big news. As neuroscientists made a detailed exploration of the cortex during the 1980s, the variety of mapping areas threatened to become overwhelming. The visual pathway alone had more than thirty maps. But it soon became clear that there was a basic division of the sensory cortex. The maps were aligned in a way which created two broad streams of processing. One stream focused on the 'what' of a sensation, extracting information about its identity and sensory qualities – the kind of information needed to recognise an object or event – while the other stream of processing concentrated on the 'where', pinning each sensation to a particular point in space.

The separation of the 'what' and the 'where' streams was most obvious with the visual hierarchy. Processing started on V1, the primary map right at the back of the brain. Then all the filters most closely connected with answering the question 'What is it?' ran down from V1 along the curving flank of the temporal lobe. The pathway started with the basic shape and colour filters, areas like V4, and then ran into the temporal lobe areas, such as IT, which specialised in object recognition and memory associations. Heading in the opposite direction from V1, up over the back of the cortex towards the parietal lobe, was the 'where' pathway. This started with the cluster of filters that could extract information about binocular depth, motion, and other positional details, then ended in the spatial mappings of the posterior parietal cortex (PPC), where cells fired whenever some significant object crossed a point in mental space. So, to put it crudely, V1 captured the basic picture, then, using one pathway, the brain created a set of labelled visual objects; using the other, it tied these objects back to a spot in three-dimensional space.

Identity and position are the two key facts that the brain has to extract in representing any element of the visual scene. It is also clear that quite different styles of processing are needed to produce such information. Making the memory match to identify a sensation is a narrow, focal kind of population vote, while placing that same sensation within a sense of space needs a more global style of representation. It is not the objects themselves, but the distance between objects that has to be captured.

As neuroscientists looked more deeply into the what–where split in the visual pathways, they found that the differences showed not just in the way the various mapping areas aligned themselves with each path, but even in the tuning of the neurons used in the processing. The split, in fact, started right out in the retina. The retina has two kinds of output, one swift and crude, the other slow but detailed. The fastest-reacting nerves – called magnocellular because of their larger size – have wide receptive fields and so gather information more quickly. This makes them good for picking up the sudden changes in light intensity that signal a movement. The slower cells – known as parvocellular because they are smaller – have tight receptive fields and also fire in a more graded, less all-or-nothing fashion. By taking their time to gather a reading, they react to the actual intensity of light rather than the mere fact of a change. This makes them the natural choice for collecting information about wavelength, shading, and the fine detail of what the eyes are seeing. So an area like V4, the colour specialist, would be living off mostly one kind of retinal output, while a motion-mapping area like the middle temporal cortex, or MT, would be living off the other.

A reductionist would look at this careful anatomical arrangement and marvel at how evolution had built an edifice of processing based on an early distinction in the tuning of retinal cells, but a dynamicist would turn the explanation on its head and say that the brain's need to represent two contrasting kinds of information would have produced the tuning difference in the neurons. Neurons are basically plastic and would have adapted themselves to fit the job the brain was trying to do. The demands created the structure, rather than the structure allowing a certain style of processing.

Thinking about the organisation of the brain as something that flows to fit makes it easier to understand the layout of the sensory cortex as a whole. The anatomy of the brain is difficult to follow because of the deep folds and tucks needed to cram it into the skull. The ever-increasing size of the primate brain has meant that the back half of the cortex has even become bent double so that it curves round on itself in the shape of a C. However, if the rumpled cortex sheet is imagined ironed out, it can be seen that each of the senses lies roughly in a row. More importantly, their position follows a clear what–where logic.

The senses most concerned with location – the somatosensory senses of touch, balance, and body position – are clustered together near the posterior parietal cortex at the end of the 'where' processing stream, while the senses more concerned with the identity of an object, like smell and hearing, are grouped at the temporal pole, standing at the end of the 'what' stream. It appears that vision, being pretty much equally torn between questions of location and identity, has found itself in the middle. Of course, each of the senses has to indulge in a bit of both styles of processing. There have to be bridges that connect the sense of touch to object recognition areas so that our hands can guess what they hold, and the ears have to have ways of placing sounds in space. But it is as if, during the course of evolution, the senses had jostled for a place on the cortex like rowdy schoolkids pushing for chairs in a classroom, each seeking a seat as near or as far from the blackboard as they preferred to be.

Looking across to the front half of the cortex sheet, the part of the brain responsible for higher thought and motor control – for a reaction to the events of the moment – researchers discovered that a what–where division ruled there, as well. The sensory cortex has to feed into the prefrontal lobes at the very front of the brain to alert them to what is happening. With simple logic, the 'what' stream of the temporal lobe turned out to connect with the ventral flank, or lower half, of the prefrontal lobes – a part of the brain known to be responsible for focusing on the detail of an action, thinking about its meaning and consequences. Meanwhile, the 'where' stream of the parietal lobe crossed over to emerge again on the dorsal flank of the prefrontal lobes. Experiments have shown that these upper half areas light up whenever people think about the direction of an event or the combination of movements needed to bring about an action.

The idea of a general what–where split in the brain's design was hugely significant because it was the first example of how a processing need could dictate processing structure. Soon after the possibility was first put forward by Leslie Ungerleider and Mortimer Mishkin at the US National Institute of Mental Health, other researchers began to suggest that the same kind of approach might explain another still more obvious division in the human brain: the split into a left- and right-hand side with apparently different mental abilities.

It has become part of popular myth that the brain has two sides, the left cerebral hemisphere having language and rationality while the right has imagination and intuition. However, especially since the advent of brain scanning to show the large networks of areas that light up whenever we speak or imagine, it has become clear that the brain does not work in this kind of compartmentalised way. All our mental faculties are represented on both sides of the cortex, and if there is a difference, it is one of processing style, the left being adapted for a narrow, excluding type of attentional focus while the right is more open and inclusive in its mapping response. Like a lens, sometimes we need to zoom in and concentrate on the sharp detail of a moment. At others, we will want to step back and take a wide-angle view on what is going on – to see the big picture. These are contrasting processing demands and so, like the what–where split, it would not be surprising if this fact were reflected in some subtle division of the brain's neural machinery.

Such a division is certainly suggested by the way the human brain handles language. The idea that language is an exclusively left brain activity arose from the observation that a left-side brain injury, such as a stroke, caused the most obvious damage to a person's powers of speech and comprehension. Yet, more careful study has shown that matching areas on either side of the cortex actually share the job of speaking; the difference is that the left does the 'foreground' tasks, such as drawing together actual sentences and recognising individual words, while the right-hand cortex deals with 'background' jobs, such as putting the emotional colouring into what we say, or making creative associations between words. So on one side of the brain we are focusing in close, getting the detail of the grammar and the choice of words just right, but over on the other we are taking the big picture view, managing the overall tone and picking up on any broader nuances of meaning.

Good evidence for this comes from cases where people with right-side brain damage have perfectly fluent speech, yet can no longer tell the difference between angry and sad voices. Or even more tellingly, they cannot get jokes or metaphors. They understand the words well enough, but can only respond to the face value of what they hear. So if asked to give someone a hand or advised to turn the other cheek, such a person would interpret this quite literally. This shows that both sides of the brain are needed for normal speech, and if we think of the left as being the dominant half, it is only because a tight focus on sequence and meaning seems the essence of language. We value this side of what the brain does and tend to neglect the difficulty of what has to take place in the background of our minds for speech to make real sense.

Exactly the reverse is the case with more spatial or imaginative styles of thought. Again it was first believed from observations of brain damage that humans depend on an intact right brain to be able to juggle mental images or manipulate a sense of space. But more careful work has shown that both sides of the brain play a part in generating imagery. The right brain does appear to deal with the more global aspects – the sense of where everything is and how things broadly connect – which is, of course, the part of non-verbal thought that we value the most. But the left brain is good at focusing in on particular locations or events, at isolating elements within the total picture. In the classic illustration of this contrasting attentional style, patients are briefly presented with a target picture of a large shape made out of smaller shapes – it might be a triangle drawn with tiny squares, or a capital M traced out by a series of z's – then asked to copy what they saw. Those with right brain damage report seeing a jumble of small squares or z's, while those with left brain damage notice only the global shape, the M or the triangle. It takes a normal brain to catch both the detail and the overall shape of an experience.

Such evidence suggests that we have a bifocal brain. With a lateralisation in attentional style, simultaneously we can see both more deeply and more broadly into the moment than we could with a single attentional setting. And it seems revealing that a lateralisation of the brain is the most obvious neurological difference between humans and other animals. Monkeys, and perhaps some mammals, show a slight degree of hemispheric specialisation. But in humans, there appears to be a very clear division of labour, giving us an exceptionally penetrating or high-contrast appreciation of what each instant holds.

There has been much speculation about how such a lateralisation could have come about. Many anthropologists believe that the brain's lateralisation began with tool use. About five million years ago, our hominid ancestors developed the posture for walking upright, thus freeing their hands to become specialised for carrying, handling, and fashioning tools. To make a tool is especially demanding. A hand axe is made by chipping away at a lump of flint with many precisely judged blows. The brain of a hominid would have had to be able to plan a sequence of steps while also retaining a global idea of the intended result. This would have put the brain on the path towards lateralisation. But once it began to develop, it would not have been long before such a capacity began to be exploited in a more abstract way. Certainly, speech depends on a very similar ability to organise a complex chain of small actions within a broader context of expectation – to keep spitting out words until a meaning has been expressed.

So there are plenty of candidates for new kinds of pressures on the ancestral human brain. It has also been argued that reorganising the brain's circuitry to fit this need would not have been nearly as difficult as it might seem. In terms of genetic adaptation, all that might have been required was a general increase in the brain's size to cope with the extra processing load, coupled with the development of a slightly different pattern of branching in neurons on either side of the brain.

As specialists in brain asymmetry, such as Georg Deutsch of the University of Alabama, have noted, the idea is still largely speculation because there has been no easy way of showing such a fine-grain difference. But it is plausible that cells growing in the left cortex might connect more locally, talking only to their closest neighbours and so having more narrowly defined receptive fields, while those on the right might connect widely and so reflect a more general picture of activity. As with the parvocellular and magnocellular streams of neurons, the distinction would not be huge, just an adjustment in tuning on a spectrum between coarse-coding and a more picky style of coding. But the dynamics of the resulting population votes and neural competitions would look quite different.

With language, for example, the power of speech might take up residence on matching areas on both sides of the brain. To the casual glance, words would seem to exist only over on the left because that would be where the most narrowly focused reactions would occur. In population voting terms, the neurons involved in representing a word like 'cow' would have mostly local connections, and would respond very weakly to any idea that was not directly related to a sharp sense of cowness. This would have the advantage of rousing a cow-shaped response and nothing but a cow-shaped response. But the price would be that the left brain's reaction would be extremely literal. If asked to make a word association, it could travel only a short distance, coming up with highly stereotyped responses such as 'milk' or 'butter'.

Over on the right brain, however, the story would be very different. The word 'cow' would not provoke such a specific reaction – the feeling of knowing the word's meaning – but there might be a loose population vote that conjured up a kind of background 'farmyardy' feel. Our minds might stir with a general sense of atmosphere or emotional context – an intimation of tractors, mud, and chickens running loose. So the left brain would do the sharp identification while the right would produce a more general state of orientation that would allow us to make creative leaps of thought or spot hidden nuances of meaning.

Thus, through some minor tinkering, the human brain may have added a complex twist to the way it reacts to the moment. We can take a sharper, almost digitally bounded view of some foreground detail while at the same time keeping in mind a broader background framework of thoughts, associations, intentions, and memories. However, before getting too deeply into the story of what lies at the end of the processing trail – the state of focus that completes a moment of awareness – it is worth considering the crucial question that the brain must ask itself an instant or two before that. As it is busy mapping the what and the where of the wash of sensory information that floods in during every waking second, the brain must start to think, 'So what?' It needs to make some kind of evaluation that enables it to choose which part of the moment should be escalated to full attention. And again, it might be expected that this evaluation step would somehow show through in the imprint it leaves on the greater organisation of the brain.

When we get a joke or a riddle, there is a definite feeling that comes with the moment. A friend might try out a brain-teaser on us, saying this person went up to bed, turned off the light, and slid under the covers. The problem was the light switch was right across the room, yet he still managed to make it into the bed before the room went dark. So how did he do it?

The answer, if you have not guessed, is that it was still daylight when he went to bed. But the point is that when we have pondered something for a second and then get the solution, we feel a sudden clap of insight. We do not just see the silly logic of the solution, saying to ourselves in dry computer fashion, 'Well, yes, that adds up.' We feel an emotional jolt, an 'aha!' of surprise or pleasure. And a similar sharp stab of emotion occurs in many situations, like when we find a set of door keys we have been looking for, when we suddenly remember yesterday was our mother's birthday, or when a dog dashes out into the road in front of our car. There is a shock that seems to grab our attention and tell us something significant has happened.

Aha! feelings come in many different flavours and vary in strength. Some lead to a feeling of elation and delight, as when we hear a funny joke or discover we have won a lottery prize; others can lead to a sinking feeling or even distress, as when we note heavy footsteps coming up behind us on a dark night, or spot a hairy spider on our shirt sleeve. The scale of an aha! can also range from the mildest buzz of interest or familiarity right up to wrenching, heart-pumping alarm. This variety is confusing, but a big clue about the origins of all such feelings is that they appear tied to the escalation of an event into consciousness. Virtually by definition, the feelings of significance, certainty, recognition, surprise, or alarm are connected to whatever it is that has just caught our attention rather than some background event of which we will not take any further notice. The feeling seems bound in with an assessment process – a decision of 'so what?' – that leads us eventually to focus on one sensation or thought out of the many that may be swirling on the edges of our minds.

To see how vital a part the aha! feeling plays in our mental lives, just try to imagine consciousness without this constant feeling of judgement. It tells us what is important. It even tells us what is familiar and what is novel, or right or wrong. If asked whether Mombasa was in Africa or South America, we might find either answer equally plausible unless we had a positive flicker of recognition to tip us in the right direction. And the way we know that we don't know an answer is when there is no click of familiarity for either alternative – such as, perhaps, if we were asked whether Mombasa was in Uganda or Kenya, or whether its population was half a million or three million.

Predictably enough, modern psychology has had relatively little to say about the aha! feeling. The emotion interested turn-of-the-century thinkers like William James and featured strongly in the theories of the Gestalt school of psychology that flourished in Germany between the wars, but behaviourism and cognitive psychology had no good framework for dealing with even the more overt emotions such as anger and fear, let alone feelings as delicate as states of familiarity or uncertainty. And yet some surprisingly revealing work was carried out in the 1950s by the Russian physiologist Evgeny Sokolov, a pupil of the great Ivan Pavlov.

Pavlov was famous for showing that if a bell was rung every time a dog was given food, soon the sound of the bell alone would be enough to make the dog salivate. This kind of simple reflex learning was one of the major inspirations for the behaviourist movement in America. But Pavlov was also interested in the more intelligent reactions of his charges, and one phenomenon that intrigued him was the way his animals would often pause, looking a little startled or even interested, whenever something caught their attention. He dubbed this sudden halting of activity the 'what is it?' or orientation response, then passed the problem on to Sokolov to investigate.

Using his physiologist's paraphernalia of blood pressure gauges, heart monitors, and other equipment to test human subjects, Sokolov found that there is a whole cascade of motor and metabolic adjustments that take place whenever we are struck by a surprising or novel event. The first thing that happens is almost too obvious to mention: we immediately stop whatever it is that we are doing and glance to bring the event into focus. Yet it is worth noting the implications. The decision to halt and inspect must be made before the event itself has entered full consciousness. The perhaps tricky disengagement from whatever it was that we had been about to do, and the reorientation of our senses to a new location, must be organised at a reflexive, or pre-conscious, level.

Furthermore, Sokolov showed that this orientation response was a complete, whole-body reaction. A person's entire emotional and physical state could be retuned within about 200 milliseconds to optimise it for dealing with the moment ahead. When we hear a knock at the door or feel an unexpected touch on our shoulder, not only do we automatically stop and turn, we also begin to sweat a little, our mouths go dry, our heart rate and blood pressure increase, extra sugar is released into our circulation, our blood vessels dilate to raise the blood supply to the muscles and brain, our breathing deepens, and our air passages widen to bring in more oxygen. Our brains go through a process of arousal as well. There is a rapid change in neurotransmitter levels to make us more alert. Substances such as noradrenaline – the brain's equivalent of adrenaline – flood key areas, lowering our sensory thresholds and causing us to react more vigorously to whatever it is that our eyes eventually land upon. In short, the orientation response makes sure that we hit the ground running, both physically and mentally.

What is particularly impressive about the orientation response is that these many adjustments are not crude, on/off reactions, but graded to deal with the situation we expect to face. A minor surprise, such as the drone of a lawnmower starting up, might produce only a fleeting change in breathing, blood flow, or brain neurotransmitter levels. However, the sight of a dog stepping in front of our car would be ushered in on the back of a full-scale, heart-wrenching jolt, sufficient for us to produce an equally dramatic level of braking and swerving. Sokolov also noted that the reactions varied to fit what was likely to be the most sensible course of action. As well as making positive orientation responses – turning towards something – we can make a defensive response. If something flutters past our face, or if we taste an unexpected bitterness in our food, reflexively we will shut our eyes, screw up our face, tighten our throat, flinch, and even throw up our hands as though to ward off danger.

The more Sokolov studied the orientation response, the more surprises he found. In some circumstances, the heart decelerates rather than speeds up. A frightening event may set the heart racing, readying us for a quick getaway, but when an event is interesting or exciting, we catch our breath and our hearts literally skip a beat, as if our bodies are quietening for an instant so that we can concentrate more fully on what is happening.

However, most importantly of all, almost any moment of consciousness seemed capable of producing a faint pre-conscious level of adjustment. It did not appear to cost the brain anything extra to keep fine-tuning a person's emotional and metabolic state. Indeed, people even reacted to the impact of their own thoughts. Some American researchers, bemused when reports of Sokolov's work reached the West, gathered together a group of students from the university's fishing, surfing, and chess clubs and showed them a series of slides. Mixed among a random selection of pictures were just a few connected to their own hobby. But by taping electrodes to their hands to record the rise in skin conductance that comes with a slight nervous sweating, it was found that simply seeing a photo of a handsome fish or a thunderous wave could cause a small startle in the students with a special interest in such things. Other such experiments showed that just thinking a pleasant thought or imagining a nasty smell could produce a measurable reaction. And in fact this was the research that eventually produced the polygraph or lie detector. The alarm at not telling the truth, in subjects bothered by such things, would show through in a fleeting adjustment by the body.

The link between Sokolov's orientation response and the aha! feeling is easy enough to see. As part of orienting to a significant event, or even a significant idea or thought, we experience a rich variety of physical changes. Our heart may lurch, our stomach heave, our face blanch, a flush of neurotransmitters may leave us feeling elated or alarmed. There is nothing ethereal about such reactions; they can be measured with voltmeters or blood-pressure gauges. And we will experience them just as directly. So the feeling of knowing that we know – of familiarity and recognition – is just a muted version of this orientation reaction. When we ask ourselves if Mombasa is in Kenya, the reason we trust the answer is because it brings a confirming twinge. It is the only pairing that triggers a slight halting – a catch in our attention – and a just-measurable quickening of the pulse. A pairing with Uganda, on the other hand, should draw a metabolic blank – the empty feeling we call unfamiliarity.

A second important source of clues about what might be going on in the brain when it is making a 'so what?' response comes from EEG recordings and, in particular, a late breaking wave of activity known as the P300 response. As mentioned, the P300 is a massive, electrically positive shift in brain activity that comes about a third of a second after a person has witnessed something surprising or novel. Researchers nicknamed it the 'oddball response' because its strength seemed to match the intensity of a subject's feeling that something strange or noteworthy had just happened. But in fact, it did not take too much of a surprise to produce a large P300.

The standard experiment involved sitting with headphones and listening out for two kinds of tone. One tone would be heard often, while the other would be the rare tone designed to trigger an aha! response – although rare in this case could mean about once every five or six trials. In some P300 experiments, researchers would try to throw in something more truly off-beat, such as using a variety of animal cries as the oddball event. But even then, because they needed to average hundreds of trials to isolate the P300 response from the brain's background noise, the level of shock would have been pretty low by the end of a recording session. Yet this, of course, merely made it all the more remarkable that the subjects produced such a strong, and late, reaction.

The actual timing of the P300 was much more variable than its name implied. With simple stimuli like tones or different coloured lights, the brain could react to an oddball event within as little as 250 milliseconds. But when the stimulus was more complicated, like an occasional boy's name popping up in a stream of girls' names, or a famous politician mixed in with a stream of entertainers, the reaction could stretch right out to 500 milliseconds or more. And fatigue or inattention could create even longer delays. The 300–millisecond figure was just an average based on alert subjects being tested with the kind of untaxing stimuli used in standard psychophysics tests.

The P300 was first reported in 1965, and for several decades a small but dedicated band of 'P3–ologists' pored over their charts, cataloguing its many variants. However, because of the difficulties in interpreting any EEG trace, there were almost as many ideas about what brain event might underlie the P300 as there were P300 researchers. There was not even general agreement about whether the electrically positive nature of the signal meant the brain was experiencing a massive bout of excitation or inhibition, whether its circuits were switching on or shutting down.

But a few researchers, such as Eric Halgren at the University of California-Los Angeles and Niels Birbaumer of the University of Tübingen in Germany, felt the P300 must be a sign of inhibition. More than this, it would be a brain-wide sweep of feedback suppression, damping cells to make one aspect of the moment stand out in sharp focus. In other words, it was Desimone's attention effect writ large – the sound of the whole brain hushing.

The evidence from Desimone's single-cell work, and also from other research such as experiments on 'pre-attentive' processing by Anne Treisman of Princeton University, was that the brain begins the moment by mapping everything. It spends 100 to 200 milliseconds arriving at a new balance of firing that best represents the latest batch of sensory input. With good anticipations to warm up the right pathways, much of the moment will fall automatically into place. But then comes the time to focus on whatever does not fit, or whatever for some prior reason has been tagged for promotion to the centre of consciousness when it eventually arrives. At this stage, the brain would need some quiet. With 100 milliseconds or so of inhibitory cross-talk, it would damp the representation of the parts of the moment in which it was not particularly interested, so throwing into sharp relief the bit that it did want to make the lingering focus of the moment. The P300 was a sign of the contrast knob being turned up in the brain.

A number of P300 researchers noted that such an intensification of the significant part of a moment would serve a double purpose. Obviously, the tighter our focus, the more targeted would be the thoughts and anticipations that flowed from the resulting stab of activation. But it would also leave us with a nicely cleaned-up memory trace to file to storage. As a living, processing surface, the brain has to be incredibly careful about exactly what it remembers of each moment, about which bumps and hollows it wants to add to its processing landscape. To flatly record every detail of every moment would be a disaster. The brain's aim must be to make a change to its existing pattern of connections only if the new connections are guaranteed to improve its processing of moments in the future.

By definition, any moment of awareness that has turned out much as expected should mean that the brain's existing habits of thought and processing had worked pretty well. So, rather than upsetting the delicate tuning of its pathways by having a strong reaction to such a moment, the brain would prefer simply to remember it as a bit of a blur – a vague mix of what was anticipated and what actually happened. There might be a gentle strengthening of many connections, but no sharp etching of a specific response. Conversely, if the moment did turn out to contain a surprise, then the brain would want to do the opposite. It would want to zero in on the event, taking extra time to shrink any population voting activity down to its bare bones. As well as quashing any peripheral events, the brain would need to rinse out any priming activity. If, through having listened to a series of low tones or watched a series of red lights, the brain had built up a slightly raised expectation of experiencing them again, then this anticipatory firing would have to be washed out of the eventual memory of the moment, allowing the brain to appreciate the slight surprise of hearing a high tone, or seeing a green light, in all its pristine splendour.

Like all ERP phenomena, the P300 wave could only ever be suggestive, never conclusive. Yet the idea of a competition-based, contrast-enhancing step tied in with a dynamic view of the brain rather neatly. And if nothing else, the P300 was certainly proof that major processing events were occurring in the brain long after it was supposed to be conscious of the world. Taken together with the research into the orientation response, the evidence was that the brain had to put quite some effort into organising a 'so what?' decision about the many events of the moment.

The next question is whether this evaluation pathway is visibly stamped into the brain's circuitry? If it is, already it can be appreciated that it is not likely to be as a simple feature involving just the cortex. For a start, the slapping of a value on a passing train of thought and experience must involve the emotion and metabolic control centres of the lower brain, so any 'so what?' path would have to take in many sub-cortical organs.

But also it is becoming plain that the brain can have quite contrasting reasons for finding an event worth escalating. In some cases, the brain will seize on a sensation or thought because its arrival has been eagerly awaited. When we finally spot a lost set of keys, or get the answer to a problem, there is a jolting aha! as we realise that the sensation or thought fits with a state of mental expectation. It is only the precise moment of arrival, or the exact form of the experience, that is unpredicted. But at other times, such as when we notice a beetle crawling across our shoe or hear our grandmother's voice, the brain will have to react to the truly unexpected. It will have to catch the significance of an event in a part of the world that it was not really watching. Given such differing processing demands, it would be expected that the brain might need two quite separate paths to handle both varieties of decision.

As will be seen, the search for the brain's 'so what?' pathway does indeed help make sense of much of the anatomy of the brain. It turns it from a collection of parts into a system with a purpose. And clearly, knowing how the brain escalates an event to the forefront of attention will explain much about the nature of consciousness. But thinking about the brain in terms of its processing pathways again leads us back into a more computational way of looking at what it does. It is focusing on the brain's structure at the expense of its wider plasticity. So as an antidote, it is worth first broadening the perspective for a moment and considering two further examples of what it means to see the brain as a fluid bag of circuits, a system drawn into shape by the pressure of its needs.

The plasticity of the brain is itself something that is carefully graded. The mammalian brain is not evenly adaptable across all its circuits. Instead, it has a hierarchy of fluidity in which the most ancient parts of the brain – the bits we share with fish and reptiles – forms a working core that evolution cannot tamper with too much. Then, draped over this is the cortex, a uniquely unstructured sheet of material that seems almost specialised for a lack of specialisation.

With its homogeneous six-layer design, the cortex is remarkably easy to grow. Our genes could never have coded for the precise placement of the billions of cells that make up the adult cortex – there is just not enough room on a DNA molecule for so much information. But the genes could cause our bodies to throw up a vault of general-purpose nerve tissue and then leave it to our experience of life to do the detailed organising.

As Donald Hebb showed in pioneering work with infant rats and chimpanzees, there are certainly a few genetically determined design features to start the cortex off on the right footing. The place where the nerves from the eyes, ears, and other senses connect with the cortex help establish a rough initial layout. The genes can also exert a tremendous influence by controlling the moment when different parts of the cortex undergo bursts of growth, or settle down and mature. They can do things like hold back the development of higher levels of the processing hierarchy while the lower levels – off which they must feed – are becoming properly established. However, the actual construction of the cortex is largely a matter of learning. Pathways are carved out in response to the sights and sounds to which we are exposed during the first few months and years after birth.

While we are in the womb, the cortex is busy sprouting a promiscuous tangle of connections. At its peak of growth, the foetal

'The brain is turning out to be the most complex system that science is ever likely to encounter. It has not just complexity, but complexity of an almost alien form . . .'

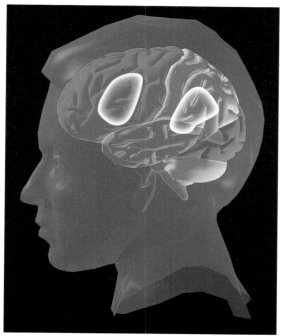

1 The major divisions of the cortex. Frontal lobe (red): for organising a reaction to the moment. Parietal lobe (orange): the head of the 'where is it?' sensory processing stream. Occipital lobe (green): the site of visual mapping. Temporal lobe (blue): the head of the 'what is it?' sensory processing stream. White shaded areas are the classic language centres: Broca's (forward) and Wernicke's (rear). Cerebellum (yellow) is a lower brain centre for refining motor output.

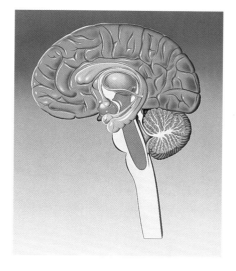

2 The clutter of hidden sub-cortical organs that are essential to the brain's processing of each moment: the brainstem (purple); the thalamus (pink), which acts as an input focusing organ; the hypothalamus (orange), an arousal and metabolic control centre; and, in a C-shaped sweep, the basal ganglia (top yellow), the amygdala (middle yellow) and the hippocampus (bottom yellow).

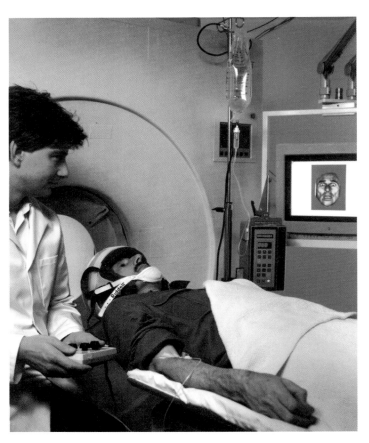

3 A PET scanner at the Wellcome Laboratory in London. The subject lies with head inside a ring of detectors as radioactive tracers are injected into the bloodstream. Blood flow changes of 30 per cent or more show which parts of the brain are having to work hard in carrying out a mental task — in this case, interpreting an ambiguous facial expression, a test in which the amygdala glows hot.

4 The EEG, electroencephalogram — an old technique making a comeback. Electrodes placed on the scalp pick up the faint rustle of firing brain cells. Computers now allow researchers to collect data from up to 126 electrodes, so making it possible to track the development of an individual state of consciousness over the course of a third of a second or so.

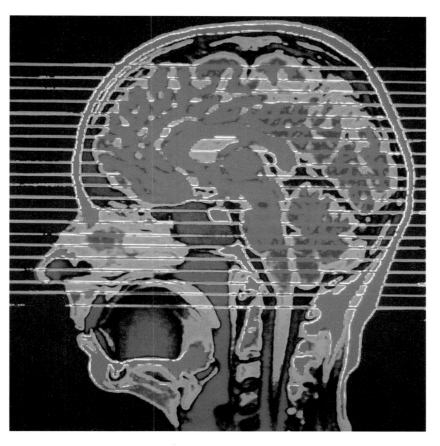

5 The awesome clarity of brain scanners. Most imaging is functional — the machines are used to track the changes in bloodflow or electrical potential produced by the brain's mental activity. But these have to be matched against structural scans — images of a subject's actual brain — to show accurately where the changes are occurring (every brain having a slightly different shape and pattern of folding). This MRI scan (which uses an intense magnetic field to 'line up' the atoms of the brain, so revealing their differing concentrations) shows how images are normally built up from a dozen or so horizontal 'slices'. Some people's brains show up more clearly than others, leading them to be used repeatedly in experiments — the so-called neuronauts.

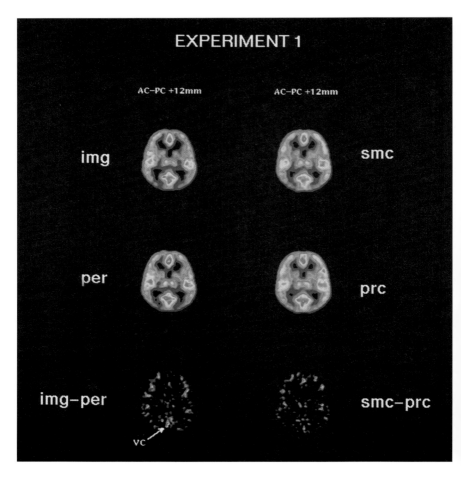

EXPERIMENT 1

AC–PC +12mm   AC–PC +12mm

img

per

img–per

vc

smc

prc

smc–prc

6, 7, 8 and **9** Stephen Kosslyn's famous experiment to prove that mental images are like a projection of a memory back across the circuits of the brain's sensory areas. These rainbow-coloured smudges might not mean much to the uninitiated, but they settled one great debate only to spark another.

Kosslyn's aim was to show that imagining a target letter would trigger a matching activation of brain cells in V1, the primary visual mapping area of the cortex. The initial set of scans (above: picture 6, experiment 1) illustrates the general principle of using subtraction to isolate mental effort. In the left-hand set of results, subjects were asked first to imagine a letter

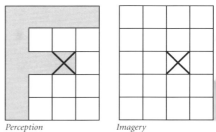

Perception   *Imagery*

(img) and then scanned again whilst actually perceiving a computer display of that letter (per). After subtraction (img minus per), only the extra effort of forming a mental image

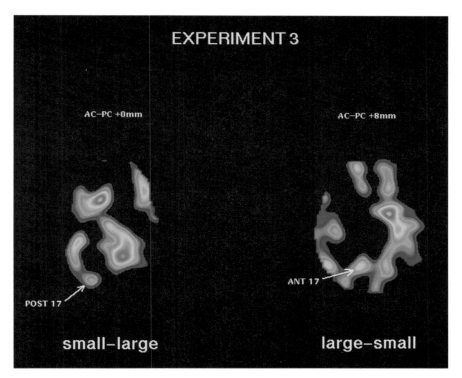

EXPERIMENT 3

AC–PC +0mm

AC–PC +8mm

POST 17

ANT 17

small–large

large–small

should remain. The right-hand set of results is a second type of control in which subjects first made a sensory judgement and foot-pedal response (smc), then again simply viewed the display (prc), so revealing the separate effort involved in making conscious decisions. The experiment showed that imagining did cause extra activity in the visual cortex as predicted — but also surprisingly widespread activity in the rest of the brain.

Note that all these scans are horizontal slices through the brain, taken at about the level of the eyes, with the visual cortex (vc) at the bottom. The block capitals that subjects were supposed to imagine (and actually saw in perception trials) can be seen opposite below (picture7).

The second experiment (above: picture 8, experiment 3) contrasts the effort of imagining a letter as small as possible, then as large as possible. As Kosslyn predicted, the larger images stirred a greater surface area of V1, which showed in the scan as activation reaching to the

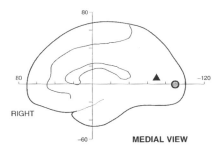

MEDIAL VIEW

▲ LARGE minus SMALL
◉ SMALL minus LARGE

forward or anterior edge of area 17. For smaller letters, activity was concentrated in the posterior region — just the centre of the map. The sketch (picture 9) summarises this result, with a triangle to mark the forward spread of the larger images.

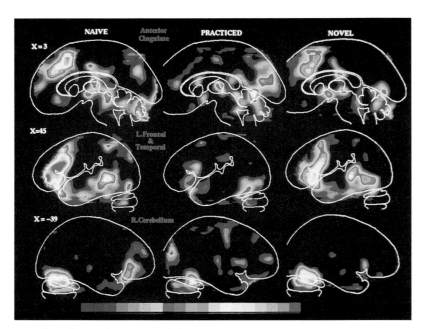

Note that scans show vertical slices first through the centre of the brain (top row: front of brain to left), then left side (middle row), then right side (bottom row).

10 Perhaps the most important, and also most unsung, result to come out of the pioneer PET work was Washington University's demonstration of how the brain automates its actions. Focal consciousness demands a global-level brain response, but much of our mental lives can be lived at a more routine level with thought and behaviour being executed 'locally'. These scans (above) show subjects thinking of verbs to match a list of nouns (e.g., the target word 'hammer' might suggest the response 'hit'). In the naive condition, a subject's first attempt, there is bright activity throughout the frontal lobes, particularly the anterior cingulate (top left) and major language areas on the left side of the brain (middle left), and even the cerebellum (bottom left) glows hot. But with just a little practice on the list, answering requires barely any more effort than the control task of simply reading out the words. When a new set of words is introduced (end column), this novel condition sees the focal level of effort return — but only weakly as, by now, even the general act of making associations is becoming automated. Such a finding changes the view of what it means for a brain event to be 'in consciousness'.

11 Further proof that the brain has a fluid organisation comes from scanning experiments of people as they mentally categorise objects. In this PET experiment at NIMH, subjects were first compared when viewing nonsense objects and featureless static (top results) to discover which areas of the brain were called into action during attempts to recognise an object. As expected, the temporal lobe became busy, but so did a host of other brain areas including high-level regions of the prefrontal lobes and low-level organs like the cerebellum. Then seeing real objects like saws and bears was compared to seeing nonsense objects (bottom results). Now still further areas, like the thalamus, kicked in, while others, like the hippocampus, fell out. The brain seemed to develop its processing logic on the fly, tailoring its response to the task. The problem for science was to marry such dynamism with the traditional computer-like view of the brain.

A. NONSENSE OBJECTS - NOISE PATTERNS

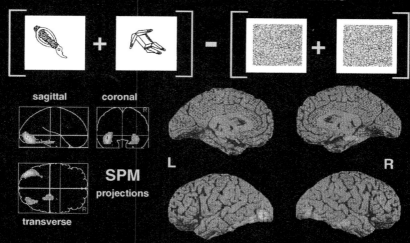

sagittal   coronal

SPM
projections

transverse

L   R

B. OBJECTS - NONSENSE OBJECTS

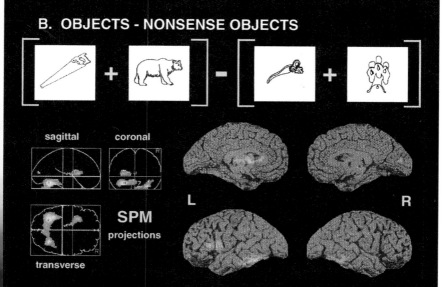

sagittal   coronal

SPM
projections

transverse

L   R

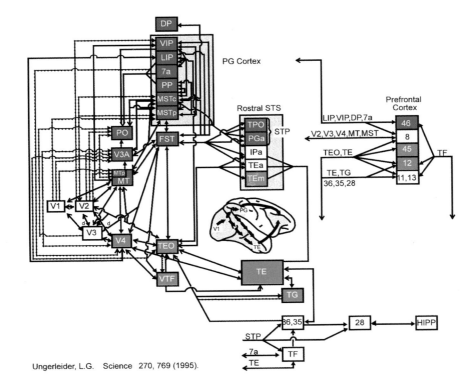

Ungerleider, L.G. Science 270, 769 (1995).

12 The culmination of thirty years of using implanted electrodes to study the animal brain: a blueprint of a monkey's entire visual processing hierarchy. Everything starts from V1, the primary visual mapping of what the eyes see. Then activity heads in two directions — the parietal or 'where' processing stream (green boxes), concerned mostly with questions of movement and location, and the temporal or 'what' processing stream (red boxes), dealing mainly in questions of object identity. More than thirty separate mapping areas have been identified. Key ones include V4, the brain's 'colour' centre; TE, a high level inferotemporal area; the hippocampus (HIPP), which is the final destination for much sensory information; and the collection of prefrontal cortex maps which decide how to respond to what is seen. But this stack of mapping turned out to work as well top-down as bottom-up.

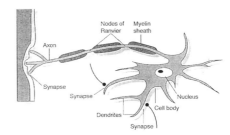

13 The textbook-simple view of a neuron. Input comes in at one end, where axons from other brain cells synapse on its bushy set of dendrites. Electrical charge creeps over the cell to accumulate at the base of its axon, eventually causing the cell to discharge and produce its own output spike. But the 1990s showed there was something deeply wrong with this idea of the neuron as a simple, transistor-like processing gate.

brain produces new neurons at the incredible rate of 250,000 a minute. These cells attach to their neighbours almost at random so that at birth, the cortex is a wild mass of wiring. But then during the first year of life there is a ruthless pruning of these connections. Neurons are forced to compete – to show that they form part of a useful circuit – and any cells that find themselves in the wrong place without a real job are programmed simply to shrivel up and die. Eventually almost half of all the neurons with which we are born will disappear, sacrificed to leave behind a sharply etched set of cortex pathways.

The fact that the cortex is self-organising in response to childhood experiences, rather than genetically designed, has many advantages. As said, it was really the only way of extending the brain in the first place, but it also means that the brain can grow closely fitted to the world in which we find ourselves. The mammalian brain has areas of circuitry in places like the IT cortex which come to wrap themselves around the contours of some extremely precise sensory experiences. Reptiles and fish have to make do with less plastic mid-brain areas to do their sensory mapping, and it certainly seems as though their mental responses are rigid. But mammals have a softness of circuitry that moulds to fit the fine detail of life.

For humans, the plasticity of the cortex has extra implications. It means that the brain must be far more sensitive to environmental and cultural influences than traditionally assumed. It is normal to think of the special human abilities, such as self-awareness and rational thought, as innate – genetic additions to the standard furniture of the brain. But all that might have happened was that our hominid ancestors grew more cortex, then relied on a childhood exposure to grown-up patterns of language and thought to shape these empty circuits. So the higher mental abilities could turn out to be culturally imprinted rather than genetically evolved.

This is yet another point to which we will need to return. But to complete the grand tour of what dynamism means to the brain, it would be useful to mention one further dimension to its fluidity, an example that also happens to count as the first really major discovery to come out of the brain-scanning revolution.

In the mid-1980s, the PET team at Washington University in St Louis was ready to begin serious testing. It was an exciting moment, because now neuroscience could tackle the workings of the human brain directly rather than having to make do with animals. A large team had been assembled, led by Marcus Raichle from the Mallinckrodt Institute of Radiology, who brought the imaging skills, and Michael Posner, a cognitive psychologist drafted in from the University of Oregon to give the work its theoretical backbone. Because of the newness of the technique, it was felt that the first experiments ought to concentrate on language, a mental process that already seemed reasonably well understood and so would act as a benchmark for more ambitious studies in the future. Yet from the very first, the scanning sessions produced shocks.

The standard cognitive model had led the Washington team to expect that the brain's handling of speech would be highly compartmentalised. Flowchart models of language production did not seem to require a huge number of steps. And this appeared to be backed up by what was known about the neurology of speech. As early as the nineteenth century, doctors examining patients who had lost the ability to talk through stroke or brain injury had concluded that there were just two key cortex regions responsible for language. They identified a spot on the left-hand side of the frontal lobes, known as Broca's area, as being in charge of speech output – for assembling sequences of words in grammatical form. Then a matching spot on the upper bank of the temporal lobe, known as Wernicke's area, was believed to deal with the meaning and sound of words – for speech decoding.

This organisation certainly made sense as Broca's area was close to the motor areas dealing with mouth and throat movements while Wernicke's area was a high-level auditory processing area. But as soon as they started scanning, the Washington group found areas of activity lighting up all over the brain. There was activation on the right side of the cortex whenever the emotional tone of speech was being processed; there was activity in many spots on the frontal cortex, including a region known as the anterior cingulate; there was even activity in the cerebellum, a large bump on the back of the brain stem which was thought to be just a lowly motor centre, involved only in reflexive control of fine muscle movements. As Kosslyn found with his mental imagery experiments a few years later, the hot spots were so widespread that it was almost harder to list the parts of the brain that did not fire during the act of speaking.

When these results were published in *Nature* in 1988, the apparent extent of the brain's language areas took researchers aback. Many just flatly refused to accept the results, feeling that the PET method must be flawed and better-controlled work would show much of the activation to be a mistake. For others, it was a first hint that brain processing depended on dynamic networks of activity. While an ability like speech or colour vision might appear to be concentrated in certain specialist mappings, the memories and response routines stored there could only be exploited as part of a whole-brain effort. Yet although the PET images gave graphic evidence of this fact, it was not the real surprise, as much the same conclusion was already implicit in the past decade of single-cell research. Something else that showed up during the scanning was the genuine revelation.

The language experiments had been designed around a verb-generating task which would allow the Washington team to subtract away the supposed elements of making a speech act, stage by stage, until just the mental effort of thinking up something to say remained. Subjects would first be scanned as they read aloud a list of words, a set of common nouns like 'chair' and 'hammer'. This created a baseline reading by activating many parts of the language system, including word articulation and word recognition areas, but presumably not any areas used in creative thought. Then, in a second scanning session, the subjects were given exactly the same list, but this time they were asked to come up with a matching verb. Reading the word 'chair', they would have to answer it with another word like 'sit'. By subtracting one scan from the other, and so removing everything to do with the act of reading and speaking, the result ought to reveal which bit of the brain was used purely in making the linguistic association.

The task could hardly have been easier, but the responses certainly needed to be quick. The verb-generating task was supposed to 'fill' a subject's mind for the several minutes it took to collect each PET reading, so target words were flashed up at the rate of one a second, giving little time to think about anything else. One of the volunteers for the experiment became a little worried that he might not be able to cope with such a pace, and asked if he could have a few practice runs first. It was felt that rehearsing with the actual test list should cause no problems because the same network of brain areas would still have to be used in dragging up the associations during testing. The act of matching 'chair' to 'sit' should look the same no matter how many times the brain had to do it. Yet when the person's results came back, a strange thing had happened. There was almost no trace of any extra brain activity. The language centres were still firing, of course, but only at a level required for reading verbs instead of having to make associations.

This practice effect staggered the Washington researchers. It was enough of a surprise that the brain burnt so brightly the first time it had to make an association between a noun and a verb. After all, the task was not really all that taxing, yet the brain showed metabolic swings of 30 and 40 per cent. However, then to find that the extra activation evaporated with just a little rehearsal was utterly unexpected.

The team began exploring the effect more systematically. In a new experiment, a group of people were again scanned while doing the verb-generating task, except this time they were recorded under three conditions: without practice; after fifteen minutes' practice with the same set of words; then finally with a completely fresh list. The outcome was clear-cut: there was bright activity during the first run, low activity after practice, and then a degree of rebound – but only a degree – with the new words. The practice effect stood up, and the testing with a second list of words suggested that the brain could even begin to adjust to the effort of turning nouns into verbs generally.

The experiment also showed what was going on in more detail. In particular, it revealed that when the task was novel, one part of the brain known as the insula cortex, on the lower flank of the frontal lobes, was actively being switched off. The area was being stopped from doing things, and was only allowed to return to normal levels of firing once the task had become practised. The reason for this was a complete puzzle, as the insula cortex was believed to be a low-level motor control area with no recognised role in language processing.

Seeking an explanation for these results, the Washington researchers realised that quite simply they were seeing a shift from focal-level processing to habit. The first time a subject had to find an association for a previously unseen word, a vast area of the brain was being called into action. Both high-level areas like the cingulate cortex and low-level areas like the cerebellum glowed hot as hundreds of millions of cells were thrown into the fray. Some of this activation was needed to rouse the core parts of the language system, places like Broca's area and Wernicke's area which contained actual memories about word meanings, but much of the activation would be produced by regions with a more supervisory role. These areas would be holding in mind a representation of what needed to be done, then checking the answers to make sure they were suitable. It would have been easy, especially given the time pressure, for a subject to slip up and simply read out the target word seen on the screen, or perhaps reply with an adjective instead of a verb. So the other parts of the brain would have been maintaining the temporary framework, the scaffolding of thought, needed to keep the language areas on track.

But once the brain had made a connection between two words like 'ball' and 'bounce' a few times, it could afford to relax a bit. The association would quickly become automatic – more a memory of a prior response. On a scan, this relaxation would show as a shrinkage in activity. All the high-level supervisory areas would drop out of the picture, and even the activity in the vocabulary areas would become dimmed. There would be no need for a general rousing of memories to look for a link between two words as the recent activity would ensure that a suitable pairing already lay close at hand. The role of the insula still needed explaining, but it might be an area which specialised in a more automatic level of motor control and so would have to be shut down to stop it interfering with the brain's attempt to deal with a novel situation. Left on, the insula cortex might have 'intercepted' the processing of the moment, leading the brain to make a more habitual response such as simply reading out the target word rather than persisting in the search for an association.

This was a lot to read into just one experiment. But a similar practice effect popped up in other early PET work, such as an IQ study by Richard Haier of the medical school at the University of California-Irvine. Haier wanted to use the school's new PET scanning system to find what made a brain more intelligent. Some psychologists had speculated that a high-performing brain would be one that had a lot of energy to burn when tackling a mental task, while others thought the opposite and said it would be efficiency that mattered. Haier first found that the general level of the brain's metabolism did not seem to have any particular bearing on a person's IQ score. Then he tested a group of subjects as they competed at the kids' computer game Tetris. Again, there seemed no relationship between a person's IQ score and whether the brain ran particularly hot or cold during the performance of the task, but what did correlate was the rate at which brain activity dropped as the game became familiar. To Haier, this suggested that a smart brain was one that could start out with a powerful attack on a problem and then, having discovered what was involved, shrink down its execution to the most skeletal, low-level set of circuits.

These PET studies, published in the early 1990s, were stunning for the sheer dynamism of the changes. Outwardly, the subjects were performing exactly the same mental skills, although perhaps a little more proficiently. Yet huge changes had taken place in the brain. When a task was novel, wide areas of the brain were lighting up. It looked just like Baars's idea of a global workspace swinging into action. But with just a few minutes' practice, the same results could be produced with hardly any visible effort at all. A skill that had been learnt or explored in the full limelight of consciousness had become downloaded to create a low-level action template.

Or, to put it another way, the process of escalation also worked in reverse. The dynamic model said the brain used a fringe of pre-conscious habits and abilities to filter the moment, doing what it could in a local, unthinking way so as to leave the brain free to react more globally and thoughtfully to whatever proved difficult or significant about an instant of awareness. But the PET studies suggested that having organised a global-level reaction, the brain would immediately try to turn it into a localised routine or skill. The brain discovered the network of connections needed to do something and then shrunk it as hard as possible so as to add yet another habit to its periphery of pre-conscious processing abilities.

Of course, escalation and automation are not quite two faces of the same coin. For a start, the cycle of processing needed to evolve a state of focused consciousness takes about half a second – according to Libet, anyway. The downloading of new thought patterns and physical skills appears to take more in the order of minutes or hours. Nor does this automation seem likely to use the brain's 'so what?' pathways. Some of the same centres and bottlenecks might well be involved, but not in the same decision-making, choice-forcing way.

Again, however, it is the astonishing fluidity of the brain that is important. Viewed in some lights, the brain may seem to have a fairly robust, computational-looking structure. But if it is an instrument, it is one that can be played in many ways. The flow of processing activity is free to run in more than one direction. Yet having said this, it is time to get computational once more – to look closely at the logic of the brain's anatomy and see if, indeed, its processing intentions are written into its processing circuitry.

CHAPTER NINE

Consciousness's Twin Peaks

Every moment of consciousness includes not just sensation, but also a reaction. Many people talk about the mind as if it were essentially a passive display. Whipping up a state of subjective experience is what counts. Yet the whole point about having a brain is to act. Evolution has no use for an organ that just wants to sit and watch the world go by, regardless of how rich or beautiful the flickering parade of images might be. The job of a brain is to turn sensations into behaviour.

So if we are looking for the needs that shape the brain's organisation, then the need to form an intelligent reaction to the moment must be the ultimate. And certainly, the importance of action to the brain seems to be suggested by the simple fact that the entire front half of the human cortex is devoted to it. Each cycle of processing may begin with a journey up the flanks of the sensory hierarchy, but it is not until the front half of the brain has taken notice and decided 'Now what?' that a moment of consciousness can take on real overtones of meaning and purpose.

As mentioned, the general organisation of the frontal cortex is like the sensory cortex in reverse. Just like the sensory cortex, the frontal cortex is a patchwork of mapping areas tied into a dynamic hierarchy. But instead of processing beginning at the bottom with raw patterns and then ascending to highly abstract representations, the processing in the frontal cortex starts with an abstract level of mapping in the prefrontal lobes and then washes down through a series of mid-level motor filters until it hits the primary motor area. An inkling of what ought to be done gets turned into a mapping of actual muscle commands by a process of decomposition.

Between the prefrontal planning zone and the primary motor output map stand two main rungs of mapping, the supplementary motor area (SMA) and the premotor area. Experiments have revealed that, broadly speaking, the supplementary motor area takes the prefrontal's urge to do something and creates a plan. It holds memories of what kind of motor acts suit what kinds of occasions. The premotor area deals more with the nitty-gritty detail of co-ordinating a movement. In scanning studies, when subjects simply thought about making a finger movement, only the SMA lit up, but when they actually went ahead and made the movement, the premotor and the rest of the motor hierarchy began to fire. This division of labour is supported by the connections the two regions have to other parts of the brain. The supplementary motor area is linked to the sensory hierarchy, making it well placed to generate planning imagery. The premotor cortex also talks to the sensory hierarchy, but at a much lower level, connecting mainly to body-centred maps of nearby space.

It is a tempting idea to think of the SMA and premotor cortex as 'commanding' a muscle movement. The simple story would be that the two areas plan an action, then pull on the strings of a topographical mapping in the primary motor cortex to jerk the body's muscles into action. But as we saw from the evidence of sports psychology and ERP recordings, the explanation is far more complicated. Conscious-level thinking is too slow to control an action as it is happening. So while the cortex areas might help prepare the plan in advance, the fine detail and late adjustments would have to be left in the hands of a network of evolutionarily more primitive motor centres, including the brain stem, the spinal column, the basal ganglia, and the specialist bulge on the back of the brain stem, the cerebellum.

The cerebellum, in particular, picks up an action plan in the final few hundred milliseconds and uses a host of subconscious habits to smooth its execution. It is the cerebellum that does things like tighten a muscle at the precise moment needed to start applying a brake to our arm as we reach to the back of a cupboard, or judging when we have a tight enough grip on a bag to begin lifting it up. The ponderous planning of the motor cortex and nimble adjustments of the cerebellum meet up on M1, the primary motor cortex map. As said, like the somatosensory cortex next to it, the primary motor cortex is organised as a topographical mapping of the body with the feet represented at the top end, hands in the middle, and then head at the bottom. Cells in M1 get input from both the cortex planning areas and the cerebellum, so stand at the end of the line for a steadily sharpening set of commands as the time for making a movement arrives. There will be a dynamic gradient which starts with a general anticipatory warning of the kind of action that might be needed in half a second's time, running down to the delicate, last instant adjustments organised by the cerebellum.

The generation of a motor act has a simple, self-assembling logic. During the first half of a processing cycle, the brain extracts a focus from the moment. An event or thought is promoted to the highest levels of mapping. This focus then feeds a top-down cascade of activation. In sensory areas, the top-down flow creates associations, images, and states of priming, but in the motor hierarchy, an instant of focus will automatically be decomposed to form a detailed action plan. Noticing a neglected cup of coffee will put us in mind of taking a sip; hearing our grandmother's voice would turn our feet in the direction of the living room rather than the kitchen, where we were originally headed. We cannot help pay attention to some fact or occurrence without beginning to think about the inherent possibilities for action – what the psychologist James Gibson dubbed a situation's 'affordances'.

Not surprisingly, generating thoughts about action and manipulation takes up a large chunk of the frontal cortex sheet. But running in parallel to the motor hierarchy lies an almost equally important hierarchy for organising orientation movements – for planning and executing shifts in attention. Every moment brings the need to get the senses in place to deal with what is about to happen next. And this will not be just as part of a Sokolov-type orientation response where we are automatically taking a closer look at something which has caught our attention. We also need to shift our attention to where we anticipate things are about to happen as part of some ongoing activity or goal. If we are reading, we have to organise our eyes to hit the next line of the page; if we are playing cricket, we have to be looking at where we predict the ball will bounce. The orientation response is about getting there after the fact. But as much as it can, the brain will try to ensure that our ears are cocked, our bodies braced for the feeling of contact, and our eyes looking at the right spot, in time to catch life as it happens.

The planning of eye shifts is, of course, the most demanding task as the fovea – the sensitive pit of cells in the centre of the retina – is so small. Foveal vision is hundreds of times more sharp, but accounts for just a thousandth of the visual field, so to bring its power to bear, we have to move our eyes about in a series of snapshot fixations. Each of these movements, or saccades, requires a full 200 milliseconds to plan; the actual movement takes between twenty and fifty milliseconds, depending on the distance travelled; then the eyes need to stay locked in place for a further 200 milliseconds to allow time for the information to register. Research has shown that the brain's saccade planning decides not only where our eyes will eventually alight, but also anticipates the necessary depth of view and pupil settings so that the eyes arrive ready adjusted for the likely lighting and focusing conditions.

Single-cell recording experiments on monkeys have revealed just how much of the brain has to be devoted to the organisation of saccades and related orientation movements, such as turning the head. Just like the motor hierarchy, the orientation hierarchy runs through many rungs of mapping so that a general intention to move becomes translated by stages into an actual eye movement. At a high level, there is an area equivalent to the SMA which uses a map of surrounding space to mark out a target spot. Cells firing at this point would put a location in mind. Then an equivalent to the premotor area takes this target position and begins calculating what would need to happen in terms of eye movement to bring them into line. Again, the cerebellum and other lower brain centres, such as the superior colliculus, chip in towards the end, making the fine-scale adjustments needed to smooth the execution of the action. Finally, this snowballing wave of planning all comes together as a specific set of eyeball and neck muscle instructions on a patch of cortex known as the frontal eye field (FEF), the orientation hierarchy's equivalent of M1, the primary motor output map.

Giving over so much of the brain to organising eye movements seems more than a little perverse. It would appear simpler just to line the eyeball with extra light-detecting cells. However, the saccade is actually a sign of evolutionary success. An animal that knows where it needs to be looking during the next moment is an animal with a sharper mind. Consciousness is not about forming a dull even view of the world but a high-contrast picture. A gimlet gaze ensures that only the events that matter most enter the limelight of awareness, and so provoke a strong frontal cortex reaction.

Recent scanning experiments have revealed that this trend has reached its extreme in humans. By 1995, Roger Tootell – who did the original autoradiography experiments to reveal the topology of a monkey's V1 mapping – had joined the flood of neuroscientists into the new field of neuro-imaging, teaming up with a world-leading f-MRI group at Massachusetts General Hospital. Using the MRI camera, Tootell was able to take pictures of the human primary visual cortex in action, and to demonstrate that the representation of foveal vision was still more exaggerated than it was in his monkey experiments, taking up three times more space than it did on the V1 mapping. So the smartness of the human brain showed in how much anything outside the extreme centre of vision was pushed into the margins of processing. Of course, we have to have some level of representation for what lies on the periphery, just to catch the occasional small flickers of movement that might spell danger or signal an event of interest. But the ability to predict where to look had become far more important than the actual business of looking.

For animals, moving muscles and moving the focus of attention sum up the possible reaction to a moment, but for humans, the frontal cortex sheet offers one further response to whatever has been escalated to the centre of attention. It has a third output hierarchy for the generation of speech acts, so each moment becomes the jumping-off point for something to say.

As the Washington University PET group found, speaking involves a whole network of areas apart from the two classic language centres of Broca's area on the left frontal cortex and Wernicke's area on the cleft of the left temporal lobe. As already outlined in the previous chapter, the traditional idea had been that Broca's area was the brain module that formed sentences, while Wernicke's area – sitting next to the brain's auditory maps – was the decoding module which understood them. But as scanning was used to explore the brain's organisation in much more detail, it became clear that language shared much the same kind of hierarchical design as the other two output streams.

The parallels were almost uncanny. For a start, speech had its own equivalent of the SMA sited right alongside the motor hierarchy's supplementary motor area at the top of the brain. Then Broca's area itself looked like an extension of the motor hierarchy's premotor maps. Its position on the lower flank of the frontal sheet also made sense as it put Broca's area hard by the part of the primary motor map responsible for making throat and mouth movements. The arrangement of speech areas made it seem that to develop the ability to speak, our ancestors simply added a bit of extra cortex to the edge of each level of the existing motor pyramid.

With a hierarchy of frontal cortex language areas, the brain would be able to take an inkling of an idea – whatever was the focus for the moment – and decompose it into a grammatically structured sentence. A speech intention could be turned into a fully fledged speech act. Wernicke's area would then fit in as the sensory cortex area with the job of remembering the verbal form of words, and also tying them to a sense of meaning. Or rather, Wernicke's area would stand at the head of its own pyramid of processing in which the high-level areas would map for speech elements, but the actual sound of a word, or the kinds of mental imagery it might conjure up, would have to emerge from a 'projection' of this information back through the rungs of the sensory-processing hierarchy. Wernicke's area would act as the brain's window into a fleshed-out state of word meaning.

So, as with the other two frontal cortex reaction streams, the language hierarchy's response to each moment could be highly variable. The cone of activation running back through the many levels of processing in both the frontal and sensory cortex might be wide and shallow, or narrow and deep. Hearing the doorbell ring, we would feel the urge to leap out of our chair, or at least look towards the door. Our mapping of the event would immediately unlock ideas about physically what to do. And as part of the overall outflow of planning activity, we would probably also be moved to make some comment – perhaps a barely voiced 'Now, who's that?' amounting to little more than a general sense of questioning. Or we might go further and develop a specific reaction such as, 'Better not be those kids next door again – they're hopeless at keeping that ball in their garden.' Every moment comes with a potential departure point for a sentence. But it is only if the reaction is pushed all the way down the language hierarchy that it will be developed into a concrete string of words, a speech act so vividly realised that we can 'hear' the anticipatory image of what we intend to say.

There is an elegance to the brain's design. Sensation flows in and condenses to produce a focus. Then, this focus becomes the spark for a wave of activation which floods back down the hierarchy to create a spread of motor plans and sensory expectations. And the more specific the focus, the more specific the mental reaction.

The part of the brain that seems to be responsible for turning the actual flow of processing around – the place where the escalated residue of each moment's processing becomes explicitly mapped – is the prefrontal cortex. It is only quite recently that neuroscientists have begun untangling the complex organisation of the prefrontal cortex. At one stage, it was believed that it might have no particular structure at all. But it has since become clear that the prefrontal cortex is probably split into at least six major processing regions, and each of these areas in turn is finely divided into an assortment of maps and maplets.

As already mentioned, the prefrontal cortex shares the same basic what–where division of the sensory cortex. The upper flank of the prefrontal lobe, the dorsal prefrontal cortex (DPFC), sees the re-emergence of the location- and motion-mapping stream of the parietal cortex – it thinks a lot about questions of place. The lower half of the prefrontal lobe, the ventromedial region, then takes the output of the object identity stream of the temporal lobe – it focuses on events and what they might mean.

This what–where split creates a basic topographical order. Then, within the two halves, there are further divisions. The ventromedial prefrontal cortex is separated into an area which deals with the senses of sight, hearing, and touch, and another which responds more to tastes and smells. The two areas are also distinguished by having different kinds of output connections. The sight, hearing, and touch mappings feed into the parts of the motor and orientation hierarchies that deal with more physical movements, while the smell and taste mapping areas feed into motor areas that handle eating and chewing behaviour. So, if our current focus of attention happened to rest on a chicken leg, part of the ventromedial would see it as an object to be manipulated while the other part would be engaged in an earnest debate over the urge to eat it.

Of course, the ventromedial prefrontal cortex is also responsible for more sophisticated trains of thought. There is evidence that we use the ventromedial to think about complex social situations – thoughts about whether people really mean what they say, or how to turn some difficulty to our advantage. But the same general logic applies; it is just that the 'object' in focus may be some nuance of body language or a familiar example of office politics.

While the ventromedial region is busy mapping different varieties of sensory object, the DPFC has its own sub-divisions for mapping different categories of space. The distinctions are not so well understood, but it seems that some parts of the DPFC focus attention on points in the real space about us while others help us navigate our way around an associative space, steering us through a mental space containing thoughts and meanings. However, the principle is the same. It is clear that the DPFC takes the spatial output of the sensory cortex and turns it around to fuel a reaction in the action and orienting areas of the frontal cortex.

The prefrontal cortex is still the least well lnown part of the brain. Single-cell recording research started at the other end of the chain – the study of primary mapping areas – and only gradually worked its way up. Besides, single-cell recording is not much use for studying the human brain, and the human prefrontal cortex is the most expanded part of the brain's entire structure. However, enough has been discovered to say that the key to the prefrontal is that it maps the focus of the moment and does so in a topographically generous way. The sensory cortex spends some reasonable fraction of a second extracting a recognised, meaning-imbued representation of the world – a quivering stack of detail. Then this stack becomes focused with one chosen aspect of the mapping being emphasised and the rest damped to form a dim, yet still organised, periphery. Whatever emerges as the focal event finally rises to fill the circuits of the prefrontal cortex. The sensory experience tapers to a point and then is mapped across a wide expanse of neural territory with the connections to explore every avenue of response. The prefrontal cortex takes the core of the moment apart to see it in terms of its eating, manipulating, associating, socialising, and orienting possibilities.

Of course, the prefrontal is not the seat of consciousness. It does not experience the fruits of the sensory cortex's labours. The sensory details of the moment stay where they are, distributed across the circuits of the brain. The firing of a prefrontal cell merely refers back to this detail. It is a pointer to the fact that a certain kind of experience is unfolding across the many levels of the brain's processing hierarchy. But because of their finely tuned connections to different parts of the brain's output areas, prefrontal cells can cause this spread of representation to be turned around intelligently. They can guide the focus of the moment back down the right paths to produce the most useful and timely response.

This is only a bare-bones description of the organisation of the frontal cortex, but it should give an idea of why each moment of consciousness is flavoured not just with a sense of understanding, but also intention. Some sort of reaction – even if it is simply a sense of wanting to see more, and so mostly making plans about an act of orientation – is the natural destiny of every cycle of processing. There is no point to awareness unless it is leading somewhere.

The prefrontal cortex caps the moment, taking what matters and guiding the consequent wave of output. This role seems to put it at the very pinnacle of the brain's hierarchy of processing. But in fact the brain has at least two peaks of processing. As well as the prefrontal cortex, there is a very different kind of concentration of information going on in a specialist memory organ known as the hippocampus.

The hippocampus is a carpet roll of cortex that has become curled up to lie inside the curve of the temporal lobe (hippocampus is Latin for 'seahorse', the roll in cross-section supposedly resembling a seahorse's tail). And, as is usual with all parts of the brain, there are actually a pair of hippocampi – one to serve each hemisphere.

The most obviously significant feature of the hippocampus is that it stands at the end of the line as far as sensory processing is concerned. The sensory cortex is split from a very early stage into its two broad streams of processing, the 'what' and 'where' paths. These two paths reach their separate peaks on the tip of the temporal lobe and the hump of the parietal cortex respectively. From there, information feeds forward to fill the various buffers of the prefrontal cortex. But it also heads down for the hippocampus where, instead of being kept distinct, the many aspects of a moment's experience begin to converge.

On the way up through the sensory hierarchy, the various kinds of sensation – such as touch and vision – are being mapped separately. Even in high-level regions like the IT cortex, there is no overlap between the sight of an object and its sound or taste. But finally, in an area of cortex surrounding the hippocampus called the entorhinal convergence zone, the many lines begin to cross so that individual cells are reacting to conjunctions of sensory information – the voice of our grandmother gets connected to her look and touch. Her representation as an object also starts to become connected to her position in our mental representation of space. Cells fire only when some target object crosses a recognised point in the greater landscape.

Eventually, all this rapidly converging voting activity feeds the hippocampus. The hippocampus is an evolutionarily primitive part of the mammalian cortex. Most of the brain is made up of neo-cortex, a six-layer sheet of material, but the hippocampus still has the simpler three-layer structure of the ancient cortex. The hippocampus also has a far more regular pattern of connections than the cortex proper. Where the neo-cortex is a tangled jungle, the hippocampus looks almost a tidy grid by comparison. But perhaps the biggest clue to the role played by the hippocampus is that its middle layer of cells has a special memory-trace-trapping reaction known as long-term potentiation (LTP).

The capture of a memory trace is actually a long-drawn-out affair. It takes only a moment for something memorable to happen in our lives, but it takes hours, and even days, for our brains to react with the growth changes that will code for a new memory pathway. To bridge the gap, the brain needs some mechanism to trap a sensory pattern and keep it jangling while the fixing work is being done. Long-term potentiation is a complex response involving a whole cascade of neurotransmitter and cell membrane processes. The fastest level of response happens in under a second. When a neuron is hit hard – or, more particularly, when it is hit by spikes arriving at several synapses at the same time – a special class of receptor pore, the NMDA receptor, will react by letting calcium ions flood into the cell's interior. This influx leads to an almost immediate change in the balance of the neuron's electro-chemistry, priming it to fire much more vigorously if the same pattern of input comes its way again. Simultaneously, the calcium acts to switch on the neuron's genes so that it begins to make more permanent growth changes.

One fast physical adjustment can be a simple swelling of the dendrites responsible for bringing the news about a significant event. By pumping up their tips like a balloon, extra synapses can be brought into play, so strengthening the connection with an input cell. This swelling appears to subside after about six hours or so, but by then more time-consuming steps, such as the sprouting of new dendrite spurs, will be cementing the pattern. Over the course of a day or two, many kinds of change will take place. And all this time, an LTP-stimulated network of cells will continue to fire spontaneously at regular intervals, as if to keep the original experience alive while it is gradually being built into the fabric of the brain. They jangle to remind themselves what they want to remember – or perhaps, as some researchers believe is more likely, to integrate the experience with the many other memories that are being fixed during the same period.

The machinery of NMDA receptors and an LTP response are actually found in neurons all over the brain. It is also being discovered that there are probably many varieties of LTP-like reactions – including its exact opposite, a long-term depression in which a neuron learns not to fire if a certain event takes place. This is only to be expected, as every part of the brain needs to be able to make connection-strength changes. But the hippocampus is notable for having a triple layer of cells with a strong LTP reaction. Still more significantly, there is a heavy overlay of acetylcholine inputs, a neurotransmitter that can damp prior activity patterns. James Bower, a neurobiologist at the California Institute of Technology, has led a group of researchers whose work suggests that a blast of acetylcholine might be used to clean up a memory trace, suppressing the background anticipations and associations that may have been part of processing so that an event can be remembered with pristine clarity. If this were true, the hippocampus would have the circuitry both to shrink down and preserve what mattered most about a moment of activity.

There is much about the detail of the hippocampus that can only be speculation. But as neuroscientists began to think about the business of sensory representation in a more dynamic way during the 1990s, a story started to come together. The hippocampus's position at the very end of the sensory trail, coupled with its memory-trapping machinery, suggested that its job was to take a snapshot of each moment of processing. This is not to say the hippocampus stores the actual record of a moment. In the brain, nothing moves between two rungs of mapping except a pattern of voting, and so, as with the prefrontal cortex, all that the hippocampus can ever 'see' is a highly abstract stippling of inputs. But because of the feedback connections between levels of mapping, each cell in the hippocampus effectively points a finger back down through the sensory hierarchy to the cells supplying it. When the hippocampus freezes a pattern of activity, it thus creates a template from which a moment of processing can be reconstructed. A stack of mapping activity converges on the hippocampus to leave behind a dry crust of memory. Then, by rousing this trace and causing it to tug on its feedback connections, the original spread of sensation can be re-ignited right back through the cortex hierarchy, reaching down even as far as V1 or other sensory mappings, if given sufficient encouragement.

Being a dynamic process, the reconstruction of some prior moment of existence would not necessarily be exact. The hippocampus trace would stir in the form of a population vote that then triggered a succession of further votes back down through the sensory hierarchy, eventually producing a fleshed-out state of sensory detail. As with other kinds of top-down flows, the cone of activation might run narrow and deep or spread wide and shallow, and sometimes a particular area of circuitry might prove more difficult to rouse. So no two 'viewings' of a remembered event would be the same. But there would be obvious advantages in the brain trapping its memories as a high-level template rather than trying to preserve the full stack of mapping that came with the original moment. It would be economic, for a start, as just a handful of hippocampal cells could signpost the way back to millions of lower-level cells. It would also mean that the lower-level cells would not be tied up while memory fixing was taking place. They would be left free to get on with the next moment of representation.

But perhaps the most important advantage would lie in the opportunity to tune the brain's reaction to a moment. As said, memories are the brain's processing pathways and so it has to be choosy about what it remembers. By having a specialist memory area straddling the flow of processing, it would be simple to control the level of our response. A dull moment with little surprise would cause a very faint reaction in the hippocampus. But a shock, such as hearing of the death of a public figure or witnessing a road accident, could trigger a flush of arousal – a release of neurotransmitters like noradrenaline and acetylcholine – leading the hippocampus to preserve the moment in fine detail. We would remember the scene with a flashbulb intensity, even down to the way a shaft of sunlight was catching the back of the TV, or a shrill screech of car brakes was followed by a crunch and tinkle of glass. Because the hippocampus stands at the end of the line, a powerful LTP snap would trap the most inconsequential details, even ones to which we would not normally pay attention.

This is not to say that the rest of the brain cannot form memories. The ability to adapt is fundamental. Even synapses and membranes show memory properties. But the difference is that by comparison with the split-second snapshot reactions of the hippocampus, change has to be drummed into other brain circuits. It will be slower to form, and will look muddier when it does. So trailing a finger on a spinning disc will eventually change the mapping properties of an area of a monkey's primary somatosensory cortex. Being set the same search task day after day may retune the visual feature representing responses of neurons in the inferotemporal cortex. However, the hippocampus is the brain's specialist organ for catching a crisp memory trace, the kind that seems to refer to a specific life event rather than a generality of such events.

So, the hippocampus and the prefrontal cortex give the brain two peaks of processing, but the style of convergence is very different in each. In the prefrontal cortex, there is a broad display. The prefrontal has the luxury of space to take the focus of each moment apart and map it onto a potential array of responses. The hippocampus, by contrast, is not for exploring the moment but for recording the results, for fixing what ended up mattering the most for future use. In one direction, the meaning of the moment is fluidly expanded; in the other, it becomes almost digitally concentrated – a frozen frame of information. The next question is how these two peaks might actually interact during a cycle of processing. The best insight, perhaps, comes from research into the phenomenon of working memory.

Working memory, or short-term memory, is the cognitive science term for the small collection of thoughts, experiences, and memories that seem near at hand, circling the current focus of attention. This buffer zone is not 'in' consciousness, yet what it contains is easy to access. And while it forms a kind of mental periphery, it is not the periphery of the conscious–pre-conscious distinction – the division of awareness into a habitually processed background and an escalated foreground. To make it into working memory, a thought or event has to have been the focus of an earlier cycle of processing. So, working memory is a lingering backdrop of all the things we noticed, or the plans we made, during previous moments of consciousness, a periphery of what has mattered to us but is gradually fading, rather than a periphery formed by things that never really managed to matter in the first place.

Experiments have shown that working memory has its limits. The most famous of these is that when people are given a list of things to remember – a set of numbers, letters, words, objects, or even tasks to do – they can only keep about six or seven items fresh in their minds. Any extra and some will inevitably be forgotten as if working memory were a vessel that can contain just so much information before the excess begins to slop over the rim. However, this idea of a fixed-capacity mechanism is misleading. Seven or so items is certainly about the maximum that can be carried over from one moment to the next, but remembering less is much easier – and the items also tend to persist in memory far longer. Catching a telephone number in one go takes some mental effort, and it is likely to disappear at the first distraction. However, remembering three digits – for example, 392 – is hardly any strain and the numbers will linger. If we took a moment to note them, there is a good chance the figure would remain mentally close at hand for the rest of the day.

The idea of working memory as a sort of cache, a fast-access memory to offset the much slower access times of the brain's bulk memory mechanisms, suited the computational outlook of cognitive psychologists. But when neuroscience actually began searching for this cache at the start of the 1990s, it soon became apparent that, like everything else, working memory was a process that involved the whole brain. There was no unique cortex module or class of neurons that acted as a holding tank. The maintenance of a backdrop of recent thoughts and experiences involved exactly the same hierarchy of areas that processed them in the first place, although the prefrontal cortex and hippocampus did have special roles to play in keeping working memories flying.

Again, Robert Desimone's single-cell recording studies using awake monkeys provided some early clues. The search image experiment in which a monkey had to hold in mind the picture of something like a boat for three seconds could also be looked on as a kind of working memory test. It was an idea that the monkey had to keep mentally close at hand until its reappearance. That particular experiment showed both an attention effect – non-target neurons in the temporal lobe area were dampened at the moment of making a choice – and also a priming effect when there was a slightly raised tick-over firing of target-coding neurons during the three-second wait. In a further experiment, Desimone's team put electrodes into the prefrontal cortex to see how cells behaved there while performing the same task. This study showed first that the prefrontal had neurons that mapped the fact that IT was mapping something. Cells in the prefrontal cortex would fire when the sensory hierarchy was representing the experience of seeing a boat picture. Then, during the three-second wait, these prefrontal cells continued to burn brightly. They fired at a high rate, not just a raised tick-over rate.

Even more tellingly, when a distraction was deliberately intro-duced to the task, the prefrontal cells held on to the thought of the target while the IT cortex was being wiped clean. As a distraction, the monkey would be given a second visual task to complete before the target came back. The need to concentrate on something else had the effect of flattening activity in the sensory area. However, the prefrontal cells kept the intention to pick a certain picture quietly glowing. Desimone's group found that having an interruption slowed the monkey's response by about 200 milliseconds. Robbed of an explicit state of sensory priming, it took a little longer for the attentional bias to work its way through the system and tilt the brain towards one of the two images being presented. But the role of the prefrontal in hanging on to the kernel of a plan or memory looked clear enough.

By 1995, the result had been confirmed in a scanning study of human working memory – in fact, by an f-MRI team run by Desimone's wife, Leslie Ungerleider, based in a neighbouring laboratory at NIMH. In this experiment, subjects had to remember a photograph of a face over a series of delay periods. When the delay was short, the activity was strong in both the prefrontal cortex and high-level areas of the temporal lobe. But as the delay stretched out, gradually the temporal firing disappeared, leaving just the prefrontal to hold on to the memory. And while the result sounded trivial enough, like Kosslyn's imagery experiments, the effect was galvanising. The monkey work could always be dismissed because nobody could be sure how relevant their patterns of brain activity were to the human experience of working memory. But with the scanning study, the level of activation could be matched to the extent to which the target photographs seemed to be in consciousness for the subjects.

These results made it plain that working memory was just one further example of the way the brain's processing hierarchy could operate in both directions. It could work bottom-up to extract a focus from the moment or top-down to reinstate a pattern of activity. And the ability to juggle a number of memories or intentions at one time was made possible by having a high-level area like the prefrontal cortex which was remote enough from the fray to keep the gist of an idea going. The sensory and the motor hierarchy might have to drop their representation of a particular state to make room for mapping new states, but the prefrontal cortex could maintain a template – a set of pointers – which could be used to stir the intention or memory back to life.

Once neuroscientists began to understand the ease with which the brain could collapse and reinflate patterns of activity, they could make sense of some earlier work which had tried to disentangle the role played by the hippocampus in working memory. In a series of lesioning studies during the 1980s, researchers had used heated electrodes or radioactive pellets to destroy either the hippocampus or the prefrontal lobes in monkeys. The monkeys were then tested to see what damage was caused to their powers of memory. It was found that a monkey with a lesioned hippocampus could hold the memory for a target for about ten seconds, but no longer. It also tended to lose the thread if there was any distraction during the delay period. A monkey with a lesioned prefrontal could not hold a working memory for a target at all – intentions vanished almost immediately.

It was clear from such experiments that the prefrontal cortex was essential for forming states of intention, and could also maintain them for a short time. However, its grip on a memory state was extremely fragile; better than the temporal lobe, but still, a major shift in attention would wipe even the prefrontal circuits clean, preparing them to deal with a fresh event. The hippocampus was the brain's real memory trap, taking the snapshots that allowed moments to become permanently woven into the fabric of the brain. So in the seconds and minutes following an event, the hippocampus must be the place where the prefrontal lobes would have to go to dig out information about its own recent history of thoughts and plans. The prefrontal lobes could do the manipulating, but it was the hippocampus which acted as a general backstop to consciousness, recording our ideas as well as our experiences.

For the hippocampus a thought – a fleshed-out state of imagery or expectation – would be seen as just another pattern of activation running back through the sensory cortex. So there was no particular mystery about how it would be able to record a moment's mental activity. But the way the prefrontal might fetch back a moment from the hippocampus was another question. Certainly, everyday experience tells us that the process is not always easy. Often we have a faint inkling that there is something we meant to do but cannot remember exactly what – as if the prefrontal representation of our intention has faded to the barest flicker. For a few seconds we will have an irritating tip-of-the-tongue feeling, then the memory will pop back into our heads. If waiting does not work, we may have to jog our minds by retracing our steps. We might have walked into our bedroom with the intention of getting something, find we have forgotten what it was, and have to go back to where we were sitting to see if the context will remind us.

As will be seen, the subtle tricks of memory recall are a tale in themselves. But as a last aside, a couple of cruel stories reveal what our mental lives might be like without the snapshot services of the hippocampus to back up our thoughts and experiences. Cases of damage to the hippocampus are rare because, of course, it would be unusual for any natural event to cause injury to such a defined area of tissue on both sides of the brain. However, in the 1950s, Canadian surgeons carried out a disastrous operation on a young factory worker who was suffering from epilepsy. The surgeons cut large chunks out of each temporal lobe in the hope of removing the likely focus of the fits. The result was that the patient, known in the scientific literature by his initials, HM, could form no new memories. He could have thoughts about the moment he was in, but as soon as an intention or experience slipped out of focal consciousness, it was gone for ever. He could meet the same doctor and answer the same questions ten times in a day, yet still react as if he had never encountered the person before.

In another more recent case that was the result of natural causes, an English music producer, CW, lost both hippocampi through a viral infection. Speaking to a film crew visiting the hospital to make a documentary about him, CW said he did not know how he had arrived in the room or what he was supposed to be doing there. Starting to get agitated at their questions, he exclaimed that the only thing he knew was that he had been dead for many years and must have returned to life just a few minutes earlier. 'Why the bloody hell can't they do anything to make me conscious?' he shouted in despair. Then, minutes later, he had forgotten who the film crew were. 'I can hear traffic sounds now,' he remarked to himself in a puzzled voice.

These two cases show that the hippocampus is essential to a normal state of mind. Without it, there can be no running sense of personal history. But brain damage cases can also be looked at the other way round. It is perhaps equally revealing that consciousness does not collapse even with the loss of so critical an organ as the hippocampus. Unlike a computer, the brain makes do with whatever circuitry is available. Consciousness may be degraded, but it is not easily destroyed.

Examples of prefrontal cortex damage make the same point. Even with massive injuries, or the deliberate destruction of large areas with the frontal lobotomy operation that was a common treatment for mental illness even into the 1960s, the result is a subtle loss of planning depth and self-control rather than a dramatic loss of awareness. People with prefrontal damage tend to have poor concentration, fickle short-term memories, impulsive behaviour, and an inability to anticipate the consequences of their actions – in other words, all the traits that might be expected of someone missing the final stages of their response organisation. However, they remain aware and sometimes even act happier because nothing in life really bothers them that much any more.

The prefrontal cortex makes a big difference to the abilities of the brain, but in the end it is just another rung of mapping. There is nothing special about its structure which makes it the seat of consciousness. The prefrontal merely offers the brain a further level of abstraction and planning before the descent into a reaction to the moment must begin.

This highlights yet another crucial fact about the brain's processing hierarchy. The ripples of activity can flow up and they can flow down, but they can also flow sideways. They can take shortcuts through the maze of pathways so as to produce a reaction to the moment without having to go all the way through slow and laborious prefrontal thinking. Indeed, as has been said, a strand of processing is escalated to the focus of awareness only by default. A lifetime of learning and shrinking down skills such as opening doors, picking up heavy bags, or striking cricket balls – and also more intellectual skills such as speaking sentences and making plans – equips our brains with deep strata of habit. There are well-trodden paths connecting the sensory hierarchy directly to the motor hierarchy so that there is no need to wait for instructions from some further level of processing such as the prefrontal cortex. The motor hierarchy has the action templates to produce a reaction all by itself. This is why destroying the prefrontal lobes does so remarkably little damage. It robs the brain of planning depth, and there is much less room for self-supervision and exploratory thought – for juggling ideas – but the brain can still do processing with the fat layers of more habitual reaction that remain.

This explains another recent neurological puzzle: why motor areas have turned out to be riddled with sensory neurons. For a long time, it was assumed that a frontal motor area should contain nothing but neurons directed towards the control of muscles. But gradually, single-cell researchers like Jean Requin of the National Centre for Scientific Research in Marseille, France realised that the motor maps they were studying were full of cells that played no apparent part in the population vote needed to program a muscle movement. In the premotor cortex, for instance, it was found that less than a tenth of the cells were classic motor neurons. Instead, about half the neurons were sensory cells that reacted to movement-related information such as the sight of an approaching target or feelings of pressure and contact. The remainder of cells in the premotor cortex had a mixed response, firing only to conjunctions of sensation and reaction.

Sensory cells proved to be abundant in the primary motor cortex and down in the cerebellum as well. What this said was that even the lowest rungs of the motor hierarchy had a direct connection to the view of the world unfolding on the sensory cortex. They did not have to wait for news of events to travel all the way up and back through the prefrontal cortex, but could intercept 'their kind' of sensory detail to drive their own more automatic level of response.

Further research gave a more precise idea about the sort of awareness of life enjoyed by a lowly area like the premotor cortex. In 1994, Michael Graziano of Princeton University, New Jersey tested a monkey while a small white ball was being moved steadily towards some part of its body, such as a cheek or elbow. The premotor sensory cells only began to fire when the monkey could see it was about to be bumped. The same cells would also fire if the body was actually bumped at that point while the monkey was not looking. So what the premotor cortex was producing was a body-part-centred sense of nearness for the world. It had no interest in objects out of reach. It was a collision detection map which saw the world only in terms of a potential for contact. The sensory cortex produced a more general panorama populated by objects and surfaces. The premotor cortex would then look into this pool of representation to extract a feel of what could be grabbed and fiddled with, or ducked and avoided.

A brain damage disorder known as the Dr Strangelove syndrome, after the crippled scientist in the Stanley Kubrick film, gives a thought-provoking glimpse of what consciousness might be like if we had only this limited view of the world – if the premotor cortex were our top rung of processing. The syndrome is caused by the SMA and cingulate cortex on one side of the brain having been damaged by stroke or disease. Victims find that the hand normally controlled by that side of the brain takes on a life of its own. Spookily, it will snake out and clutch anything within reach. Patients have to prise off their own fingers with their other hand, then either hold the misbehaving hand down or even sit on it. The explanation is that, robbed of the higher-level planning areas on one side of the brain, the patient's premotor cortex is free to launch simple grappling and handling actions whenever a likely object hoves into view. To the patient, the hand feels paralysed – there is no planning stream to will a movement – but the premotor cortex remains a dim island of purpose, peering out at the world and reacting impulsively to anything that looks at all grabbable.

Our idea of a cycle of processing has now been stretched to include a frontal cortex reaction. The moment is not complete until the brain has also settled its representation of a response – the path out of one instant and into the next. This need to generate a state of intentionality might have something to do with Libet's half-second results. The evidence seems to be that it takes about 100 to 200 milliseconds to sort out a pre-conscious spread of mapping in the sensory hierarchy. It then takes somewhat longer to draw the moment into some kind of sensory focus. Perhaps embellishing this focus to produce a fully developed frontal cortex reaction might require still more time. And certainly, until we have some idea of what a moment means to us, there is not much about it that is worth remembering. The brain only wants to add useful connections to its circuitry so there would seem little point in the hippocampus taking its snapshots until the brain had evolved the representation of a moment far enough to include an intelligent response.

At least the right kinds of questions seem to be coming into view. In the 1980s, neuroscience threw up a lot of baffling detail. Much was being discovered, but the findings did not slot neatly into a framework. During the 1990s, with new ways of thinking, and also new kinds of evidence like scanning, the way the parts of the brain might hang together during a moment of processing started to become clearer. The shape of the hierarchy, and the many paths that could be taken by a wave of activity, could be seen.

It also became possible to appreciate how the brain was an organ that built itself – how today's thoughts became tomorrow's circuits.

The brain has areas like the prefrontal lobes in which it can really explore the meaning of a moment. The broad expanse of prefrontal maps could feel out every potential avenue of response, looking for the optimal reaction to whatever had proved itself the most noteworthy event of the instant. Then, as the PET studies of verb generation and Tetris-playing showed, solutions developed in the bright glare of attention would be pushed back down the hierarchy just as fast as the brain could manage. It did not want to – indeed, could not afford to – go through the same elaborate evolution of a response every time it faced the same set of circumstances. Once an answer had been found, the brain was quite happy to slip into unthinking routine.

The way the brain automates skills is obvious from watching a baby. A new-born begins life from scratch. It has no habits or expectations. A child even has to learn that things still exist after they are lost from sight. When a blanket is dropped over a toy, a six-month-old has no expectation that it will be seen again once the blanket is lifted. But gradually a baby learns to make these very basic predictions about the qualities and properties of the sensory world. It also begins to see life in terms of possibilities – the kinds of things it can do in response to what it is experiencing. The first grabbing or kicking movements of a baby are little more than spastic expressions of excitement. However, it soon discovers that it can do things with its hands. Through trial and error, its motor and sensory areas begin to connect in meaningful ways. Eventually, an infant finds out not just how to move its hand smoothly to a point in space, but also how to shape its grip in advance so that it arrives prepared, ready to pick up something small or large, hot or sharp, soft or sticky.

In the same way, the skills of speaking, imagining, and thinking are learnt through trial and error and then added to the circuitry of the brain as compacted layers of habit. So over a lifetime, the frontal cortex sheet silts up with automatisms. We become be able to deal with almost every situation with a minimum of effort. High-level thought is reserved for managing the wider picture: for choosing the moment to act, for guarding against impulses, for hanging on to a longer list of intentions, for maintaining the general context within which all actions must unfold, and for supervising the quality of the results.

So the brain's hierarchy of processing is a living structure. It provides a landscape for processing the moment and is, in turn, shaped by the flow of that activity. However, as talk about the prefrontal choosing the moment and supervising results should remind us, the brain also seems to have some measure of active control over its flows. Somewhere along the line, the idea of being able to will things to happen has to come into the picture. The higher levels of the brain would need such a system of control simply to time the release of an action or to block an undesirable impulse. As we sit in our car at the lights, our motor and sensory areas will be revved up and ready to go. Our visual centres will be primed to see the lights turn green and our hands and feet will be itching to get the car racing off the line.

So somehow the brain must have a safety catch that stops these urges from immediately becoming fact – some mechanism for thinking without acting. There has to be a gateway or bottleneck in the flow of messages running between cortex maps so that the brain does not simply run away with itself, erupting into action as fast as it senses the possibilities. The explanation of how the brain can assert an often digital-like grip on the generation of thoughts and the release of behaviours lies in the sub-cortical machinery of the thalamus and basal ganglia.

CHAPTER TEN

Of Sub-cortical Bottlenecks

For years, David LaBerge, a mild-mannered professor of cognitive psychology at the Irvine campus of the University of California, felt rather out on a limb. LaBerge learnt his psychology at Stanford University in the 1950s and got his first job at the University of Indiana – both hotbeds of behaviourism where researchers were expected to devote their lives to watching rats pressing bars or scurrying about mazes. But LaBerge always believed psychology should offer more. So, as the opportunities presented themselves, he eased himself into the study of attention, and from there into the still more unfashionable subject of how expectations pave the way for states of focused attention. To cap it all, LaBerge thought it was essential that he should be able to say something about the neurology underlying these processes. There was not much point in a model of anticipatory mechanisms unless it applied to the workings of real brains.

His serious research began in the 1960s when he started doing reaction-time experiments which demonstrated that the brain was exquisitely sensitive to the biasing effects of anticipation. Groups of students would be hauled into the lab to do tests which had hardly changed since the era of mental chronometrists such as Wilhelm Wundt. A series of different coloured squares would flash up on a screen and the subjects would have to respond to some colours but not others. LaBerge showed that people could react up to a tenth of a second faster if they were given reason to expect a particular target, such as by simply showing it more frequently than the alternatives. On the other hand, if the subjects had come to expect to see mostly non-targets, or some second target that needed a different key response, then this mental set would delay their processing of what

218

was now a surprise event, skewing their reaction time by about 100 milliseconds the other way.

In itself, the result was no shock. But LaBerge was trying to make the point that anticipation was something that had to affect every moment of processing one way or the other. The normal approach of researchers doing such tests was to look for the shortest response times and treat those as the real measure of brain processing. They deliberately made sure the stimulus was as predictable as possible and that the subjects were fully alert so as to get rid of any variability. But the variability was what impressed LaBerge. Expectations made a measurable difference. It even turned out that a state of expectation had its own volatility. Experiments showed that it took about a third of a second for a subject to rouse a state of anticipation and that the effect of any priming peaked at between half a second and a second, then faded rapidly. Anticipation was quick to build, but also quick to decay.

Such details may have fascinated LaBerge, but it put him well outside the mainstream even when cognitive psychology arrived to get researchers thinking about mental processes again. The cognitive approach put the emphasis on the broad generalities of a process like attention rather than the particulars. LaBerge paid no heed, however. By the early 1980s, he had already moved on and was avidly absorbing all he could about the latest findings coming out of neuroscience. Single-cell recordings were beginning to reveal the full richness of the cortex's hierarchical design, and LaBerge was quick to seize on the need for states of attention to evolve through a brain-wide neural competition. His work on anticipation also made it plain to him that priming could be used to bias the mapping activity right from the start. The only question was, how did the cortex manage its feedback flows so the right kind of biases were in the right place at the right time to focus its activity? The pathways connecting one rung of cortex mapping to another did not seem to offer any obvious mechanism of control.

There was certainly no shortage of actual connections between areas of the cortex. Most of the traffic up and down the cortex's hierarchy of processing has to travel outside the thin rind of grey matter itself, passing through a thick, fat-insulated layer of trunk cabling known as the white matter. Inside the cortex, the axons linking cells are naked and good only for very short-distance connections. They can carry signals no more than a millimetre or so, and even then the spikes crawl along at just a few miles per hour. For long-distance traffic, messages have to be routed through the myelin-jacketed pathways of the white matter. And the amount of talking the cortex does with itself can be judged by the fact that white matter makes up more than 40 per cent of the entire brain. It has been estimated that, stretched out, the wiring from a single brain would form a strand long enough to reach twice around the world.

However, while there was plenty of white matter cabling to carry a huge weight of feedback and cross-talk messages between cortex areas, it seemed an unmanaged connection. The open pathways would allow for the rapid development of a brain-wide state, acting like sluices to equalise feedback pressure and let a stack of maps subside into some kind of coherent balance. But the white matter did not seem to offer any way of actively steering a competition towards a desired outcome.

But during the 1980s, yet another advance in research technique began to suggest that the thalamus might be a strong candidate for the job of being the cortex's feedback-focusing bottleneck. For a hundred years, neuroanatomy had been carried out with tissue dyes that had not changed much since Camillo Golgi, an Italian doctor working by candlelight in his kitchen, discovered the nerve-staining properties of the compound silver nitrate. Then, suddenly, a whole range of biologically active stains were developed that could be injected into the brains of living animals. There they would be absorbed by target cells, and over a matter of days be transported back along nerve fibres, and even across cell junctions, to show exactly how different areas of the brain were coupled. These new stains started to reveal a great many unexpected patterns of connection, the thalamus especially throwing up some surprises.

As said, the established view of the thalamus (or thalami, as there is one serving each hemisphere) was that it was simply a gateway for new information heading up to the mapping areas of the cortex. All nerves from the sense organs and body relayed through the thalamus, their messages 'changing trains' by synapsing on a thalamus neuron that would then complete the journey to the cortex. The thalamus also handled the traffic coming from other major brain regions such as the cerebellum, the amygdala, and the hypothalamus, the body's metabolic regulation centre. Many neuroscientists suspected that the thalamus might have some role in concentrating the input traffic, and perhaps also in shutting off the flow at times, such as when the brain needed to sleep. But no one believed that it would play an intimate part in the cortex's processing activity.

However, as the 1980s progressed, the whole picture changed. It was realised that the thalamus was tightly woven into a network of cortex cross-talk. The anatomical studies showed that only the smallest part of the thalamus actually dealt with raw sensory input at all. The thalamus was not that big anyway – about the size of a little finger joint. But the junction of the optic tract, the auditory nerves, and the somatosensory bundle formed no more than small bumps on its surface. Even the inputs from lower brain organs like the amygdala and cerebellum accounted for a fraction of the traffic passing through the thalamus. The great majority of lines came down from cortex areas, and then returned to cortex areas.

Looking closer still, researchers were surprised to discover that the thalamus had a topographical organisation that seemed precisely to mirror that of the cortex above. The bulk of connections with the prefrontal cortex were grouped together in a forward section of the thalamus known as the medio-dorsal nucleus (MD). The various levels of the motor cortex connected to a series of nuclei along the side of the thalamus. The sensory hierarchy was then connected to the pulvinar, a large nucleus taking up the whole back third of the thalamus. The divisions kept on going with the pulvinar, for example, having at least four sub-regions. Some anatomists even felt they could make out the match for individual cortex maps, such as V4 or the IT area.

There was no question that the thalamus had ties with all corners of the cortex. Some neuroscientists even started calling it the seventh layer of the cortex. But the mystery was, what was the purpose of these connections? The picture was further clouded by the discovery that there appeared to be a very strong feedback loop built into the lines coming back out of the thalamus. Whatever input was arriving by these lines, it looked like its impact could be massively amplified – and possibly sustained so that it lasted longer than usual – by this excitatory circuit.

A further finding was that between the thalamus and the cortex stood an equally strong inhibitory mechanism. All the lines heading for the cortex first had to pass through a thin sheet of cells known as the reticular nucleus which wrapped itself around the outside of the thalamus. The reticular nucleus did not touch the message running through it, but instead responded with a sideways wave of inhibition that suppressed activity on any neighbouring lines. So this created a double mechanism for intensifying a signal. The turning up of the volume on a thalamo-cortical loop would simultaneously dampen the activity of surrounding traffic streams.

The pattern of circuitry seemed to put the thalamus in a position to do something powerful, but the scantiness of the data and the complexity of the connections made it difficult to be sure exactly what. For a while, many researchers were taken by the idea that the thalamus might act as the pacemaker for consciousness. Wolf Singer and Charles Gray's discovery of synchronous firing rhythms being used to bind mapping activity in a cat's brain led theorists to think that conscious states might have a characteristic frequency. It could be a beat of forty Hertz – forty pulses a second – which defined the areas of cortex mapping that made it into focal awareness. The thalamus, with its amplifying loops, then seemed a conveniently central place from which to drum out an entraining beat. The idea was heavily promoted, especially by Francis Crick. However, it soon became plain that consciousness had no magic frequency. Synchrony might well be important in coding, but it was most likely a dynamic, self-organising phenomenon, a coherence that evolved rather than needing to be dictated by a central brain clock.

A second in-vogue idea was that the powerful feedback circuit between each bit of cortex mapping and its matching spot on the thalamus might be a reverberatory loop that kept a chosen pattern of input in consciousness. Again this was a theory championed by Crick, who continued to seek the one neural trick that might turn awareness on. Crick felt that an echoing thalamo-cortical trace might linger in a way that ordinary cortex traffic did not. But the reverberatory loop hypothesis also soon fizzled out as it became apparent that the search for simple neural tricks was a doomed exercise. The loop might certainly sustain activity, but this alone would not make such activity conscious. It was the way the brain responded as a whole to the disturbances of life that produced states of awareness, not the workings of any individual part.

A third – and, for LaBerge, much more likely – line of speculation was that the thalamus was simply a processing bottleneck. It offered a route by which the cortex could focus its own mapping activity. The trouble with this idea of the thalamus as an attentional lens was that it presented most theorists with a logical conundrum. How did the brain know what should be brought into focus before it had been brought into focus? They could see how the amplifying circuitry might be used to turn up the volume of an area of mapping, but not how the thalamus could know which bits to turn up.

LaBerge had the advantage that his years working on anticipation meant that he had already begun to think about attention in a more dynamic way. He could see that states of focus would have to evolve through a balance of bottom-up and top-down pressures. Competition within a mapping area would sort out some basic winners and losers – brighter stimuli drowning out weaker ones so that a pattern of activity would emerge – but the brain would also have to have ways of steering this competition if it was ever to attend to faint or more meaningful events. Feedback from higher rungs of processing like the prefrontal cortex would be needed to ensure the escalation of aspects of a spread of mapping it had reason to find interesting. And this feedback could come either during the moment of processing or, better still, the higher brain could make sure the competition was biased from the start. The thalamus now seemed a strong contender as an organ that would be able to concentrate such a top-down expression of interest.

LaBerge still saw a priming and competition-settling role for the direct connections between levels of the cortex hierarchy, but he felt that the backward ripples of activity would be rather weak and diffuse by comparison. A spreading stain of activation might rouse an area of cortex circuitry much as Desimone's work with monkeys and search imagery suggested, so setting up a fine-grain pattern of bias. But the thalamus would be the route to a much blunter, more forceful guiding of a moment's competition. By reaching down to prime some input channels and suppress others, the higher levels of the cortex would make sure that the competition started on an uneven footing, with whole areas of potential experience blocked out. LaBerge likened it to a hi-fi system in which the brain's planning areas used their direct connections to select the record to be played, and then used the thalamus to turn up the volume setting for that bit of cortex to make sure the record was played at full blast once inputs actually started arriving.

This would give the brain a great advantage, because not only did it provide a means of preventing naturally noisy neural events from drowning out fainter patterns of activity, but a bias at the level of the thalamus would also be more open-ended in nature. The brain often cannot predict what it will hear, or what it is about to see, just that it expects to hear or see something in the next few moments. The thalamus would be able to prime the right input channels so as to give a boost to the raw information, ensuring its rapid escalation to the centre of attention, without needing to know in advance exactly what shape the input will take. So the brain would have a gradient of anticipatory mechanisms ranging from precise working-memory-type states of priming to the broad tuning of attentional settings.

And importantly, as with the cortex, the thalamus would keep tightening its settings right through the moment of processing, adjusting the volume levels to fit with how the moment was actually panning out. The act of focusing would be a tapering cone of activity in which the thalamus was prepared as accurately as possible in advance of a cycle of processing, but then had the dynamism to go with the flow of events. A weak block on surrounding activity might become a strong block as expectations were confirmed; alternatively, on occasions the thalamus might have to change tracks entirely, relaxing its grip to let news of some surprise event fill awareness. As LaBerge's reaction-time experiments had shown, wrong expectations could cause a split-second extra delay. But the thalamus would not be a static device. If the volume setting was wrong when the record started to play, the frontal cortex could quickly snake out a hand to turn the knob up or down.

LaBerge had his theory. But as a psychologist, he never thought he would be able to test it. The circuitry of the brain was the preserve of neuroscientists with the surgical skills to stain pathways or do single-cell recordings. Then, in 1988, the medical school at his university – not wanting to be outdone by its rivals like the Hammersmith Hospital and the University of Washington – hastily installed a PET scanning system. Suddenly there was a need for psychologists with big ideas to test. Richard Haier was recruited to do his IQ studies, and LaBerge was able to check to see if the thalamus really did play an active part in the focusing of a state of attention.

LaBerge rather groans at the memory. One advantage of doing his reaction-time experiments was that it was small science. The gear cost next to nothing and he could run the tests more or less by himself. But PET was science on an industrial scale, with all the committees, turf battles, and funding wrangles that went with it. 'The medical boys and the physicists were into the politics, but I spent too much time stuck in meetings at great expense to the health of my stomach walls,' says LaBerge. However, despite the grumbles, LaBerge was pleased with the eventual result.

Thinking up an experiment was simple enough. LaBerge adapted a visual detection task that he had used in his reaction-time studies. In this test, eight student subjects were asked to watch a computer screen and report if they saw the letter O appearing as part of a brief display. On some trials, just a single large letter was shown, making it easy to spot the Os. This was the control condition where the brain did not have to work hard to focus on an anticipated event. But on other trials, the O was placed at the centre of a grid of similar-looking letters, such as Gs and Qs. And the alternative letters used inside the square – either a C or the symbol Ø – were equally easily confused. With these letter arrays flashing up for just a fifth of a second, it was a tricky job for the subjects to catch the Os as they appeared. The subjects would have to keep the O target firmly in mind and also concentrate in a way that blanked out the distracting letters surrounding the centre spot. If LaBerge was right about the thalamus's role as part of an expectation-driven focusing circuit, then it should light up brightly on the trials when the subjects had the extra distractions to filter out.

The Irvine medical school's PET scanner was actually of quite a primitive type, using only a small number of detectors and so only able to record from a tiny bit of the brain rather than the whole of it. It also used radioactive glucose rather than labelled oxygen as its tracer. This meant that it measured the gradual rise in sugar consumption with mental activity rather than the almost instant surges in blood flow. But despite the poor resolution of the system, it was plain that when the subjects had to concentrate, the pulvinar – the sensory part of the thalamus – was working overtime. And evidence that the pulvinar was actively blocking out the distractors surrounding the target came from the fact that the letter arrays actually produced a slightly lower glucose consumption in the cortex. It might be assumed that areas like V1 would need to burn more energy to represent the many components making up the letter grids, but V1's activity appeared to be lowered by the thalamus's attempts to narrow the cone of vision.

This experiment was reported in 1990. In 1993, LaBerge was able to do a follow-up experiment using a much better PET camera recently installed at the University of Texas in San Antonio. This time he could see what happened as the difficulty of the task was varied from slightly hard to very hard, and also look for the other parts of the brain which might be involved in the attention circuit.

The level of concentration needed could easily be changed by surrounding the target square with a much milder distraction, such as an array of slash marks rather than the Gs and Qs. Satisfyingly, LaBerge found that the activation of the thalamus matched the subjective difficulty of the task. The Texas PET camera also showed the whole brain and LaBerge found his triangular pattern of active areas. The thalamus lit up in concert with the prefrontal cortex and letter-recognition areas of the temporal lobe. This did not prove that the prefrontal lobe was using the thalamus to focus the mapping activity of the sensory cortex, making certain that the right target event got escalated to the forefront of attention when a slide was flashed up for a fraction of a second. More detailed anatomical or single-cell work would be needed to show that the actual pattern of circuitry existed. But the result was certainly consistent with such a triangular set of connections.

Buoyed by the result, LaBerge could still see there were a few problems with the idea of the thalamus as a focusing bottleneck. It seemed to give the higher brain a way to manage the evolution of a state of sensation. He felt it was likely that the frontal cortex would also use exactly the same mechanism to switch the spotlight onto areas of its own thinking and planning. The medio-dorsal nucleus of the thalamus, which connected to the many mapping areas of the frontal output hierarchy, was almost as well developed in humans as the pulvinar. So if the prefrontal cortex wanted to find the solution to a problem, it could turn up the volume on the parts of the motor and planning hierarchy where it most expected to discover the answer. But the control of the execution of output plans was another matter. The thalamus by itself did not look like a safety-catch mechanism that would allow the brain to hold plans in mind without immediately moving to execute them.

However, the rapid progress in neuroanatomy was revealing that there was another set of sub-cortical organs, the basal ganglia, that appeared to be woven into the web of cross-cortex traffic. And some researchers believed that the basal ganglia might do for the output side of the equation what the thalamus did for the input – act as a bottleneck by which the cortex could focus its own processing activity and so harness the dynamics of its information flows.

Dick Passingham, a reader in cognitive neuroscience at Oxford University's department of experimental psychology, is a specialist in the brain's motor control system – which for a neuroscientist is about as far away from the study of consciousness as it is possible to get. During the 1970s and 1980s, any researcher with a hankering to discover something about the nature of the mind would be studying the sensory cortex, or if they really wanted to push it, the process of attention. To be content with the field of motor control was not even to be in the race. While scientists working on the senses were at least forced to admit there might be a subjective side to the representational processes they studied, motor control scientists could as well be dealing with a vast system of ropes and pulleys. The driving metaphors were not even computational but electronic: switches were thrown, relays closed, energies discharged. However, in 1989, Passingham – a slight figure with a gruff demeanour – found himself being drawn into the new field of scanning, and from there into serious questions about consciousness.

Passingham's own research had involved mostly lesion studies. A monkey would be trained in a simple task, such as learning whether it had to turn or pull a handle in order to get a food reward. Then an area of the brain, like the premotor cortex, would be cut out – or rather sucked out using a fine vacuum cleaner-like probe. After the monkey had had a few weeks to recover from the surgery, it would be tested again on the task. The effect on its performance would provide an argument for what job that chunk of motor tissue might have been doing.

Like all other animal research, progress was gruellingly slow and there was no reason to expect sudden insights about consciousness. But the meticulousness of the approach could not be faulted. Then one day – as with many other top researchers at that time – Passingham felt a tap on his shoulder from someone wondering if he had any good ideas for a scanning experiment. And although Passingham is still quick to growl about the many shortcomings of brain imaging – he says that maybe some time in the twenty-first century the method will become truly rigorous – very soon he was making the trip down to the Hammersmith Hospital in London at least once a week. Furthermore, the question he ended up asking in his scanner work was about as direct about the nature of awareness as it was possible to be.

Passingham's first PET experiment was a simple study to see which cortex areas would light up when a person made a movement such as tapping a finger or hitching up a leg. Colleagues joked that the only reason he was interested in the results from a human subject was that it might confirm some long-held suspicions about the organisation of the brains of laboratory monkeys. However, poring over the rainbow-coloured PET images, it did not take long for Passingham to be struck by something that his animal work had not even hinted at. When he asked his subjects to do a task such as rapping out a slightly intricate sequence of finger movements on a keypad, the activation would start out as a bright and widespread network of brain hot spots. But as the sequence was mastered, the activity shrunk rapidly to almost nothing. Just as the Washington group had found with their verb-generating task, there was a visible difference between the thinking and unthinking performance of an action.

More detailed experiments followed. In one test, subjects had to discover a sequence of eight finger taps through trial and error. They would lie in the scanner with one hand resting on a keypad, then tap each key in turn to find the first movement in the sequence. The beep of a tone would tell them when they were correct. Having found one, the subjects would have to keep going, remembering what they had already learnt while searching for the next. Then, once all eight had been discovered, they would keep hammering out the sequence until it became automatic. By the end of an hour's scanning session, most of the subjects managed to get a rhythm going so that their fingers were skipping through the sequence almost of their own accord. The subjects were still aware of what they were doing – there was a background level of supervision waiting to step in if the fingers got tangled up – but they no longer needed consciously to focus on the detail. In fact, with enough practice, even a vague supervision was not really necessary. Like car drivers motoring along with their minds occupied by shopping lists and day-dreams, the subjects could drift away so that they forgot their fingers were still going.

Passingham is almost apologetic about the initially demanding nature of this particular test. 'You've got working memory, attention, absolutely everything going at the same time, which makes it a rather impure task,' he says. But over a series of experiments, Passingham began to see just which areas marked out a brain that was focally conscious – and where the activation seemed to go as a skill shrank to become a matter of habit.

A move with the Hammersmith team to their new laboratory at the Institute of Neurology helped greatly. Passingham could now use an f-MRI scanner to film the shrinkage in action. A drawback with PET was that it still took about twenty minutes to record the series of two-dimensional slices needed to produce a full three-dimensional image of the brain. But f-MRI could capture a slice in just 100 milliseconds, and the whole brain in about six seconds. With this kind of speed, it became possible to string together a sequence of three-dimensional snapshots and produce something four-dimensional – an actual film clip of the brain changing as it learnt.

Passingham's f-MRI tests showed which areas were involved in fine detail. When the subjects were first working out the sequence of eight taps and juggling a set of possibilities in their heads, most of their frontal cortex was firing. The prefrontal areas leading into the motor hierarchy burned bright, as did the cingulate cortex. Parts of the thalamus and parietal cortex – the area dealing with the 'where' information for targeting movements – also ran hot. It was plain that the brain was throwing every circuit of any relevance at the problem. But by the time the sequence was mastered and had become a rhythm rather than a series of individually remembered taps, all the higher levels of processing had dropped out. They had served to establish a program and now the lower rungs of the motor hierarchy, such as the premotor cortex, cerebellum, and, in particular, the straggling gaggle of nerve centres making up the basal ganglia, could perform the action at the level of a habit.

Passingham's experiment was an advance on the Washington University language study not only because he was using the latest generation of scanner but also because his background in monkey research meant that he knew where activity should be expected to occur. The impact of the Washington University finding was rather diluted by the fact that the original intention had been just to identify the language areas of the brain, and it was a big enough shock simply to see so many different parts of the brain lighting up during the generation of a speech act, especially such apparently unlikely spots as the cingulate cortex and cerebellum. However, the motor hierarchy was already well defined, so immediately it was the sheer dynamism of the changes that was taken as the significant finding.

Before such scanning evidence, it was possible for motor researchers to ignore the issue of how much conscious effort was being made in the execution of a task, or what actually might change as the task became subjectively easy. Passingham admits that most researchers probably had some simple learning effect at the back of their minds; the same brain circuits would organise the output, but with strengthened connections or lowered thresholds to make triggering the action pattern less trouble. However, after seeing the striking shifts in brain activation, the issue of levels of attention could no longer be ignored.

The scanning studies also brought home the fact that there was a lot of overlap in the way that the brain handled what seemed to be completely different kinds of mental output. The same high-level areas, such as the prefrontal cortex and cingulate cortex, lit up when people were having to concentrate on making a novel response, no matter whether the response was verbal, a motor act, or even an act of orientation. And when the activity shrank down, it was the same low-level areas that seemed to take up the slack, areas like the insula cortex and, especially, the basal ganglia.

The basal ganglia are not much to look at. Buried deep inside the cerebral hemispheres, they form twin streaks of grey matter which curve round, following the general C-shaped sweep of the cortex above. The front end of each streak is a grape-sized blob known as the putamen, which is tied by looping pathways to the motor cortex. The two-inch-long tail, the caudate nucleus, is connected by more loops to the prefrontal and sensory cortex. Indeed, as with the thalamus, anatomical tracing studies carried out during the 1980s showed that the basal ganglia sit at the centre of a web of cross-cortex traffic and have a careful topographical organisation. The putamen, for example, maps parts of the body, while the caudate nucleus is split into separate areas to handle 'what' and 'where' information. And there is even a region, the nucleus accumbens (NA), which deals specifically in traffic from the emotion centres of the brain.

One of the reasons the basal ganglia have confused neuroscientists is that the looping connections with the cortex appear to have a back-to-front logic. There were many reasons to think of the basal ganglia as a lowly motor output station whose main job was to add a few final embellishments to smooth the execution of a cortex-planned action. So it seemed that the basal ganglia should mostly be delivering motor instructions to the primary motor cortex – the M1 mapping that acted as the brain's final launch pad for muscle commands. But the anatomical studies revealed something quite different.

The standard plan of a loop was that it collected together lines from a great variety of cortex areas – both sensory and motor – funnelled them through a point in the basal ganglia, and then returned to a single point on a frontal cortex mapping. Some of these loops did indeed feed into M1, but most were spread generously across the frontal cortex, connecting even to the many maplets of the prefrontal cortex. A further surprising fact was that the returning lines came back via the thalamus. Rather than being an end-of-chain motor area, the basal ganglia looked more like some kind of bottleneck in a cortex-wide focusing of information.

Apart from their looping paths, the other notable feature of the basal ganglia is that they have a special relationship with the brain neurotransmitter dopamine. Most neural traffic depends on fast-acting neurotransmitters such as glutamate, which tends to be used in excitatory connections, and GABA, which is preferred for inhibitory connections. But the brain also has a range of substances, often called neuromodulators, which can produce more complex and long-lasting effects when released from a nerve ending.

The most familiar of these is noradrenaline. In the body, the release of the hormone adrenaline sets the heart racing when we get a fright or become excited. As mentioned, noradrenaline has a very similar effect within the brain, boosting mental arousal and making us more jumpily aware of anything that might be happening at the time. Noradrenaline is produced by just a few thousand gland-like cells deep in the brain stem, yet it can be delivered immediately anywhere in the brain through a thickly branching network of axons. If the brain has decided something exciting is happening – or has even anticipated that it might, as when we turn down a dark alley in a strange part of town – a signal is flashed to the noradrenaline reservoir and in a hundred milliseconds or so, about the time it takes to organise a lurching orientation response, the axon tips deliver a wake-up dose to the required processing areas.

If noradrenaline is the chemical for producing jumpy alertness, then dopamine seems to be the neurotransmitter used by the brain when concentrating its thoughts on output plans. At a cellular level, the effect of dopamine often seems contradictory: in some cases it can act as a powerful inhibitory signal, switching a neuron off; at other times it can produce a rise in the excitability of a cell that lasts for some hours. The picture is further complicated by the fact that there are at least four versions of the dopamine receptor, suggesting that dopamine might produce quite different reactions in different classes of neuron. However, the places to which dopamine is delivered seem to tell a story.

Dopamine has two main delivery pathways, one thick band of fibres which serves the basal ganglia and another set of fibres that spreads more diffusely to blanket the frontal cortex planning and motor areas. The reason for a dopamine supply to the frontal cortex has been revealed by some extraordinarily delicate single-cell recording work by Patricia Goldman-Rakic at Yale University's school of medicine in Connecticut, who reported in 1995 that dopamine appeared essential to keeping a state of working memory burning in the higher brain. Goldman-Rakic's experiment involved using a hair-thin glass tube to squirt microscopic doses of a variety of amphetamines and tranquillisers – drugs known to work by interfering with dopamine transmission – onto individual neurons in the prefrontal cortex of a monkey as she recorded their firing behaviour during a standard working memory task.

In this task, the monkeys watched a computer screen on which a light flash would appear in one of eight places. Then, after a three-second delay – during which they had to use their memories to keep the location in mind – the monkeys would point to the spot for a reward of a sip of juice. Goldman-Rakic knew from earlier experiments that cells in the 'where' mapping areas of the prefrontal cortex would code for the location of the target, firing brightly during the delay period to maintain a sense of place.

By using drugs alternately to block or exaggerate the effects of any dopamine signals reaching a cell, she was able to show that dopamine activity was necessary for prefrontal neurons to keep an activity pattern or memory trace flying. In simple terms, dopamine kept the cell on. However, the balance of dopamine input had to be just right because flooding a cell had the reverse effect, shutting down its activity completely. So the dopamine system allowed the higher brain to pick out the precise arrangements of cells needed to keep a thought or intention alive. However, the same system could then deliver a flush of dopamine that simply knocked activity flat, wiping the slate clean of the very memories that were being preserved.

Dopamine plays an equally powerful, yet delicately balanced, role in the workings of the basal ganglia. The most straightforward evidence comes from the degenerative disorder Parkinson's disease, in which the brain's dopamine-producing cells gradually die and the supply to the basal ganglia dries up. The most obvious symptoms of Parkinson's disease are physical: sufferers find they can no longer move smoothly. In fact, they begin to seize up as if a hidden brake has been applied to their muscles. They have to fight with their bodies to make an action happen, and when it does, it often comes out in an abrupt and clumsy rush. For the victims, it is like a disease of the will; they know what they want to do but cannot make it happen. The safety catch on their planning seems stuck, and when it comes off, it does so in an unpredictable way.

Movement problems are the most prominent symptom of Parkinson's disease because the motor output areas of the basal ganglia are particularly dependent on a healthy supply of dopamine, and so get hit first. But as the condition worsens and more dopamine-producing cells die, the basal ganglia areas connected with the sensory and higher-planning parts of the cortex also begin to fail, causing mental symptoms to appear. Sufferers feel as if the same stuttering brake has been applied to their thinking, stopping them from forming plans, retrieving memories, or switching lines of thought with normal smoothness.

Parkinson's disease suggests that the basal ganglia do indeed do some kind of execution-smoothing service for the frontal cortex. The frontal cortex has the ideas and then calls on the basal ganglia to manage their actual timing and release. At the crudest level, the basal ganglia can act as the brakes for voluntary action, doing the stopping and starting so that the higher brain can think without acting. But other evidence suggests that the basal ganglia also have a lot to do with a more general sense of timing and the sequencing of planned movements. For example, an f-MRI study by Warren Meck at Duke University in North Carolina, where subjects were scanned as they squeezed a ball for what they estimated to be eleven seconds, showed intense activity in the basal ganglia. Meck then demonstrated that damaging parts of the basal ganglia in rats caused them to lose the ability to judge even short time intervals. It is also known that the loss of certain key areas of the basal ganglia, such as the subthalamic nucleus, can lead to a syndrome known as 'ballism' where movement becomes wild and random. The urge to wiggle a finger becomes translated into an uncontrolled flailing of the whole arm as if the brain can no longer focus on a single component of an action.

So, as in the cortex, dopamine seemed to have a dual effect: it could be used in a fine-grain way to hold down the precise pattern of connections needed to represent an action sequence, or a blast of the neurotransmitter could be used simply to throw out the anchors and lock up all activity. But despite learning about the safety-catch and release-timing abilities of the basal ganglia, many neuroscientists felt this did not really get under the skin of the chain of nerve centres. Again, there was the problem of the loops: the connections between the basal ganglia and the cortex still seemed too diverse. And then there was the puzzle of the scanner studies which showed mental skills shrinking in the direction of the basal ganglia.

A flurry of single-cell recording experiments in the early 1990s suddenly suggested a much more dynamic view of the basal ganglia's execution-smoothing role. A particularly revealing study was carried out by Ann Graybiel, a neuroscientist at the Massachusetts Institute of Technology.

In collaboration with a number of Japanese researchers, Graybiel put recording electrodes into the basal ganglia of a monkey while it was learning a simple mental association – that the sound of a click signalled the availability of a sip of juice. Obviously, during such conditioning, a connection between hearing the stimulus and thinking about the reward soon becomes automatic. What Graybiel found was that at first, the activity within the basal ganglia simply reflected what was going on in the cortex above. The looping pathways through the basal ganglia hummed with news of the click, news of the monkey's licking response, and then news of the pleasant taste. But soon a special class of basal ganglia interneuron – spidery-armed cells that spanned the throughput lines as if to eavesdrop on their traffic – began to fire in rhythm. The interneurons started reacting sharply to the sound of the stimulus themselves.

To Graybiel, this finally made sense of the apparently back-to-front logic of the basal loops. Each line passing through the basal ganglia concentrated news of activity across a wide range of cortex mapping areas and channelled it back to one spot in the response-organising areas of the frontal lobes. The various sensory aspects that went with the moment – the sound of the click, the rewarding taste, and even the feelings of the licking action – would be tied together by a convergence of inputs and then fed into the bits of the brain having to discover the connection. Although, importantly, what the converging lines would carry was not the sensory picture itself, only a report of significant conjunctions. The firing of the loops simply told what arrangement of cortex circuitry was active when the brain finally decided to do something.

This bottleneck design would have the effect of reinforcing the trial-and-error learning of the higher brain. In a rather abstract way, it would focus information about what cues were significant to a moment of response and also what level of success or failure resulted, so helping the cortex feel its way to an optimal solution. But having supported the higher brain in evolving its pattern of reaction, Graybiel's experiment showed that the basal ganglia were perfectly positioned to start making the same connections themselves – to produce exactly the same response on auto-pilot.

The interneurons straddling the looping pathways would see the pattern of sensory cues that led up to an action such as turning and licking a sip of juice. By mirroring the firing rhythms and learning to react to the same cues themselves, the interneurons would be ready to step in and short-circuit the frontal cortex's decision-making. They could begin pulling the same strings, but at the level of quick, unthinking habit. They might not know why a particular output response was appropriate, just that in certain contexts it was the best thing to do.

Other single-cell recording work, much of it carried out by Wolfram Schultz at the University of Fribourg in Switzerland, was also beginning to make the part played by dopamine in this basal ganglia learning a lot clearer. One of the puzzles about dopamine was that early research had often suggested it was the brain's neuro-transmitter for feelings of significance, pleasure, and exhilaration. It was known that many drugs, such as cocaine and amphetamines, worked by flooding parts of the brain with dopamine, particularly the nucleus accumbens, the area of the basal ganglia most closely connected to the brain's emotion and metabolic control centres. This fact caused some neuroscientists simply to treat dopamine as the brain's 'joy juice'. However, in the 1990s, the more detailed studies of Schultz and others made it plain that dopamine was just the common messenger of the parts of the brain dedicated to arriving at the right response to whatever has proved the most noteworthy event of the moment.

In his experiments, Schultz recorded the firing of the dopamine-producing cells supplying the basal ganglia of a monkey. Schultz had been expecting that the cells would release small doses of dopamine whenever the rats made any movement, but instead, the cells fired only when there was a significant conjunction of events in the moments leading up to some physical action. As so often is the case in single-cell work, this unexpected result was discovered by accident. After days of baffling silence, an assistant happened to hand a monkey a small piece of apple and its dopamine cells went wild.

This suggested that dopamine was the signal used by the brain when it realised something significant was happening and it wanted to prod the output hierarchy into taking proper note. In the frontal cortex, a fine drizzle of dopamine would be used to help fuel a working-memory state, making sure there was a lingering picture of the event to drive a wave of planning and thought. Then down in the basal ganglia, dopamine would be used to fix a more permanent memory of what kind of situations should evoke what kind of response. The interneurons would be learning just when to fire from pattern-reinforcing jolts of dopamine.

In yet another experiment to check this part of the tale, Graybiel destroyed the dopamine supply to one side of the basal ganglia of the monkeys doing the click test. The result was that the interneurons on the damaged side no longer fired on hearing the familiar signal. They needed a dopamine input to produce a response.

The basal ganglia recording studies appeared to confirm the idea that the brain was cleverly set up for automating its thinking and acting, for turning solutions evolved at a global, focally-conscious level into quick, reliable, and unthinking habit. Like a composer, the higher brain would invent the tunes, but a web of interneurons in the basal ganglia was always waiting, eavesdropping as the cortex plonked away at its keys and ready like the hippocampus to trap a memory template. Then, in the same way that a scroll drives a player piano, the basal ganglia could robotically produce the same runs of notes.

Of course, the basal ganglia are not the whole story of automation. Parts of the cortex, like the insula, appear closely involved with the remembering of habitual responses. The cerebellum is also crucial in the learning of automatic actions – indeed, the cerebellum is probably the storehouse of all the really fine-grain components of any motor activity. So the process of automation is itself a hierarchy with rungs of processing both above and below the basal ganglia.

However, what was key about the basal ganglia was that they were in a special position to bring the mind's output – both its voluntary and its automatic actions – into sharp focus. If the thalamus was the brain's bottleneck on the input side of processing, helping it focus on one chosen sensory event, then the basal ganglia appeared to be the bottleneck on the output side, helping the brain focus on doing just one well-chosen action at a time.

The evidence of Parkinson's disease showed that the basal ganglia were essential to smoothing the path for voluntary, consciously initiated behaviour. The maze of loops and dopamine connections were a brake that allowed the higher brain to think about the right moment to release an act, or even to catch and veto an impulse it did not like. The basal ganglia would allow intentions to be executed in an orderly, timely fashion.

And because the brain goes into each moment with some kind of prevailing, consciously appreciated context, the basal ganglia's store of habits would slot into place in an equally orderly way. As pointed out, automatic actions normally unfold as part of some greater state of intention, or at least orientation. We head into a moment perhaps with the goal of getting down the corridor to see our grandmother, or in expectation of flexing our wrist in the spontaneous way demanded by some researcher's experiment. So there will always be a global framework, a wider scaffolding of thought to guide the basal ganglia. With the hand flexing, there will have already been a decision to let a rising impulse go. With getting down the corridor, the basal ganglia will be using a lifetime's store of accumulated habits to intercept all the familiar and easily anticipated parts of the moment, such as avoiding rucks in the carpet and reaching for door knobs, without the need for a specific conscious say-so on each step. The basal ganglia would do what they could that was compatible with the brain's overall goal state, unthinkingly channelling activity down the right paths, with the higher brain only having to come back into the picture whenever there was a stumble or some other novelty that demanded the evolution of a fresh mental set.

With the twin bottlenecks of the thalamus and basal ganglia for focusing the cortex's activity, most of the machinery for evolving a response to the moment now seems in place. But there is still one last basic problem for the pathways of the brain. As may have become obvious, every instant of awareness actually offers a forking path. There are actually two quite different kinds of event that the brain must be prepared to escalate to the forefront of attention: events that are significant because they are sought for, and events which are significant because they come as a surprise. So, during each processing cycle the brain has to make an almost binary choice about which way to head. It can either stick to whatever it had planned, or turn to respond to something that has come as an interruption. These decisions are so distinct that they ought, once again, to be reflected in the architectural plan of the brain.

CHAPTER ELEVEN

The Brain's Forking Pathway

Jeffrey Gray, chief of psychology at London's Institute of Psychiatry, swung a pair of sandalled feet up on his desk, propped his hands behind his head, and mused about the rather wayward journey both he and his field have taken over the past fifty years. A tall, rangy figure with leonine hair and a fatherly twinkle to his eyes, Gray admits he arrived where he is by accident. He started out thinking he might be a lawyer, then got a first at Oxford University in languages. Unsatisfied with this, at the age of twenty-three he went off to a kibbutz in Israel. After a few backbreaking weeks picking stones out of a field, Gray decided he might as well return to university and give something else a go.

Psychology took his fancy, but this was in the 1950s when it barely existed as a field. Even Oxford took on just twelve new students a year in a course that was mixed in with philosophy and physiology. 'There definitely weren't any career prospects when I started. We were stuck away in this old condemned Victorian house. Experiments were done on a shoestring. At Oxford, our lecturers weren't even accepted as fellows of the college – they couldn't be dons – which made a big difference to the pay and prestige. As a field, things hadn't developed much since the days of Wundt and Helmholtz,' says Gray. But luckily for him, psychology blossomed almost as soon as he had got his Ph.D. A combination of military funding to support the new discipline of cognitive science and a sharp rise in student interest led to a demand for professors to teach the subject. To his surprise, he had a future as a researcher.

Consciousness was, of course, what interested him. But he found that it was not on the agenda. Behaviourism was never as entrenched in Europe as it was in North America, but still there was a feeling that psychology needed to prove itself as 'good science', so the c-word was taboo. Even the arrival of cognitive science did not help. Gray says the hype about computer models and artificial intelligence actually seemed to hinder creative thought about the problem. For stimulating talk, he used to spend his lunch hours sitting with Oxford's philosophers. Gradually Gray began to find his own direction, steering away from the highly abstract approach of cognitive psychology and beginning to think about the neurological mechanisms that must underlie awareness, especially the processes by which the brain must make an emotional 'so what?' evaluation of each new moment.

A crucial stimulus to Gray's thinking was Evgeny Sokolov's work on the orientation response. Sokolov's findings were important because they showed just how much the brain must be doing at an early, pre-conscious stage. We react to a startling or significant sensation with an automatic shift in attention and a matching adjustment in the body's arousal levels. To do this, the brain must be making at least a rough, habit-level assessment of the value of an event. It would seem pretty hard not to respond automatically to some happenings, such as the bang of a car backfiring or a hand grabbing our shoulder. In the competition to map the moment, the sheer sensory energy of such an event would seem to push it into the centre of awareness. But the whole point of Sokolov's results was that the brain showed the tell-tale twinge of an orientation response even to the most trivial surprises. Just hearing an oddball high tone in a sequence of low notes, or seeing a picture related to a hobby such as camping or fishing, could cause a brightening of the eyes, a prickle of sweat, and a flutter of the heart.

Then there was the fact that the response can be just as strong for the sudden absence of a sensation as a sudden onset. The silence when a fridge goes off, or our abrupt noticing that a vase is missing from its usual place on the mantelpiece, can be equally as intrusive on consciousness. Such negative instances forced Sokolov to conclude that the brain must maintain a 'neuronal model' of the world, a running backdrop of anticipation against which it actively matches events as they happen. To feel surprised, the brain would already have had to build up a state of expectation.

There were many immensely important clues buried in Sokolov's work. His findings suggested that consciousness was a cycle of processing – a structured competition with stages – and that anticipation paved the way for what took place. Yet his writings failed to have much impact in the West. Psychophysiologists who did evoked response potential recordings certainly knew about his work, and extended it through their study of the P300 and other EEG peaks, but mainstream psychology did not tend to take too much notice of the findings of psychophysiologists either. The dimension of time, whether it was the anticipating that led up to a moment of awareness or the lagging processing that followed, just did not fit with the abstract, contextless approach that cognitive scientists wanted to take.

However, it just so happened that Gray was a fluent Russian speaker. Like many bright conscripts doing their national military service in the 1950s, he had been sent by the army to language classes because it might be useful for intelligence work. Then, when he first began his research career, he started scraping together a little extra cash by translating Soviet articles, and this brought him into contact with Sokolov's writings. Reading the originals, Gray admits they may have made even more impact.

Gray's own research began to follow the same path. His early studies involved the biology of anxiety – what made some brains more jumpy than others. Part of this involved breeding rats that were especially anxious. Put in the classic behaviourist apparatus of boxes and mazes, these rats would be slower to explore a new environment, and would take longer to forget the effects of a shock. Eventually this line of work led him to the Institute of Psychiatry where he could get deep into the neurology of anxiety, and still more extreme conditions such as schizophrenia. And as he learnt more about the brain and the role played by emotion centres, such as the amygdala and hypothalamus, and also arousal-producing neurotransmitter systems, such as noradrenaline, he began to see the forking pathway which might actually be responsible for Sokolov's orientation response.

The brain has to make plans, yet also leave room for surprises. The fact that representations and reactions take time to develop means it has to prepare well in advance of the moment, making guesses about what it expects to be central to its eventual conscious state. The brain goes in with an objective, whether it be the intention to open a door, read the next line of a book, or continue day-dreaming peacefully by the fireside. But shocks and distractions happen all the time. While in the middle of moving to the next line of a book, or image in a day-dream, for instance, a dog might bark or a curtain flap in the breeze.

So the brain has to run two kinds of processing side by side, making a mapping response to both expectations and potential interruptions to at least a pre-conscious level of organisation, before finally deciding which track to follow. Sometimes the flapping curtain may remain ignored, staying on the outermost periphery of our awareness as we stay concentrated on our book; at other times its flapping might win the competition. Our concentration will be broken and all activity that was part of book reading become frozen as we look away to take in the unexpected movement catching our eye.

In fact, thinking more deeply about this quicksilver switch at the heart of consciousness, Gray realised that there was a four-way logic at work. The basic decision was between sticking with an anticipated train of events or diverting to attend to the unexpected. However, there were actually two kinds of possible surprises: ones caused simply by a plan taking an unexpected turn, and then ones which were true shocks, events like a car backfiring which came completely out of the blue. The difference was that with the first, our senses and a lot of our thinking would already be in place waiting to deal with the situation.

When we are searching an essay for spelling mistakes or listening to a series of tones as part of a P300 recording experiment, we have already chosen the type of event we want escalated into consciousness. Our intention during the next moment is to be reading a word or staying awake to a train of sounds. The job of the brain is to make a simple match–mismatch decision. If the word reads correctly or the tone is the common one, we want a quick OK so we can move smoothly on to the following word or tone, but if the brain picks up a mismatch, we want a tightening of the focus so that the spelling mistake or rare tone becomes brightly appreciated. Our consciousness will skip along in a blur of successful matches and occasionally contract around the more particular contours of a mismatch. Any surprise would be relative because both its position and consequences would already have been anticipated. Our brains would even be primed about the correct reaction. With the tones, we would know we were just supposed to keep on listening and let the instruments record; with spelling mistakes, the word-processing areas of our brain would already be roused to make changes. The surprise would unfold within a well-organised context.

A real surprise would be one that caught the brain completely off-guard. Our senses would be looking elsewhere, and our thalamus would be primed to promote a different stream of input. The memory circuits needed to interpret the event would be dormant, if not actively suppressed to avoid their interfering with the expected cycle of processing. This sort of surprise would need a sharp orientation response. There would have to be a halting and a flushing to rid the brain of its existing attentional set-up, and a redirection of the sense organs to allow the surprise to be fully analysed. And as a surprise is also always going to be a potential threat, the brain would want to throw in a few swift changes in arousal levels, gearing up the body and mind in case of emergency.

A match, a mismatch, and a true surprise were three possible outcomes for a moment of processing. A fourth kind of moment would be produced by the deliberate use of the brain's surprise-detecting mechanisms to produce an open-minded state of vigilance. There are times when we expect something to happen, such as when we are walking down a dark alley, but we do not know what form it will take or from which direction it might come. In such a situation we will want to shut off our usual planning and thinking activities so as to maximise our alertness. The settings of an organ like the thalamus would be left wide open so as not to add any delay. We would also pump up our noradrenaline levels and heart rate well in advance, not waiting for something nasty actually to occur. Vigilance is about blanking the mind and learning to expect the unexpected.

In truth, many different blends of expectation and surprise seem possible. A radar operator's state of open-minded vigilance forms a cone no wider than the span of the radar screen. The operator will be alert for the random appearance of any blip, but will actually be trying to suppress the temptation to react to stray movements in the control room. As might be expected, the dynamic nature of the brain would mean that it can easily tailor a mix of planning and open-mindedness to suit the coming moment. Yet still, there seems to be a basic difference between arriving at a planned point of focus and dropping those plans so as to be able to deal with an interruption – a difference that should be reflected in the architectural blueprint of the brain.

It took Gray many years of poring over experimental results and anatomy books, but eventually he felt he could see the outline of these two pathways. And while they began separately, it turned out that they intersected on the 'emotional' part of the basal ganglia.

Joseph LeDoux, the researcher most responsible for defining the interruption-handling pathways of the brain, is another psychologist with an unorthodox background. Bearded and cropped-haired, LeDoux was studying marketing during the early 1970s and had completed his Masters at Louisiana State University before he made the sudden switch to a Ph.D. in psychobiology because he wanted to pursue the relationship between the brain and the mind. By the age of thirty he was a professor at Cornell University Medical College, and eventually got his own neural science lab at New York University. While he was never able to chase the problem of consciousness quite as directly as he might have liked, he again was one of the few workers in the field with the breadth of vision to ask some big questions about the organisation of the brain.

LeDoux saw that the threat-detecting, alarm-bell-ringing side of the brain took in many structures, but the two key ones were the brain stem and the chickpea-sized bump of the amygdala. The brain stem is the first thickening of the spinal cord and so the earliest part of the brain to hear most incoming sensory information. It does not have much intelligence, but its position and stores of neurotransmitters such as noradrenaline and serotonin make it a good watchdog for consciousness. The brain stem would straddle the input lines, listening in on the traffic heading for the higher brain. Then, when it detected an abrupt shift in activity of a line – whether it be a sudden buzzing or a sudden silence – it could flag the signal, sending an accompanying jolt of neurotransmitter to the thalamus and cortex areas responsible for its forthcoming processing, thus ensuring the signal got noticed.

The very dumbness of the brain stem's response would be an asset. Up in the cortex, there would be a vast, finely tuned state of intention and expectancy. The ground would have been lovingly prepared to receive some particular type of sensory event. But the brain stem would take a raw view of sensory input, looking just for evidence of sudden change. It would catch any flickerings or rustlings coming from unexpected directions, or the disappearance of some previously steady sensation, and push them through at greater strength.

The second organ on this interruption-detecting pathway, the amygdala, would then lie in wait to sound the alarm on the basis of what the cortex was actually mapping. The amygdala, like the basal ganglia, is a sub-cortical organ whose role for a long time seemed deeply mysterious. Each hemisphere has its own amygdala buried in the tip of the temporal lobe, just forward of the hippocampus. This position looks telling because, like the hippocampus, the amygdala would be well sited to view the mappings taking shape at the end of the sensory trail. The picture converging on the hippocampus would also converge on the amygdala. Another clue to the amygdala's function is that it has a fat trunk cable, the stria terminalis, which ties it to the hypothalamus, the body's emotion and metabolic control centre. This anatomy suggests the amygdala watches the sensory patterns taking shape in the cortex and organises a suitable emotional response, making it the equivalent of a motor-planning area for the feelings.

Evidence to support this idea came from experiments in which neuroscientists stimulated the brains of monkeys with an electrode. The amygdala turned out to be topographically divided so that jolting different parts produced different emotional reactions. Some of the responses were violent; the monkeys would explode with blind rage, or quiver with panic. But a lot of the amygdala also seemed to be tied to eating behaviours. Stimulation triggered reflex actions such as salivation, chop-licking, swallowing, and even retching. The amygdala was responsible for many actions concerned with Sokolov's orientation response as well. In certain places, the electrodes could make a monkey catch its breath, flinch, prick up its ears, or widen its eyes.

Of course, such stimulation experiments did not show that the amygdala did all the work itself. The actual movement of licking the lips or flinching would be organised by the brain's motor areas, while the metabolic changes would be controlled by the hypothalamus. But the amygdala seemed like the general co-ordination centre. It learnt to see threats and significance in the outside world, then leapt in with a swift, emotional response. This interpretation was backed up by experiments in which researchers cut out the amygdala of a monkey and found the animal became passive and unemotional. The monkey failed to recognise threats and could no longer learn new associations, such as that the sound of a buzzer would signal a shock.

Many of these experiments on the amygdala were carried out as far back as the 1940s and 1950s, but it was not until the late 1980s that LeDoux showed that, like the thalamus and basal ganglia, the amygdala had a subtle dynamic relationship with the cortex. The standard view of the amygdala was that it merely responded to what the cortex saw. It stood at the end of the line of sensory processing and reacted to the contents of consciousness. However, by severing the connections between the amygdala and cortex at various stages along the auditory-processing pathway of a rat, LeDoux was able to prove that the amygdala saw the pre-processed as well as the processed picture. The amygdala had its own input lines running up from the brain stem and thalamus, and so was positioned to catch new information early. Again, as with the basal ganglia, the amygdala seemed to have the circuitry to watch the cortex and learn what it found startling, then to start intercepting these events at a more automatic and habitual level. The amygdala would note the kind of things that should cause interruptions, then make sure that they did.

The final piece in the puzzle was the discovery that the amygdala could halt the actions the cortex had planned for the moment by hitting the nucleus accumbens – the bump on the underside of the basal ganglia that neuroanatomists felt was so insignificant-looking that for a long time they did not even give it a name. However, the nucleus accumbens has turned out to be a crossroads for the brain's planning and acting activity. The loops passing through it bring concentrated traffic from the highest parts of the cortex as well as the amygdala, and also the hippocampal formation. The nucleus accumbens also has a finely balanced dopamine machinery so that it can hold down either a state of precise working-memory-like intention – a crisply defined pattern of connections that helps ready the basal ganglia to follow a certain plan of action – or else just use a general blast of dopamine to wipe the slate clean. Like a contention switch, the nucleus accumbens can either give the moment's goals the green light or it can throw out the anchors, freezing the frontal hierarchy in mid-step and forcing it to take notice of something different instead.

So the interrupt story was that the brain stem tried to catch the cortex's attention by flagging any abrupt sensory change. The amygdala then sat on the sidelines, keeping an open eye on a developing field of sensation. It could use its direct inputs to see some things for itself. There is evidence that the amygdala is born with the wiring to get worried about certain kinds of sensation such as small creeping or slithering movements on the edge of vision – the kind of movements that might give away the location of spiders, snakes, and other nasties. Such alarm does not have to be taught, but appears genetic or innate. Then there are other kinds of events which the amygdala learns to recognise and grow concerned about. As PET scanning allowed researchers to begin looking at the responses of the amygdala in humans, they found that it glowed hot in any kind of emotional assessment situation, such as reacting to violent expressions on another person's face, hearing dirty words, or even thinking about unhappy memories. In other words, if something seems to nag on our attention or causes a shocked double-take, the amygdala is likely to be the brain organ making the connection.

And, of course, while most of the research tended to emphasise the amygdala's part in anxious and fearful responses, it was also involved in tagging events as good or potentially rewarding. Single-cell recording in the basal ganglia had shown that its interneurons learnt to fire to the sound of a click signal, automatically leading a monkey to think about a fruit juice reward. Further single-cell work showed that it was most likely the amygdala that actually made the initial positive connection and then used its many input lines to the nucleus accumbens and other parts of the basal ganglia to train up these interneurons. So the amygdala's job was to be the general-purpose salience detector for the brain, keeping an eye on the cortex's 'dry' mapping of the moment and supplying its own 'wet' emotional evaluation of what actually mattered most. Experiments by LeDoux and his team confirmed that the amygdala had both the organisation and the memory mechanisms, such as cells with a sharp long-term potentiation (LTP) response, to remember what aspects of life promised either good or ill. It then seemed to be connected to all the right places to make sure that the brain reacted when it rang the alarm bells.

The ability of the amygdala and other parts of the interrupt pathway to catch events of significance, even when they come from unexpected directions, is important in explaining how the brain can go Janus-like into the moment, facing each way. It can have intentions without sacrificing a more general level of alertness. With the interrupt pathway to boost the signal strength of 'corner-of-the-eye' experiences such as fluttering curtains or the drone of the neighbour's radio being turned off, we can find our attention being grabbed by things that we did not anticipate would grab it. And as it was busy redirecting our attention, the interrupt pathway would also be launching a full body-wide orientation response to make sure that our focal-level consciousness could hit the ground running when it eventually arrived on the scene.

First the interrupt path would have to call an immediate halt to whatever the brain had been intending to do. It would lock up the basal ganglia, so suspending any further progress of the current action plan (a startling experience can cause us to freeze in mid-step or mid-sentence). Then the interrupt pathway would co-ordinate the turning needed to bring the interrupting event into proper focus, or alternatively the flinching away if it seemed more like a moment for emergency evasive action. Finally, the interrupt path would also throw in some rapid metabolic and arousal-level changes to prime the body for an instant of possible danger or excitement. As best it could – given that it must work at the level of learnt habit and genetically tuned instinct – the interrupt pathway would prepare the ground for full consciousness. Of course, the amygdala and the other links in the chain could often get things wrong – the heart-stopping writhing movement on the floor might turn out to be a twist of paper bowling along in the breeze rather than a wriggling snake. But in evolutionary terms – in the predator-filled world of rats, monkeys, and other animals – it would always be better to be safe than sorry.

Just to confuse matters, however, while the amygdala is clearly a critical part of the brain for noticing the unexpected, it would also have a role in driving the brain's output. As mentioned, the electrical stimulation of the monkey's amygdala triggered many actions connected with eating, such as sniffing and chewing, as if the amygdala were part of a pathway that turned the sight or smell of food into thoughts about feeding. Likewise, experiments have shown that the amygdala is crucial to the recognition of other kinds of emotional 'opportunity', such as a chance to have sex, and also situations that demand aggression or avoidance. When the amygdala is lesioned in animals, they stop responding to the social cues that would normally inspire them to action. They see the world in a dry way and no longer see it in terms of how urges might become satisfied.

So how researchers view the amygdala tends to depend on which aspect they are stressing – its ability to recognise opportunities, or interruptions. The interrupting amygdala would want to knock the basal ganglia flat, freezing further output until the brain has reassessed the situation. The opportunity-recognising amygdala would instead be urging the motor hierarchy on, supplying a drive that would lead it to take advantage of what seemed on offer.

This mixed use of the amygdala's abilities as a salience detector should be no surprise in the dynamically organised brain. The brain has a processing structure, but actual flows of activity are free to develop in many directions. Indeed, other researchers, such as Antonio Damasio of the University of Iowa College of Medicine, were finding that higher parts of the cortex, like the ventromedial area of the prefrontal lobes, seemed to bounce their plans off the amygdala for evaluation. People used the amygdala to detect danger and opportunity in their own thinking! But in a general way, the brain does appear to have a route for making sure it maintains a background state of alertness and so can catch interruption-worthy events at the earliest possible point of the processing cycle.

As Gray spent the 1980s trying to fit all the brain's organs and pathways into some sort of wiring plan capable of making a multifaceted 'so what?' decision about what to escalate into consciousness, another knot of tissue began to seem important – the long-ignored area of the cingulate cortex.

Gray's own pharmacological research on rats was helping to show that the nucleus accumbens occupied a pivotal position in the 'so what?' pathways, acting as the intersection between intentions and interruptions. Three major streams of activity funnelled through the nucleus accumbens during the evolution of each moment: inputs from the amygdala, the hippocampus, and the highest parts of the cortex planning hierarchy. Between them, they would bring a condensed snapshot of the brain's current set of plans, and also break news about potential alerts. The delicately poised, dopamine-titrated circuits of the nucleus accumbens would then act as the battleground for ownership of the moment. The weight of the rival streams of input would tip the balance decisively one way or the other, making sure that the brain came out of the moment with one point of focus, one coherent departure point for the planning and anticipating that would lead it into the next.

But it was also plain that the nucleus accumbens could not rule the moment alone. It could not be left to a simple matter of who shouted the loudest during the pre-conscious build-up to the escalation decision. Sometimes we have to be able to concentrate despite interruptions; at other times we may deliberately want to put ourselves in a state of open-minded vigilance. We will need to be able to still our thoughts, stifling the frontal cortex's natural tendency to fill any free space with plans and expectations. Once into the moment, things will happen too quickly to control the nucleus accumbens's decisions. But the nucleus accumbens – and probably the pathways leading into it – could be primed in advance. There could be a higher part of the brain that tuned this contention switch, adjusting the level of concentration or vigilance that the brain took into each moment.

Late in the 1980s, neuroscientists began to get very excited about the cingulate cortex – a region which had been neglected because it seemed a rather primitive part of the cortex with nothing much to do. The cingulate's cause was not helped by its rather inaccessible location. It lines the central parting of the cerebral hemispheres, its two faces pressing flat against each other where the hemispheres meet. The cingulate stretches in a long C-shape arc from the front of the brain, where it merges with the prefrontal cortex, back to a rather uncharted region of the parietal lobe which may or may not be an area for processing spatial memories. The cingulate itself is not a single structure, but divided into perhaps twenty or thirty maplets.

Anatomical studies had shown that the forward section in particular, the anterior cingulate, was closely connected with emotion centres such as the amygdala and hypothalamus. The anterior cingulate also had fat links with the hippocampus and a fine tracery of paths bringing information from both the planning areas of the prefrontal cortex and the higher levels of the sensory hierarchy. These various connections suggested that the cingulate cortex might have some bottleneck role in the interactions between the highest and lowest parts of the brain, but no immediate job suggested itself.

A hint that the anterior cingulate might be more interesting came from a variety of bizarre medical conditions. It was one of the parts of the brain damaged in the Dr Strangelove syndrome, where a person's hand feels paralysed and at times takes on a life of its own. More extensive damage to the cingulate lobes seemed to paralyse even the ability to think. After a stroke in the cingulate area, one woman lay flat in her bed for weeks, her eyes open and aware, yet her face blankly expressionless. She recovered enough to be able to speak again once the swelling in the area began to subside, and when doctors asked whether she had been in some kind of coma, she replied she had been perfectly conscious, just completely lacking in any urge to act. Sensations did not provoke any thoughts or plans.

These cases piqued the curiosity, but it was with the arrival of brain scanning that the anterior cingulate suddenly caught the attention of neuroscientists. From the Washington University PET group's first studies of the brain's speech areas, it was obvious that the cingulate lit up brightly every time a subject was asked to concentrate hard, or be sharply mindful of an action. Michael Posner, the University of Oregon psychologist brought in to lead the unit's investigations, decided to do further experiments to test the idea that it might be some kind of special awareness-focusing area.

Posner – who has both the look and the brisk, purposeful manner of a corporate lawyer – set up an experiment based on a standard psychological phenomenon known as the Stroop effect. In this test, subjects are shown a list of words printed in different colours. Their task is to report the colour and not read the word. When the words are neutral ones like 'moon' or 'cow', concentrating on the colour is easy; but when the words are themselves names of colours – so that the subject is seeing the word 'orange', but it is printed in some other colour such as yellow or blue – then the task becomes a real strain. Under any sort of time pressure, it is very easy to slip up and read out the word instead.

As with the verb-generating test, subjects are being asked to suppress a habitual reaction – to read any word that falls within the line of sight – and replace it with a more novel response. This is easy enough when the words and colours have no meaningful connection, but when both tap into exactly the same area of vocabulary – that of colour names – then the very memory circuits that the brain will need to rouse are also the ones it would like to shut down. The result is that the habitual response, rather than the required response, keeps threatening to creep in. The brain needs to be extra mindful of what it is doing. And as Posner predicted, during the Stroop test, the anterior cingulate ran hot indeed.

Some PET experiments to highlight the brain's response to pain shed further light on the anterior cingulate's role. Medical researchers were keen to use scanning to map the brain's pain pathways so as better to understand ways of providing pain relief. A group of volunteers was recruited and then scanned as a hot electrode was pressed to the back of their hands. The PET images showed that when the electrode merely felt warm, brain activity was confined to the somatosensory cortex. Heat was mapped just like any other skin sensation. But as soon as the electrode crossed the threshold and became painful, the anterior cingulate lit up. In fact, it turned out that the cingulate had a cluster of pain-mapping areas, with different maps for external pains and internal ones like belly-aches.

Still more interestingly, yet another part of the cingulate became active when the subjects were asked to try to ignore the pain – to put it to the back of their minds, or at least to tolerate the heat as well as they could. Pain, of course, is not a feeling that can be easily switched off. As a message telling us that our body is suffering some form of physical damage, it is the one sensory interruption that is guaranteed a direct line through to consciousness. And the reaction it demands is always the same: immediately do whatever you can to withdraw from the cause of the hurt. Yet there are times when we need to be able to override this response, such as when we are stuck in the dentist's chair, or if we have just picked up a full saucepan with a burning handle. We have to control ourselves and hold our head still, or lower the pan gently, otherwise the consequences may be far worse. So the PET findings suggested that the cingulate had areas to notice this potent interrupt signal, and also to push it into the background as much as possible when circumstances dictated.

In almost every PET experiment, the anterior cingulate was turning out to be a key area whenever the brain was being mindful of what it was doing or experiencing. This prompted some fevered speculation about a 'cingulate hypothesis' of consciousness. It seemed that science might finally have found the brain module that switched on the light of strong awareness. But Posner took a much more pragmatic view. Like Gray, he realised that the brain needed to be able to strike a balance between its internally generated plans and any unanticipated interruptions, what Posner called the endogenous and exogenous streams of processing. And to him, it seemed clear that the anterior cingulate did a general scaffolding service, holding together the fragile states of intention and expectation that the brain often took into the moment while at the same time priming the lower brain-processing pathways for the balance of concentration that would be needed.

Posner argued that the prefrontal cortex created a broad rather than a focused idea of what should happen during the next moment of processing. It had the space to keep a general backdrop of thoughts, goals, conditions, anticipations, and other working-memory states flying. But some other area would be needed to draw together the threads and set up the brain with a firm plan, a single manageable aim, for the coming moment. The fact that so many lines converged on the anterior cingulate, and that it glowed hot whenever the brain was having to keep a tight grip on a novel state of response, suggested that it must do this job. It would be the lynch-pin of the output hierarchy. The prefrontal cortex would buzz with the ideas, but the anterior cingulate would sharpen the focus on which of them would actually be executed.

Of course, when a plan was routine – such as opening a door or naming a colour – the workload for the anterior cingulate would be light. It would only need to keep a general aim in mind, the guiding image that would provide a context for the unfolding of a series of untaxing, habit-level responses. But whenever a task was difficult, as with the Stroop task or Passingham's finger-tapping exercises, the anterior cingulate would have to burn energy, holding together a rather more tenuous state of intention. It would have to keep track of the parts of the brain that would need to remain roused, or stay suppressed, so that the demands of the processing plan were met.

And as another part of this scaffolding service for the action of a moment, the cingulate would also determine the level of interruption that could be tolerated. It would add a level of intelligence to the simple-minded operations of the nucleus accumbens apparatus, priming its circuits on how to react and telling it what would be the right degree of interruptability for the moment ahead. So scanning studies might suggest that the anterior cingulate has a lot to do with a highly conscious brain. But this would be because the area is active whenever we have to stuggle hard to hold together a frail intent to do something new or different. Consciousness would not reside in it. The anterior cingulate would be just the high-level bottleneck by which a moment of frontal planning could take on sharper organisation.

Back at the Institute of Psychiatry, the box marked 'cingulate cortex' was almost lost in the clutter of arrows, squares, and labels making up Jeffrey Gray's ever-growing circuit diagram of the brain's planning and interrupt pathways. However, there was still another large chunk of machinery to be included. Gray had the cingulate cortex to draw the frontal cortex hierarchy into a state of declared intent for the coming moment, the nucleus accumbens to act as a pivot, and then the interrupt pathway for making bids to take over the cycle of processing. But what happened when the planning side won out and the desired focus became the actual focus? Which bit of the brain did the comparison that allowed the anticipated event – or equally, a surprised mismatch with expectations – to be identified within the swelling spread of sensory mapping?

Catching the core aspect of a moment would have to be an active process. It might be thought that the winning event would select itself. With the help of advance priming to warm the right pathways, one area of activity would begin to burn bright and overshadow all others. But a little introspection shows that what gets escalated into the focus of attention is not whole events, but rather some corner of the field of representation.

As our brain closes in on the moment, our focus tightens around whatever bit matters the most. And, as said, what matters can be several kinds of things. It might be some aspect of the moment that is significant because it was sought for, like the sight of a set of lost keys which we hoped we would see, but did not know exactly when or in what position. Or what matters could be noticing the one part in an otherwise well-anticipated moment that did not go quite right. We may have hit a smash off the edge of our tennis racket, and it will be the precise feeling of the clunk that becomes the vivid centre of the moment. The brain will try to boil down each moment to its gist. The brain's aim is not to get things into the centre of consciousness but rather to keep as much as possible out so that only the most crucial details will linger to spark the next cycle of planning and expecting.

As has been seen, in a broad way the whole brain plays a part in this filtering process. Every neuron has a tonic state of firing and every mapping area hums with some balance of feedback tension, so every part of the brain can become primed for processing the moment ahead. All parts of the brain also store memories – they form landscapes of connections – and so have well-established habits of processing. The brain thus goes into the moment equipped to distil as much as possible, quickly, locally, and pre-consciously, from the arriving wash of sensory input. It is one giant predictive organ. Yet still, the brain would need some kind of pathway to bring to a head the efforts to extract a focus. There would have to be a bottleneck area which could at least point back to where the many bits of mapping activity lay, and probably also sustain these fragments of representation long enough for them to begin generating a reaction.

Gray felt that the hippocampus – or rather the hippocampal formation, which included surrounding regions such as the septal area, the subicular area, and the entorhinal cortex – was the obvious candidate for the job. Most researchers working on the hippocampus saw it as simply a passive trap for memories. As a moment came together, the hippocampus would take its snapshot record – the template that pointed back to where the most important areas of activity had been. This preserved voting pattern could then later be used to reproject a memory image back across the sensory cortex. However, Gray felt that the hippocampus must also play an active role in the processing of a moment. As the most rarefied level of sensory representation, the hippocampus would also be the natural place for the most precise comparisons between a state of priming and the events that followed.

Lower levels of the sensory cortex could recognise general qualities, such as a change of colour or a stirring movement, but the hippocampus would be the place where the brain could catch complicated conjunctions of events. If we were searching a crowded party in anticipation of seeing a friend's face, or watching ourselves make a tricky manoeuvre such as reaching to the back of the cupboard for a packet of sugar, then it would be the hippocampus where all the strands of information eventually converged. Not until processing rose to this level would the state of representation refer to a specific occurrence in a specific place, thus allowing a sharp comparison to be made.

The idea that the hippocampus did the most detailed level of matching was supported by the fact that it had a particularly heavy connection with the cingulate cortex. The cingulate sent messages down to the hippocampal formation via a fat bundle of fibres known as the cingulum. Gray presumed this was the route by which the hippocampus became loaded with the next moment's anticipatory state. The hippocampal formation then had its own bit of cabling, the fornix, to send messages back to the cingulate. To Gray, it looked like the fornix completed the loop, putting the cingulate back in touch with the fruits of the moment's processing. The cingulate would ask the hippocampus to expect certain events and certain actions; the hippocampus would then report back on what had actually happened, or rather what turned out to be least routine about the moment's happenings.

So the hippocampus would be the brain's significance trap as well as its memory trap. It would help select the focus for the moment as well as remember it. Then, through its inputs into the nucleus accumbens and the machinery of the basal ganglia, the hippocampus would also play a part in the actual stopping and starting of actions. As it subtracted what had happened from what should have happened, it would be able to give the green light for a planned action to go ahead or call a sudden halt when something seemed out of place. The amygdala would interrupt because it had noticed an event where none was expected. The hippocampus would interrupt because an event was not going exactly as expected.

As the final resting place for working memories, the hippocampus would also interrupt in yet another way. Often we will have thought about something we ought to do, and then forgotten all about it until something happens to jog our memory again. We might be walking around the shops and suddenly the sight of the cheese counter will remind us we had made a mental note to buy some Parmesan. Briefly, our frontal cortex would have become organised with a state of intention, so priming the hippocampal machinery with the idea of executing some action. Then, because of the memory powers of the hippocampal formation, this thought would have remained quietly ticking over even after it had slipped from conscious view. There would be a goal state waiting to intercept a sensory event, like the sight of a pile of cheeses, and thrust it into the limelight of awareness.

Of course, our minds would be busy thinking about something else entirely by that point, so the hippocampus would have to jolt us from our expected track. It would have to deliver a quick blast to the nucleus accumbens to flatten the current state of intention and so allow a remembered intention to fill its place. As with the amygdala, the circuits of the hippocampus would be employed in many ways, and so could itself end up looking like an interrupt organ at times.

Despite such confusions, by the mid-1990s Gray had arrived at a hugely detailed anatomical story about the brain's forking 'so what?' pathway – the processing bottleneck, or rather whole chain of bottlenecks, that selected the contents of a moment's focus of attention. Even the thalamus was part of his picture as the output of everything – the hippocampus, the amygdala, the loops of the basal ganglia – had to pass through the thalamus's controlling eye before returning to the various parts of the cortex such as the cingulate and prefrontal lobes.

Gray was also finding that the climate of opinion had so changed in science that he could now openly claim his work said something important about the brain processes leading up to a state of consciousness. When Gray first began publishing his circuit diagrams in the early 1980s, ostensibly he was talking about the brain pathways behind anxiety, conflict resolution, and other standard psychiatric problems. Even when he wrote a major review of his work in 1991, the explanations were made in the context of understanding schizo-phrenia, another defect of normal consciousness. But by 1995, Gray could declare that he had a theory about how particular aspects of a moment came into sharp conscious focus.

Of course, Gray's reading of the brain's anatomy produced plenty of debate. For a start, both LeDoux and Posner had their own pet theories that placed a somewhat different stress on certain parts of the likely escalation network. And it did not help that organs like the amygdala and hippocampus seemed to shift their role depending on whether they were being seen from the point of view of a bottom-up interruption or a top-down flow of planning. Gray's work could only be a start on a more complete characterisation of the brain's sub-cortical machinery.

But what counted was the evidence of an emerging consensus about the brain and what it had to do during each moment of processing. A number of psychologists – researchers like Gray, Posner, and LeDoux, but also Bernard Baars, David LaBerge, and several others – had begun taking a systems perspective of the problem of consciousness and realised that the brain could not be instantly and effortlessly aware of the world. There had to be a lengthy cycle of events, beginning in the laying of plans and expectations, then proceeding through a pre-conscious phase where it tried to settle as much as possible, before finally the brain flowered into a state of focus that allowed for the evolution of an equally focused state of response. It was also becoming agreed that to find its way to the best view of the moment, the brain had to be able to handle both plans and interruptions. It had to go in poised, but also open-minded.

Another point of agreement was that this 'so what?' evaluation of the contents of the moment took in many of the brain's supposedly more primitive areas. It had long been assumed that the cortex was the conscious part of the brain. The cortex spun the gossamer maps that led to a state of subjective awareness, while organs like the brain stem, amygdala, basal ganglia, and thalamus were either merely unconscious reflex centres or way-stations leading to where the real processing action took place. But the brain was turning out to be an integrated whole. There was a whole string of sub-cortical bottlenecks which acted to organise and focus the current flow of cortex traffic.

This made the tale of the brain still more complicated. There was an even greater amount going on inside each moment of awareness than anyone could have imagined. But at least the brain's intentions left some imprint on the design of its pathways. Thinking about its needs – the logic of what the brain has to do during an instant of processing – could take mind researchers a long way towards understanding the logic of its design.

CHAPTER TWELVE

Getting It Backwards

So what is consciousness? In the 1980s, people would have given you two kinds of answer, both equally unsatisfying. Many advised just to accept that it was a mystery – or, to use the preferred term, 'cognitively impenetrable'. Others said it was a lot like what computers did, except on a much grander scale, of course. The 1990s brought no great visible change either. No one stood up and said 'I've cracked the problem!' – or at least no one that anyone else took very seriously.

From the outside it even looked to be a decade of disappointment. Brain-scanning technology created a momentary surge of hope. At last we can experiment on living human minds, said the scientists. But reading the message in the glittering pictures proved to be far more difficult than anyone had imagined. The same frustrations hit single-cell researchers. The 1980s had brought tremendous discoveries. The whole processing hierarchy of the brain seemed laid bare. But the enterprise foundered in its search for the brain's neural code. In the end, researchers could find little that was stable about the way the brain did its computations. Cognitive science was another movement that was not delivering on its promises. And as the glamorous sheen began to wear off the new sciences of chaos and complexity, even they too began to be talked about as failures. Fancy maths alone did not seem likely to solve the problem of consciousness.

But change was happening, even if it was a kind of change that was difficult to trumpet. First, there was indeed an absolute torrent of new facts. The advances in technology did make a difference, as did the move of more psychologists and computer scientists into the field of brain research. And then the willingness to address the subject of consciousness directly – to be heard using the c-word – served to remind researchers just what a subtle phenomenon they were dealing with. An explanation of awareness would have to take in all its shades, all its degrees, even all its timescales. With this, the questions being asked started to become more pointed. A lot of half-forgotten facts and overlooked scientific figures also began to be talked of again.

However, personally speaking, the biggest change for me was not how much new needed to be learnt, but how much that was old and deeply buried needed to be unlearnt. I thought my roundabout route into the subject would leave me well prepared. I spent most of the 1980s dividing my time between computer science and anthropology. Following at first-hand the attempts of technologists to build intelligent machines would be a good way of seeing where cognitive psychology fell short of the mark, while taking in the bigger picture – looking at what is known about the human evolutionary story – ought to highlight the purposes for which brains are really designed. It would be a pincer movement that should result in the known facts about the brain making more sense.

Yet it took many years, many conversations, and many false starts to discover that the real problem was not mastering a mass of detail but making the right shift in viewpoint. Despite everything, a standard reductionist and computational outlook on life had taken deep root in my thinking, shaping what I expected to see and making it hard to appreciate anything or anyone who was not coming from the same direction. Getting the fundamentals of what dynamic systems were all about was easy enough, but then moving on from there to find some sort of balance between computational and dynamic thinking was extraordinarily difficult. Getting used to the idea of plastic structure or guided competitions needed plenty of mental gymnastics.

But at least I had the luxury of time. And as I began to feel more at home with this more organic way of thinking, it also became plain how many others were groping their way to the same sort of accommodation – psychologists and brain researchers who, because of the lack of an established vocabulary or stock of metaphors, had often sounded as if they were all talking about completely different things when, in fact, the same basic insights were driving their work. The problem, then, was to put some of this new understanding into words.

So, again, what is consciousness? In some sense, it must be the outcome of a moment's processing. But the very idea of an outcome is a dangerously reductionist notion. It implies a definable time when all processing is finished and suddenly the brain lights up with a state of subjective awareness. Once the work is done, the results can be surveyed.

Of course, it is undeniable that every cycle of processing seems to have a trajectory. Over the course of perhaps half a second, a state of representation will emerge from a structured competition with a broadly settled background and sharply felt centre. Yet somehow any concept of consciousness has to be inclusive of the whole duration of the cycle. Our everyday use of the word may fit the way that consciousness feels to us – instant, effortless, and uniform – but a scientific use of the word has to embrace the fact that awareness is a dynamically evolving construction. So how can consciousness be described in a manner that would do justice to both the inner experience and the objective evidence?

If there were a way, it would seem to be to define consciousness as the view we establish into a moment. After a lot of work sifting and reacting, we find a central vantage point from which to look back over all that has just happened. Every cycle of processing produces a vast mass of information, but what we remember about the moment is a carefully ordered view – a tale told from the perspective of whatever emerged as the focal event.

The advantage of this definition is that it includes both the escalated part of the moment and the pre-conscious fringe. We climb the mountain not to look at the rocks at the top but to have a panoramic view back over the slopes of our recent journey. The details nearest us may seem the clearest, the most immediate, but our very position implies all that came before. In rising up, we will have settled a whole landscape of processing into shape. So, even if we are not brightly aware of the many pre-consciously handled events, like the footsteps that took us down the corridor or the preparatory lift of a cricket bat, we know that they must have happened to get us to where we are. Consciousness is not something that follows an unaware stage of processing, it is the arrival at a point from which everything that has happened so far falls into a sharper, more coherent perspective. The ending of a cycle of processing is also, in retrospect, our entry point. It establishes the angle from which we will remember and relive the moment.

One of the benefits that follows from describing consciousness in this way is that it avoids the need to claim a fixed finishing line for a cycle of processing. Awareness is just whatever level of crisp order the brain has managed – or, indeed, wanted – to extract from the moment. During some moments, the brain may develop a tight and memorable state of focus. For example, spotting a long-lost friend across the street would cause the immediate collapse of whatever intentions or expectations had been filling our head, leaving the sensory experience and our emotional reaction to be remembered with extra vividness. But at other times, life may pass as a bit of a blur. We still develop some focal stance that orients us to the moment, but this angle points back to a broader, more entangled field of activity. When we are driving quietly down the road or day-dreaming in an armchair, the brain is still doing a job but it does not produce such sharply felt trajectories back into each moment. The landscape of processing is more one of rounded humps than vertiginous peaks. Being a dynamic phenomenon, consciousness is free to evolve as little or as far as it needs in order to reach the pitch of processing most suited to the moment. In the end, all that matters to the brain is the quality of the fit.

And, as said, the other benefit of defining consciousness as our angle into a moment is that it keeps the pre-conscious stages firmly in the picture. The tendency of a reductionist approach is to think that only the brightest part of consciousness, the escalated focus, is properly experienced and explored. The fringes of the mind – the realm of background sensations and habit-based reactions – are treated as non-conscious and unanalysed. It is as if consciousness were a limited resource, a spotlight beam that is so busy illuminating the important part of the moment that it cannot do more than throw a little scattered light on anything happening out on the periphery. At best, the fringes of our cone of awareness have to put up with a hasty, make-do kind of processing.

But with the dynamic view, the story is, of course, completely the other way round. We might remember each moment as if our minds went straight to the point of focus and ignored the surroundings. But the evidence is that the fringe of awareness was processed first. To use the spotlight metaphor, we began by illuminating the entire field of input and straight away noted all the areas which could be dealt with quickly and without fuss because they were either well anticipated or slotted neatly into some waiting routine. Only afterwards did the spotlight shrink and come to highlight some part of the moment for further processing. So while the background of awareness might not intrude on our subjective record of the moment in the same way – it does not feature as the first thing to come to mind when we introspect – it is still fully processed.

The connection between the quickly handled fringe and the focus of consciousness is actually even closer than this suggests. After all, when the brain deals with sensory input at the level of anticipated or habitual reactions, it is acting on an 'as if' basis – as if it had gone to all the time and trouble of developing an extended focal-level appreciation of what was happening. Every habit or expectation was once a prefrontally explored problem. The first few times a toddler sees a cat or fumbles to pick up a sheet of paper, it has to escalate the experience to a high level of awareness. It has to notice details about cats, such as that they generally have pointy ears, a slinky way of walking, and occasionally can flash a fearsome set of claws. In experimenting with picking up a sheet of paper, a child has to discover that a different grip is needed and other interesting facts, such as that a satisfying tearing can result from tugging at the paper in the right way. Quickly, however, the many aspects of seeing cats and manipulating various materials become automatic. The child's brain becomes silted up with a predictive knowledge of the world. So while in any one moment there might only be room for a single exploratory route in, the rest of the moment is being handled by skills and habits of association which were once focal themselves.

This strikingly beautiful fact was brought home to me during another of many snatched conversations during coffee breaks at conferences, this one at the famed 'Woodstock' for consciousness studies, the inaugural 'Toward A Scientific Basis For Consciousness' meeting held at the University of Arizona-Tucson in 1994. I was talking to Walter Freeman, the neurobiologist at the University of California-Berkeley, renowned both for his pioneering efforts to apply chaos theory to brain activity and also for the rather gnomic quality of his writings. In person, however, his ideas no longer sounded so obscure. Stroking his luxuriant snowy-white beard, Freeman made the point that being dynamic, the brain could bring the full weight of a lifetime's experience to bear on each moment as it was being lived. Every second of awareness as a child or as an adolescent would in some measure be part of our consciousness of the present. That was what being a memory landscape really meant. Lambasting the standard uninspired view of the brain, Freeman chuckled: 'The cognitive guys think it's just impossible to keep throwing everything you've got into the computation every time. But that is exactly what the brain does. Consciousness is about bringing your entire history to bear on your next step, your next breath, your next moment.'

The idea of bearing down on the moment turns out to be a useful corrective to the processing-cycle approach to what the brain is doing. This book has tried to unravel what goes on inside a single instant of awareness. It has talked about the hidden micro-structure of the mind with its various processing stages such as the generation of anticipations, the settling of a pre-conscious spread of mapping, a 'so what?' decision, and then the elaboration resulting from establishing a single point of focus. But the trouble is that this way of looking at the brain only deals with dynamism on a single timescale – the plasticity exhibited by the brain over the course of half a second or so.

As Gerald Edelman argued, the key to understanding the brain is that it is plastic on all scales of organisation. Each moment of processing is actually connected to a whole continuum of selectionist pressures and evolutionary adjustments. Behind the events of the instant lie the minute-by-minute adaptations in neurotransmitter levels, the hour-by-hour sprouting of new memory connections, the year-by-year changes of childhood development, and even the generation-by-generation changes of the evolutionary history of a species. Information is being captured on all these levels and, as said, the entire weight of this information is brought to bear on the processing of a moment.

The situation is actually even more complex because the relationship works both ways. The history of the brain has its impact on the processing of the moment, but then the processing of the moment has its impact on the history of the brain. In the very short term, one moment's focus drives the next moment's plans and anticipations; in the longer term, it adjusts the brain's arousal levels and helps build new memories and habits. Then, if the decisions made during a moment turn out to be particularly good or bad, they could even affect the survival of the individual and therefore the genes used in building the next generation of brains. So the shimmy of an instant of mapping is just the tip of the iceberg. It is all right to talk about processing cycles and stages of a structured competition so long as it is remembered that the activity depends on circuitry that is itself fluid. As a Hebbian landscape of connectivity, the memories make the moment, but the moment also makes the memories.

Or, to emphasise the point in quite a different way, consciousness began with nervous systems that adapted very slowly and which now can adapt very fast. The origins of the brain lay in an ancient need for some kind of intelligent link between sensory input and motor output. It was a simple product of the race between predator and prey. To start with, both sides were slow-witted with limited sense organs and limited manoeuvrability. This meant they could get away with a few reflex loops to slam shut a shell or jet away from danger, and some hormone systems to regulate activity levels and mating behaviour.

But steadily the pace quickened. To keep making the right decisions, animals had to get more meaning out of each moment. Simple reflex loops became knots of nerves, and then whole brains.

The pathway from input to output became tangled with extra wiring so that animals could tune their reflexes and even learn to chain them together. Eventually, animals began to represent. They made topographic mappings of the inputs arriving at their sense organs and then put them through the wringer of a hierarchy of feature-extracting filters. On the output side, the residue of the moment was decomposed by a matching hierarchy to form an action plan. And as the levels of complexity began to stack up, it could be said that the animals became aware.

What defines consciousness in this most basic sense is the ability of the brain to wrap itself around the moment. The evolutionary trend has been towards a more fluid, more precisely adjusted response to each instant. The first nervous systems were fairly stiff in their organisation. Sense organs and reflex pathways were accommodations to the world made on a genetic timescale. So while a worm may wriggle as soon as it is touched, any consciousness of the threat could be said to be stretched thin over many generations of small evolutionary adjustments.

The next step after a reflex pathway is a primitive form of learning such as habituation. A snail or sea anemone will take less and less notice of being prodded. But this slight degree of plasticity in a withdrawal reflex is still genetically circumscribed. It was only with the development of brains and hierarchies of mapping that the responses gradually became more plastic. Fish and frogs have very simple brains and can probably be said to be conscious at about the level of a habit. They map the passing parade of sensation quickly and accurately enough, yet the response is stereotyped both in terms of the meaning being extracted and the reactions that follow. A frog sees the world basically as a bug to be eaten, a looming shadow it ought to escape, or a potential mate it will want to mount. The events of each moment are appreciated in a routine, rather than a unique, way.

Then came the higher animals with a cortex that gave them a large sheet of uncommitted circuitry. The foundation for brain processing was still a bundle of time-proven programs – motor reflexes, instinctive behaviours, and neurotransmitter systems – genetically wired into the lower brain and brain stem. These created a stable platform for the brain's response to each moment. But the cortex provided a blank slate for the accumulation of habits and memories particular to an individual's own life. And with the room for dozens of rungs of mapping, the cortex even developed its own internal gradient of plasticity. Lower-level mapping areas like V1, and even V4, had to be rather general in their ways, but higher up the trail there was room for cells to fire only in response to more specific events, such as the sight of a hand or a sense of swirling movement.

Then, right at the top of the trail, in the prefrontal cortex and the hippocampus, the reactions could become extremely particular. Freed of the need to make habitual connections because all the necessary ground work was already being done by the levels below, the peak-level areas would have the privilege of being able to make one-off, hand-crafted responses to the moment. They could wrap themselves tightly around whatever was emerging as the most significant, or least habitual, aspect of the instant. The hippocampus would pick out a pattern of unique memory pointers while the prefrontal cortex would provide the space to develop a suitably considered reaction.

So consciousness, as we think of it, would have crept up slowly on the brain. It was a steady progression as new levels of plasticity were added. Crucially, what the brain was doing at each level was always exactly the same: making an evolutionary adjustment to its environment. But as one layer was built atop another, the timescales kept shortening and the adapting also became more global, until eventually the whole brain could leap from one viewpoint to another in under a second. One instant it might be wrapped around the thought of smacking away an oncoming cricket ball, the next it could be smarting at the embarrassment of having swung and missed. With a convulsive shift, the entire circuitry of the brain would become aligned to produce a coherent angle of view back into the latest moment.

The processing-cycle approach also has a problem in that it suggests we take in life frame by frame, in a rather clunky, computational way. With it needing between 100 and 200 milliseconds to shake down a basic mapping of the world, and then as much as half a second to evolve this mapping into a state of sharp focus, it would seem that the flow of experience should be uncomfortably lumpy. Awareness would come in discrete pulses twice a second; either that, or the brain would have to be juggling a number of overlapping frames of processing at the same time, all chasing each other up the flanks of the processing hierarchy and peaking one after another in rapid succession. So there would appear to be a real dilemma about how moments of processing could be chained together, or projected at a fast enough rate, to blur into the smooth stream of consciousness we actually seem to experience.

Fortunately, this is only a difficulty if it is believed that the brain is like a computer display, lighting up pixels of information against a dead background. But if the brain is a dynamic system, with its cells always active with some tick-over balance of firing, then continuity is not a problem because its circuits are always in a state of representation. Even when nothing is doing, the brain's mapping areas will rustle with a pattern of activity that reflects the strength of their many memory connections and so stand as a defocused portrayal of all the brain has ever known. Thus, at the beginning of each moment there will already be a coherent surface of representation. New input only acts as a disturbance, a provocation to send the system in search of a new state of balance.

This means that what the processing cycle times actually measure is the speed at which the brain can mount such a shift. Rather than making something out of nothing, as the computational tone of the story would imply, the brain is always conscious in a vague sort of way and then this stream is punctuated by the eruption of regular shifts in the point of view. A defocused state is tightened in a particular direction for an instant before being allowed to relax so it can be tightened in some new direction about half a second later.

So there is both change and an underlying continuity in every moment of processing. This makes sense, because in real life one moment is actually very much like the next. It is not as if the world jerks us from one situation to another, one second leaving us suspended a mile above the Gobi desert, the next pressed face-down on a busy restaurant floor. Even the most dramatic surprises, like coming out of the house in the morning and discovering that our car has been crushed by a falling tree, take place within a steady, predictable framework of experience. As we step through our front door, the look of the street, our memories of our car, and even our intention to set off for work will all be constants which can be carried over from one instant to the next. Our mental perspective on the scene might alter completely in a single gut-wrenching moment, but much of the background – the look and feel of everything else – will remain the same.

If our life is going smoothly – we come out of the house and the car is fine – then there is even less to change from moment to moment. We might still put life through the wringer to maintain the best possible view into the current stream of events, and also to keep our anticipations updated, but our consciousness would become something of a blur. As said, instead of sharp peaks, it would consist of gentle humps. So a cycle of processing only becomes defined as such in retrospect. A certain set of steps is needed to pinch up the brain's gently rolling stream of representation into something tighter, and so produce a moment in the flow of experience that is noticeably more distinct. And some moments involve a much more memorable reorganisation in outlook than others.

What, then, does this say about both Libet's half-second result and the status of the pre-conscious fringe in a moment's processing? First off, it suggests that pre-conscious processing is not tied down to set times in a repeating cycle of activity. The brain does not do pre-conscious-level work for a tenth of a second or so and then switch over to its focal mode for the next third of a second; it can take in new information and send out fresh instructions continuously. Because a habitual or well-anticipated brain process would be executed locally, there should be no interference with other habit-level activity taking place in the brain. There would be room for pre-conscious responses simply to bubble up and trigger another small adjustment in the brain's representational state at will.

But extracting a focus from a moment – drawing up the mapping activity into some brain-wide peak of response – would be a different matter. Effectively, the brain would be saying it had discovered some new angle on life and so all of its circuits had better take notice. To allow such a re-centring, some of the brain's areas of activity might have to be suppressed to make the focal event stand out more sharply, while other areas would have to be roused to discover what sort of response the focal event should inspire. The disturbance would no longer be local and so free to happen willy-nilly. The brain would only have room to mount one global shift in perspective at a time. The adjustment could have its beginning at any point – any scrap of pre-conscious activity might act as the spur for a massive reorientation – but the processing characteristics of the brain – the time it takes to evolve a coherent state of response – would determine how quickly the brain could move from one peak to the next.

This perhaps puts Libet's half-second finding in a different light. It is certainly now possible to understand a little better what was going on when his subjects turned a steady trickle of electricity into a subjectively reportable experience. The stimulation of a spot on their primary somatosensory cortex did not simply unlock some sensation that was stored at that position. Instead it set up a persistent nagging, telling the brain something was happening and that it ought to place an interpretation upon it. As signals flashed up the processing hierarchy, top-level somatosensory areas would have been dragged in to make some sort of population response, voting in a way that would help settle an initially ambiguous pattern into a more definite arrangement. A new balance of representation would be negotiated so that the subjects went from feeling that nothing in particular was happening to their hands, to a feeling that they were experiencing a pulsing pressure, a flush of warmth, or some other experience drawn from a large stock of somatosensory memories.

At the same time as the somatosensory pathway was evolving its way into a state of representation, the rest of the brain would be having to organise itself to take notice. It would need actively to suppress any other candidate points of focus for the moment, such as thoughts about the strangeness of lying on an operating table while someone poked about inside the skull, or the fact that a big toe was beginning to itch again. The brain would also be having to set up its output reaction. Libet's subjects were being asked to report the exact quality of the sensations, so the representation of the pulsing pressure or hot flush would need to be held together for long enough to generate a string of words adequate for expressing what had just been felt.

There would be plenty for the brain to do in order to become aware of the electrode buzz. But it is another question whether the brain would always spend exactly half a second in evolving to a peak of response. It would seem that the processing time should be dynamically variable – the brain would take as little or as long as was needed to get the job done. The simple physical characteristics of the brain should at least lay down a certain minimum processing period. Factors such as its conduction speeds, the time it might take to establish transient patterns of coherence between distant rungs of mapping, or the time required to mount a P300–like sharpening response, would mean that the brain would not be able to achieve a global shift in perspective in less than some fixed duration.

But with the help of anticipation to smooth the path, some events might be shoved into focus quite a bit faster than half a second. They may arrive within, say, 250 to 300 milliseconds – although being well anticipated, they might also result more in one of consciousness's rounded humps than a sharply felt peak. On the other hand, achieving a state of focus might sometimes take much longer than half a second. If life really takes us by surprise – if there is a lot about the existing state of representation that has to be wiped away and a lot of evolving to do to reach a suitable state of response – then the cycle of processing might be stretched right out. And, indeed, the evidence of P300 experiments seems to support this. The classic P300 peak can occur as early as 250 milliseconds for very humdrum forms of surprise, but can be delayed for as long as a second if a subject is not concentrating properly, or is forced into a real double-take.

Speculating even further, introspectively such a belated arrival at a settled view back into a moment might possibly make it feel as though time had been temporarily suspended for an instant. In extremely shocking moments – such as during a car crash, for example – events can seem frozen or to be taking place in treacly slow motion. The reason could be that suddenly our existing state of representation and expectation would seem very inadequate, and yet it would be equally hard to install a settled state of interpretation and response in its place. So when we eventually managed to establish a view back into the moment, it would include a record of a protracted period of confused, but intense, experience.

Or again, if we went up to smack a volleyball and something rather amusing and fluky happened – like the ball getting bounced off someone's helpless head – then the contrast between our expectations and the actual turn of events might also be so great that it took some extra time to settle a state of representation. Normally a moment of skill passes in a blur of automatic activity. Part of the training is not to think too much about any particular point in the cycle of action, but to let the moment flow – to get in and out smoothly so that you can react fast to whatever happens next. In consequence, the memory usually seems fragmented or partial. But the confusion over whether we did or did not intend something so lucky might stretch out our awareness for such a moment in a freaky way.

The idea of looking back through the moment, bearing down on the moment, wrapping around the shape of the moment, pinching up the flow of the moment to form a fleeting peak, are all ways to try to get at the dynamism behind the structure. The processing-cycle approach to consciousness and the workings of the brain hopefully gets things more right than wrong, but it should not be allowed to obscure the still more complex story that lies behind it.

Consciousness is a popping stream of adjustment. The brain draws itself tight to form a responsive surface which ripples with many quick and automatic changes. Then, on the back of this – every third or half a second or so, depending on the steepness of the adjustment required – the flow becomes pinched up more sharply. There is a global shift, and it is the fact of this shift that would create a pointed feeling of subjective experience. The brain is always representing, but it is really only the changes in representation that we notice or remember. It is the sense that something is happening, of a rapid succession of new viewpoints, that makes awareness feel as if it is going somewhere.

But if this seems like the end of the tale, well, it is not quite. The kind of consciousness we have been looking at so far is ordinary biological or animal awareness – the brain's development of states of representation. But on top of this foundation humans have added yet another level of complexity. We have evolved both language and a collection of thought habits to go with it. These give us a range of special capabilities, not the least of which being the ability to introspect, to be self-aware and not just aware. And because the brain is a fluid bag of circuits, shaped by its experiences, this cultural software becomes an intimate part of our brain's organisation. So for a science of the mind to be complete, it will have to come to embrace this social dimension. As well as psychology, neuroscience, computer science, and complexity science all being part of the marriage, a partnership will have to be forged with sociology, anthropology, and linguistics as well.

CHAPTER THIRTEEN

The Ape that Spoke

It is obvious that the consciousness of humans is different from that of animals, but the question is, in what way? Humans appear to have all sorts of added extras, such as self-awareness, rational thought, and free will, yet it is hard to tell whether the distinction is one of degree or kind. Is the human mind just a scaled-up chimpanzee brain, or is the gap so great that there can be no comparison between the mental lives of animals and humans? Do animals even have subjective states?

The place to start looking for answers is the archaeological record. The hominid family line is thought to have broken off from the rest of the African apes about five to six million years ago. The first adaptation appears to have been upright walking: early hominid skeletons have completely modern hips and legs but still an ape-sized brain – and presumably an ape-like mind. However, it was not long before the brain of the hominids began to swell. With the arrival of *Homo erectus* about a million years ago, it had trebled in size, and there is plenty of evidence that this extra brain power was put to good use. *Homo erectus* made simple stone tools, built rough shelters, and by about half a million years ago was keeping warm beside fires.

In evolutionary terms, this was certainly a swift, but not unprecedented, rate of development. The dramatic increase in brain size was achieved by the quite simple mechanism of letting the brain continue to grow for a year or two after birth. The brain of an ape normally does almost all its growing in the womb, but humans develop as big a brain as they can in the womb – large enough to make childbirth a risky, painful affair – and then keep on going. It is true that half of the neurons actually die back during the first year of life as the brain's pathways prune themselves to shape, but overall the brain gets bigger as the remaining cells swell in size, sprout their connections, and develop a fatty myelin insulation. There is also a huge increase in the number of support cells that keep the neurons healthy and fed. So within a year the brain doubles in weight. By the time it reaches full adult size at the age of six or seven, its size has trebled.

As far as the genetics go, producing the human brain could hardly have been simpler. Whatever our mental abilities depend upon, it does not seem to have been the development of any radically new brain structure. The human brain looks exactly like a scaled-up monkey brain; all that needed to be changed was the timing of the genetic clock ruling the starting and stopping of the various phases of brain growth. What makes the self-conscious human mind all the more puzzling is that until very recently, the ballooning of the hominid brain did not even appear to have much effect on the nature of consciousness. Of course, it is impossible to say anything certain about the mind of *Homo erectus*, yet there is no evidence of a huge mental gap between this ape-man and the apes of today.

Chimpanzees might not use fire or put a roof over their heads, but they can be reasonably skilled at fashioning tools. They will strip twigs to fish termites out of a nest, or crumple leaves to act as a sponge to get water out of a tree bole. One chimp was observed to make four different kinds of tool to get at honey in a bees' nest, using successively finer gauges of splintered branch as chisels to pierce the wall, then finally nipping off a thin whip of vine to collect the honey. Chimps show many other signs of intelligent behaviour, such as making nests of branches in which to sleep, seeking out medicinal plants when they have worms, and wiping their bottoms with leaves when they have diarrhoea. Chimps even hunt co-operatively. In the wild, a pair of males will creep up on a group of feeding colobus monkeys, panicking them and driving them into an ambush set up by the rest of the troop.

The fact that *Homo erectus* was a tool maker and fire user is impressive, but it seems equally significant that *Homo erectus* then hung around for about a million years without making any further spectacular advance. Having mastered the skill of chipping a flint into a teardrop-shaped hand axe, *Homo erectus* went on producing exactly the same tool for thousands upon thousands of generations. This ancestor had a mind that was intelligent, but not explosively inventive.

About 100,000 years ago, *Homo sapiens* – true modern humans – appeared on the scene. *Homo sapiens* was a puny creature compared to the Neanderthals who happened to be the dominant species of hominids at the time, and they did not even measure up in brain size, having a cranial capacity of about 1,500 millilitres compared to 1,600 millilitres for the Neanderthals. Yet despite this, *Homo sapiens* took over the world. The Neanderthals were shoved aside, the last few disappearing about 30,000 years ago while *Homo sapiens* spread rapidly to fill every corner of the planet.

From the first appearance of *Homo sapiens* in the fossil record, it is plain that we are dealing with a fully modern mind. The tools made by early humans represent a quantum leap over anything produced by a Neanderthal hand. As well as finely crafted harpoon tips and arrowheads, archaeologists have dug up bone needles for sewing clothes, and small fat-burning lamps. *Homo sapiens* made the first proper campsites, with huts and hearths rather than crude windbreaks and open fires. The dwellings were also arranged in a way that suggests a clear social pecking order.

The clincher, however, is that by about 40,000 years ago *Homo sapiens* was capable of art. Bone knives were decorated with carvings of deer; shells and bones were whittled into beads to be strung on necklaces. Most famously of all, people were crawling deep down into caves to paint the walls with awe-inspiring murals of animals and hunting expeditions.

So the archaeological record tells of a long haul to produce a large-brained ape. By resetting the brain's genetic clocks, the hominids grew into smart chimpanzees. But then something else happened. For some reason, in a blink of geological time, one line of hominids suddenly became symbolically minded and self-aware. The only possible explanation for this overnight change appears to be the development of language – or, more precisely, of articulate, rapidly spoken and grammatical speech.

*

Speech would have emerged out of the extremely close social lives the hominids were leading. Living in a group depends on good communication. Chimps and gorillas may not have a formal language system, but they are expert at reading the moods and intentions of one another from subtle signs such as facial expressions, direction of gaze, general posture, and, of course, grunts, screeches, pants, and other emotional noises. A lot of these signs are involuntary, and so depend entirely on the intelligence of the receiver to notice them and interpret them. But the apes also make a deliberate use of gestures. Chimpanzees have been seen holding out a hand as an invitation to groom, or directing attention with a flick of their eyes.

Even more revealingly, chimps are capable of deceiving. The anthropologist Jane Goodall, who studied wild chimps on the shores of Lake Tanganyika, reported how a bunch of bananas was once dropped in front of a junior chimp after the rest of its group had wandered on ahead. The chimp made excited 'food barks' which immediately brought the others racing back, so the youngster lost out. As a test, the same thing was tried again the next day. This time the chimp stayed silent, although a faint choking noise could be heard coming from its throat as it struggled to stifle its glee.

Many other examples of concealment, misdirection, and feigned nonchalance have proved that apes are not only smart enough to read one another's emotions, but also to make some predictions about how their own signals might be interpreted. An ape like the chimpanzee seems very close to language. However, two further developments were probably needed to clear the way for a symbol-based system of communication.

As many speech researchers, such as Doreen Kimura of the University of Western Ontario in Canada, have argued, the first of these would have been the gradual lateralisation of the brain as an adaptation for tool use. Making and handling tools demands the ability to plan a focused sequence of actions. A flint has to be turned in the hand as precise blows are struck to produce a sharp edge. And, as said earlier, doing this would require two kinds of thinking to be going on simultaneously. While one side of the brain was shrinking hard to isolate the separate steps of the sequence, the other would have to be taking a more global view, holding the overall outcome in mind. Having spent a million years or so developing the brain so that it could handle complicated sequences of hand and finger movement, the same ability would then have smoothed the path for language. Speaking requires both the ability to chain words together and the facility to keep in mind a general idea of where a conversation is going. With tool use, there would have been an existing motor and planning hierarchy for language to hitch a ride on.

The other, less obvious, pre-condition for the development of speech would have been the social change of stable pair bonds. Almost every other species of animal that lives in a group depends upon rivalry. The males compete for dominance and the right to father most of the children in the troop. The result of this pattern is that the males have little reason to help in the care of the young – they either have no offspring or are too busy keeping their place at the top of the pack. This tournament style of mating works fine in smaller-brained animals such as baboons and hyenas, but big brains are metabolically expensive. The young need to be well fed. It also takes longer for the young to grow up. The strain of this already shows in chimpanzee and gorilla mothers, who have to space out the production of offspring. A chimpanzee mother can only have a baby once every four to five years. A simple extrapolation shows that the apes were heading up an evolutionary blind alley: they could not grow brains much bigger before they ran into problems raising enough babies to start the next generation.

To get out of this bind, anthropologists such as Donald Johanson of the Institute of Human Origins in Arizona say the early hominids would have had to make a drastic change in their approach to parenting. The need was for a more monogamous, more co-operative, reproductive strategy. By pairing off, fathers would have a stake in looking after their children. The relaxation of tensions between males would also have allowed more group activities, such as hunting and shelter building. Food could be gathered collectively and then shared. As hominids like *Homo erectus* became more socially organised, there would also have emerged a division of labour between the sexes. Weighed down by their brood of children, the mothers would have stayed close to the safety of a base camp, spending the day collecting staple foods like roots and berries. Meanwhile, the males would have gone after riskier, but nutritionally more rewarding, food. They would hunt, fish, or climb trees in search of honey.

The advantages of these changes in food-collecting and child-rearing behaviour are easy to see in modern-day hunter-gatherer tribes. Even though civilisation has pushed such people into the most marginal lands, they can still feed themselves with just a few hours' effort a day. Chimpanzees, by contrast, have to spend most of their time foraging. And instead of children being spaced out, humans have no trouble supporting another child every couple of years.

The importance of pair-bonding was that it would have put a premium on the development of communication skills and also served to preserve the first hesitant steps towards proper speech. Obviously, to live in a co-operative group would demand a more explicit method of communication than just meaningful grunts, face-pulling, and eye-rolling. At some point, the bridge must have been crossed so that the use of gestures and sound became symbolic. Rather than things simply standing for what they were – a grimace of annoyance, or a grunt of interest – they would refer to something else.

Learning such an association is not in itself all that difficult. In experiments, chimpanzees have been taught to recognise and use hundreds of words. They can use hand signs or plastic tokens to ask for a cup of orange juice, or to go for a walk. Chimps in the wild have also been seen to invent their own personal gestures, such as holding out one arm clasped in the other as a plea to be groomed, or shaking a head to say no. Even with a brain a quarter of the size of ours, chimps seem to be capable of making the first small steps towards the use of symbols. But the hurly-burly of chimpanzee life also means that these first steps are unlikely ever to be preserved. Mother chimps have a close relationship with their infants, and so might create some shared quirk of expression; however, as adults, this bond is not repeated, and any experiments with symbolism would get washed away. With the hominids, on the other hand, the necessary closeness between parents would close the cycle and allow the advances of one generation to be passed on to the next. So, from minor beginnings, the use of symbolic communication could quickly snowball.

All these many kinds of evolutionary change seem locked together in a virtuous feedback spiral. The apes had reached a bit of a dead end, their brains being about as big as their social system would support. Then the hominids broke out. Walking upright freed the hands and encouraged the greater use of tools. Using tools demanded more brain power, and also the ability to imagine sequential actions. To afford bigger brains, the hominids had to change their parenting style. Closer bonds, in turn, fostered developments in communication. Closing the feedback loop, every small advance in the ability to communicate would have further tightened the social ties, helped to formalise the use of tools, and so paved the way for the growth of yet larger brains. Once started down this evolutionary path, the hominids were propelled forward until eventually a true speech capacity began to develop.

Language would not have appeared all at once. Words must have come before grammar. The first words would have been names for everyday objects such as firewood and antelopes. These would have made possible simple one-word sentences whose meaning was made clear with expressions and gestures. A tribal elder would only have to say 'Wood' with a commanding tone and a nod of the head towards a dying fire to tell a passing youngster to go and add a few more branches. Or a hiss of 'Antelope!', said with a quieting slash of the hand, might have been enough to warn a band of hunters to hush up and spread out.

It is not hard to teach chimpanzees and gorillas to communicate at this grammarless, single-word level, so with a much larger brain it seems almost certain that *Homo erectus* would have been using just such a proto-language for most of its million years on the planet. And for the daily round of a hunter-gatherer species, single words would have been plenty. Speech does not need grammar to be useful. But eventually, somewhere along the line, the next step was taken and a race of hominids began to talk in complex sentences.

Inventing the rules of grammar was probably more of a jump than it seems. When speaking in single words, the only choice to be made is one of topic, but as soon as words are chained into sentences, there is the problem of what to say first. We can only transmit one word at a time, so a sentence has to have a backbone logic connecting its parts. A standard formula is needed for breaking down an often complex tangle of thought into a linear stream of symbols. In other words, to be grammatical, our ancestors had to learn how to think like reductionists!

There are over 6,000 dialects and 200 language families in the world today. Superficially, the grammar of each looks very different with almost any rule of word order or declension of tense appearing to apply. Yet the first feature shared by every known language is that they all divide the flow of words into sentences, speech modules with a self-contained logic. The second is that these sentences always have the three fundamental components of a subject, a verb, and an object – a doer, an action, and a done to. The story they tell is a straight-line, cause-and-effect tale of who did what to whom.

Of course, these three essential components can be arranged in any order. The standard English order of subject–verb–object seems the most sensible because it is what we are accustomed to, but other languages, like Japanese, use subject–object–verb. Instead of saying 'the cat sat on the mat', a Japanese speaker would say 'the cat on the mat sat'. Then some languages, like Gaelic, use a verb–subject–object order, so the sentence would read 'seated was the cat on the mat'. In a few rare cases, the object can even come before the subject. This reverse logic sounds as though it would be difficult to follow, but even to English-speaking ears 'on the mat was sitting the cat' still makes sense. What matters is that it seems to tell us the complete story. The sentence has the three components needed to express a simple linear relationship.

So where did this reductionist formula come from? Tool use would have pre-adapted the brain for grammar, giving it the resolving power to think in sequences. The three-part logic then adopted would have been no more than an accurate reflection of life as seen through human eyes. We view ourselves as creatures of purpose, taking decisions and making things happen. The world we inhabit is primarily one of actors and actions, so it was only natural that, forced to find a serial logic to structure our speech, we should chose to break each sentence down into a tale of cause and effect.

The in-built reductionism of human language is clearly a problem for the would-be dynamicist. Its standard straight-line logic becomes a handicap as soon as we want to talk about systems which are the product of competition and complex feedback relationships – systems like the brain, for instance. But, for a tribe of hunter-gatherers, it was more than adequate. Their life depended not on philosophical clarity but on being able to communicate what were essentially social ideas. A grammar centred on actors and their actions would capture precisely what most needed to be said about any social situation.

The question then is, when did grammatical, articulate speech enter the picture? It is a query that has caused surprising heat among language scholars. But many researchers, such as Philip Lieberman of Brown University in Rhode Island, believe that grammar was only perfected with the arrival of *Homo sapiens*. The most convincing line of evidence for this comes from the shape of the human vocal tract. Reconstructions of the throat and mouth of Neanderthals suggest that they would have simply lacked the equipment for talking fast. They did not have the right-shaped lips or tongue to nip the flow of air into crisp sequences of syllables. The shallowness of their throats would also have greatly limited the range of vowel sounds they could make. It has been guessed that the articulation rate of a Neanderthal may have been as much as five to ten times slower than a human.

The actual changes needed to evolve a better vocal tract were all relatively slight: a resetting of the genetic growth clocks to arch the palate, thicken the tongue, jut the chin, refine the lips, and drop the voice-box lower down the throat. So, as Lieberman argues, the fact that the Neanderthals did not make such changes was a sign they had no use for them. Equally, the fact that humans did make the changes necessary for rapid speech appears proof that we were also the first to want to speak in sentence-length bursts of words.

Thus far, the general story is well enough accepted. Yet then comes the puzzle of how grammatical speech made a difference. Why was there the sudden explosion in art and creativity which heralded the arrival of the abstract-thinking, self-aware human mind?

To answer this, we have to be able to say something more about what distinguishes the mental life of an animal from that of a human. In one way, the difference is terribly simple. Animals are locked into the present tense. As numerous philosophers from John Locke to Ludwig Wittgenstein have remarked, animals live entirely in the here and now, their minds responding to whatever is currently going on around them, or to whatever urges happen to be welling up from their bodies. Humans, by contrast, have broken free of this tyranny of the moment. We have a consciousness that can wander about, thinking back to review memories from our past, or thinking forwards to imagine how life might be in the future. We can even take a step away from ourselves to contemplate the fact of our own conscious existence.

Animals have awareness, feelings, associations, anticipations, memories, self-control – seemingly the full complement of parts needed to make a mind. However, all these mental abilities are tied to the moment. For example, an animal can recognise but not recollect. If a monkey sees a banana or a picture of a ship, the mapping of the sensation will surge upwards through the hierarchies of the cortex, stirring a sense of familiarity and meaningfulness in the memory areas of the temporal lobe. And yet there is no reason to think that a monkey ever sits around mulling over the story of its life.

Of course, an animal as intelligent as a monkey can make associations, and this will seem to bring memories back. Feeling a pang of hunger might trigger a memory for the whereabouts of a tree of ripe figs; being placed in an experimenter's box would rouse expectations about the kind of events likely to follow. However, the monkey's response would still be prompted by outside cues, the pains, urges, and alarms of the body being something 'outside' as far as the brain is concerned. There is no hint that a monkey has the independence of thought to be able to indulge in fond reminiscences about the fig trees it used to clamber up as a youth, or to wonder whether it might face some new kind of experimental test for a change today. Wittgenstein summed this up neatly when he commented: 'We say a dog is afraid his master will beat him, but not he is afraid his master will beat him tomorrow. Why not?'

So the mind of an animal is caught in the flow of life. It is always adopting an intelligent view of the moment, but has no freedom to take flight. Language would have changed all this. By taking a tool developed for organising their social world and turning it around to organise what went on inside their own heads, our recent ancestors would have been able to interrupt life's ceaseless tide and begin directing their thoughts elsewhere. Words have the power to spark images, so by getting into the habit of talking to ourselves, using a self-questioning, self-prompting inner voice, we are able to transport ourselves to imaginary viewpoints. We can roam our memory banks to relive past moments or spin fantasy images. The invention of articulate, grammar-driven speech was also the invention of articulate, logic-driven thought.

The idea that speech must be responsible for the special mental abilities of humans is also a very old one. It was present in the writings of Enlightenment philosophers such as Locke and Thomas Hobbes. Charles Darwin and many of the Victorian evolutionists speculated about the idea. Then, in this century, a strong case was made by the Russian psychologist Lev Vygotsky and his colleague, Alexander Luria. Working in the 1930s, Vygotsky observed children closely and realised that the development of their capacity for self-awareness, memory, and thought went hand in hand with their learning of speech. From this, he argued that such abilities were not innate – wired into the genes – but habits which every child had to pick up during the first seven or eight years of life.

However, while the suggestion has been frequently advanced, it has been just as easily dismissed. For most people, the claim that language could make the human mind falls at the first hurdle. After all, it seemed only logical that before our hominid ancestors could have felt the urge to speak, and so put a premium on the development of their vocal equipment, they must have already had something on their minds they wanted to say. The special abilities of humans had to precede grammar and words rather than result from them. Speech would be the outward sign of what was an inner revolution.

The belief that words can only clothe thought has seemed so axiomatic that until a rediscovery of Vygotsky's work in the late 1980s, it was exceptional to find a recent Western philosophy or psychology text that even mentioned the possibility that language might make a difference. And in truth, introspection seemed to suggest that words are indeed mostly secondary. Sometimes speech clearly appears to lead our thoughts – as when we feel that we did not know what our opinion was going to be until we heard ourselves expressing one – but at most other times, thought definitely seems to come first. There is at least an inkling of what we intend to say before the words start spilling through consciousness or out of our mouths. So if this is the way we are now, then it seems only sensible to think that the same was true for our ancestors, that they must have been thinkers before they were speakers.

The realisation that the brain is a dynamic system – a fluid, adapting bag of circuits – of course changes the whole conception of the problem. There can be no chicken-and-egg dilemma because simple ideas of cause and effect do not apply. The brain is constructed according to a genetic blueprint, but it is a blueprint that codes more for degrees of plasticity and bursts of growth than actual pathways or processing circuits. As said, even the way the brain sees and hears is something that is largely learnt – or at least self-organises to fit. There is an unbending level of neural structure – some basic proportions of cells types, cell numbers, and cell branching patterns – needed to give the brain its correct shape. But the pathways are responsive to the world they find themselves in. It takes months for areas like the IT cortex to become even partially organised in a human child. Then, all our lives, we are adding extra memories, further habits of sensory processing, that allow us to see people, places, and events as familiar or novel. The circuits of the brain are tailored by experience, and although they may eventually accumulate a thick sludge of habits, they never actually settle.

What this plasticity means is that the pathways of the brain are open to being colonised by culture. Language, and the habits of thought which it supports, may have developed first in the social sphere, being the product of a cultural rather than a biological evolution. But because nerve tissue is plastic, words and grammar would immediately grow into the brain. Every tiny advance in the usage of one would show up immediately as circuit changes in the other. Speech would not be a small add-on, a symbol-processing module strapped belatedly to the side of the brain. Instead, as scanner images have so graphically revealed, the structure of language would penetrate just about every part of the brain.

The early PET studies at Washington University in St Louis, it will be remembered, dazzled those who expected language processing to be confined to the two classic coin-sized speech centres, Broca's in the frontal cortex and Wernicke's in the bend of the temporal lobe. The Washington experiments showed stains of activity all through the higher cortex, and also down in the thalamus, basal ganglia, and cerebellum. As new generations of scanners came into service with better resolution, language researchers found more and more of the brain lighting up. The traditional language centres handled core tasks such as chaining words into grammatically correct sequences and storing representations of the sounds of words – the aural mappings used to drive the muscles of the throat and mouth. But, as might be expected, the meaning of any word – the associations and imagery it might spark – were spread right across the cortex.

If the word was the name of a tool, such as a screwdriver or hammer, then thinking about it would cause a stir of activity in the motor areas of the frontal cortex. The verbal symbol 'screwdriver' was connected to memories of what it was actually like to handle such an instrument. Likewise, if the word was the name of a colour, an area around V4 would light up. And if a person was asked to think about a yellow screwdriver, then both areas would fire. Like any kind of memory, word meaning was something that was distributed back across all the mapping areas that would have to process aspects of what that word stood for. This did not turn all sensory and motor areas into outposts of the language hierarchy, but it does show how something socially invented could come to texture the whole processing landscape of the brain.

In fact, giving the plasticity story another twist, the already plastic cortex of the mammalian brain – with its extra dollop of uncommitted circuitry in the form of the ever-swelling prefrontal cortex – has taken on yet new levels of plasticity in humans so as to allow language to have its maximum impact. As is obvious, human children are born much more helpless than the animals of any other species. It takes a year before human infants show much ability even just to move about and handle objects with any skill. The reason for this is that the maturation of the brain – the insulation of its many paths with myelin to make the connections fast and efficient – is delayed quite abnormally, particularly in the cortex areas critical to language and the higher levels of thought. The main language areas do not really begin myelinating until the second year of life, and it takes six or seven years before the process approaches completion – which is why learning a second language is easy for an infant but hard for adults. In other regions of the brain, such as the prefrontal lobes and the area around the hippocampus, the delay is even more dramatic. These parts do not reach full maturation until the late teens or early twenties.

The slowness of the brain's maturation is an immense evolutionary burden. Children need total care for the first few years of their life, which would have made the caring, co-operative lifestyle of our hominid ancestors all the more essential. But it also creates a wide window of opportunity in which children can be exposed to language and can soak up their culture's rhythms of speech and thought. So our genes have set us up with not only a large brain, but a brain that patiently awaits its coming brush with culture.

What this means is that the evolution of a language system was always something that was socially scaffolded. The advances were not individual – a lucky genetic break that allowed some distant ancestor to start expressing the many private thoughts he or she had been having – but collective. It would be the group that developed new habits of sentence forming or new words. Then, because of the extreme plasticity of the infant brain, the next generation would find themselves doing with ease what had been moderately difficult for their parents. Ways of thought that were once alien would become second nature for those who grew up with them, a situation that might be repeating itself today with the operation of video recorders and home computers.

There is still a question about whether the human brain has made any specific genetic adaptation for grammar – whether it has a wired-in motor template for generating subject–verb–object-shaped sentences as so many linguists, especially the towering figure of Noam Chomsky, the MIT theorist who was one of the major inspirations of the cognitive science movement, have argued. The more distributed-looking the language system becomes according to scanning studies, and the more that is discovered about the way children can pick up the rules of grammar from simple statistical inference – from noticing patterns and regularities – the less likely it seems that there is much that needs to be hardwired.

As said, most of the genetic changes are permissive. Human brains are four times bigger than an ape's, they myelinate many years later, and they were probably already lateralised and skilled at sequencing motor operations from several million years of tool use. Once some vocal equipment changes had been made to allow faster, more distinct articulation, not a lot else seems that necessary. Babies do seem to possess some useful instinctive behaviours to get them on the path to learning to talk, such as a keenness to experiment with babbling noises. Babies also have an instinct for conversational turn-taking. They will alternate between periods of babbling and listening, and they will naturally follow another's gaze – a trick which means they can more quickly learn the connection between an object and its name. So there are a host of genetically simple adaptations which make the infant brain fertile soil for the establishment of language. But it is the human brain's sheer size and plasticity – its capacity to respond to a culturally imposed need and to develop some language-textured circuitry – that seems the main advance.

The dynamism of the brain's design means that our ancestors did not have to be thinkers before they were speakers. Cultural change and genetic change could go hand in hand. So exactly how does language make a difference? As Vygotsky and others have argued, just having words would have started the process of breaking us free of the moment. A word is no more than a puff of air, a growl in the throat. It is a token. But saying a word has the effect of grabbing the mind of a listener and taking it to some specific spot within their memories.

The animal brain is tied to the present by its very dynamism. The evolutionary story of the brain has been one of an ever-increasing ability to wrap itself around the shape of a moment. The brain is a living memory surface, tuned by learning and experience, but this means that it has no such thing as memories in the digital sense used by computer scientists. A computer memory is made up of discrete bits that can be picked up and shuffled about, but in the brain, memory is embedded. It is the processing landscape. So while an animal can use its circuits to react – to bear down on each moment with the full weight of a lifetime of experience – it has no mechanisms to fetch and replay arbitrary chunks of data.

Words, however, allow us to treat our brains as if they actually were digital warehouses. We cannot shift the data – that always has to stay in place – but we can use words to trick the brain into making a shift in its point of view, to open up an angle into an area of experience. Hearing a word like 'rhinoceros', 'camel', or 'cat' will cause an adequately trained brain to react as if it had just seen the real thing – or, to be more accurate, to react with the anticipation of seeing the real thing.

The power of words is all the greater because we can slap a label on anything. Large or small, simple or complex, it takes the same effort to speak its name. A word can take us as quickly to the idea of the universe as the idea of a rhinoceros or the colour blue. We can even give names to abstractions like love and honour. This gives the human mind another kind of freedom. The animal mind is not only trapped in the present tense, it is also stuck with a concrete level of categorisation. A monkey can recognise a banana or a fellow troop member when it sees one, but it does not classify these experiences in abstract terms, thinking about them as examples of a fruit or some relation like an uncle. An animal develops natural categories of processing, ones that lump together experiences in terms of their basic sensory qualities, but words allow humans to create artificial categories for organising and exploring memory. Things which might otherwise be very difficult to think about, such as bravery, outer space, or family relationships, get given a convenient handle. Using these, we build our way out of the world of immediate sensation and move our consciousness into a realm of culturally evolved thought. We can reach points of view that are only possible because a symbol serves to bind the meaning together.

Words are a critical first step towards taking control of the brain's memory landscape, making it possible to wrest the focus of attention away from the events of the moment. However, grammar would have been needed for any real breakthrough. Simply possessing a proto-language based on single-word utterances would not have made that much difference to the mental abilities of our distant ancestors. An isolated word could be used to force a big shift in the thoughts of others, causing them to focus on something like firewood or big game which had been far from their minds. But the speaker would still be left trapped in the here and now. The urge to utter a word would arise only because of some immediate need or circumstance the person was experiencing. There would be no mechanism for calling up a word unless something happened to jog its use. However, developing a grammar with a driving reductionist logic would instantly have put an engine into our thoughts.

As has been seen, the language centres hang off the prefrontal lobes to form a third flank of the frontal motor hierarchy. The focus of every moment is mapped within the prefrontal cortex's broad expanse and then ripples back down through the motor, orientation, and speech output areas, so sparking ideas about suitable reactions. This means that the itch to say something in response to each fresh shift in focus is automatic. Once the reductionist logic of grammar has taken root in our brains, sheer force of habit will make us look at each moment and search for the way it can best be broken down into a story with a subject, verb, and object.

Of course, as with any stirring impulse, we can always cut short the urge to speak – or, more likely, we will be interrupted, and will have moved on to something else before the urge has got very far. But having grammar means that every instant of our lives is viewed through a reductionist prism. We are always just about to launch into a sentence given half a chance. Our thoughts are always just about to head off somewhere. And then as soon as we do let ourselves speak – or imagine the same words through our inner voice – our minds will quickly be carried to far places. A sentence may start out as a comment about the moment, but it will immediately open up its own chain of thought. The words we have just used will trigger their own associations and images, creating the potential focus for a fresh sentence. The act of sentence-making feeds on itself to draw us into an entirely private world of thoughts about thoughts.

For example, seeing the glassy eye of a fish staring out at us on a trip to the supermarket might spark an inner comment about the fish not looking too fresh; its opaque gaze would be recognised as significant and our words would then serve to make explicit a link with the idea of freshness. Throwing the emphasis on the idea of freshness might next prompt an inner puzzling about how this fish got to the shop. Before we know it, our minds would be filled with images of trawler boats tossing in the waves, or crates of glistening fish being slung across market floors, together with some inner comment about it taking days or even weeks to get back to shore to unload. Do they freeze the fish solid, or just chill them? Borne along by the trick of grammar, rapidly we will find that our thoughts have travelled a long way from the here and now of standing by a supermarket fish counter.

Importantly, we do not actually have to voice full sentences for our thoughts to be pushed along like this. One of the awkward points faced by those who, like Vygotsky, wanted to claim that speech drives thought was that too often our inner dialogue seems sketchy at best. A lot of the time it is no more than a string of half-formed phrases, or even just the feeling of being about to say something. When we strike some difficult moment in a train of thought, we usually do seem to try to prompt ourselves with fully articulated questions and suggestions. We may even talk out loud to ourselves or attempt to clarify our ideas by putting them down on paper. But mostly our thoughts seem to consist of a confused jumble of inner mutterings and fleeting glimpses of imagery. Our minds appear to move along too fast to be dependent on the laborious probings of self-directed speech.

However, as has been seen, the brain does not have to have everything 'in consciousness' to do useful work. A speech intention must go through many levels of decomposition before it arrives at the primary motor surfaces as a fully fledged set of muscle instructions ready to drive the mouth and throat. Like a spontaneous flexing of the finger or any other motor action, it may take half a second or more for the brain to gear up to speak a sentence. As well as selecting the actual words, the brain has to decide the pattern of emotional emphasis and plan for any face or hand gestures that will accompany the message. Even the lungs have to be instructed to take a breath proportional to the expected length of the sentence.

Yet the part of the speech act that is crucial to thought – the extraction of a logical link from the current focus of the moment – actually happens very early in the process. So just getting to the stage of having a first inkling of what we want to say – an idea of what should be the subject, verb, and object – will set our thoughts moving along quite nicely. Going the whole way and voicing the fully formed sentence might well produce a greater impact on our thinking. The more clearly we state an assertion, the easier it is for us to notice its implications or shortcomings. But long before this stage, the key step of establishing a logic connection and rousing the relevant areas of memory will already have been achieved.

Exactly the same is true of the mental imagery that is the other half of the equation. Given time, our brains can respond to a word like 'rhinoceros' with a whole succession of rhinoceros-related views, each fleeting mental picture taking about half a second to generate. But long before a vivid, fully fledged sensory experience comes together, our anticipatory state will already be doing work. Merely having the sense of being in the right spot to start seeing rhinoceros images would be enough for us to feel we understand the meaning of the word. We do not need to unpack the mental pictures that go with each word of a sentence for its meaning to carry us along. Letting the imagery blossom will always give a sentence greater impact. But we only tend to pause to allow this to happen at critical stages in a train of thought.

As usual, the brain prefers to operate at the lowest, most habitual level that it can. Like any other kind of action, much of our thinking is actually rather stereotyped. The things we say and the imagery we rouse will tend to repeat what went through our heads the last time we faced a similar situation, so unless we have good reason consciously to want to check the links in a chain of thought, there would be no need to slow down and make each step explicit. Yet what matters is that the full structure of grammatical speech and reductionist thought is always there at the back of whatever we do. We will either be using it to deal with the moment, or it will have been used successfully to deal with an almost identical moment in the past, establishing the necessary connections to do the same job swiftly and pre-consciously. So even when speech does not seem to be acting overtly, our thoughts will still be moved along on an 'as if' basis – as if each word and image were making the individual, full-blown trip through the spotlight of consciousness.

Grammatical speech put a motor into human thought, allowing our minds to break free of the tyranny of the present. We could then start going places in both our imagination and our memories. Our brains are still basically the same as an animal's, always reacting to whatever has just been placed in front of them, but with words we can start feeding our brain fake moments – pseudo angles in. And this control over our state of mental representation gave us two new powers in particular: recollective memory and self-awareness.

Memory is a confusing term because it is used to cover such a wide spectrum of the brain's activity. Any measurable change in the brain's organisation – even the fleeting pattern of a working-memory state – is seen as a type of memory. But what we really mean by memory in humans is recollection. The animal brain has memory in that it can accumulate pathway changes and sensory habits; however, as said, these circuits only show themselves during the processing of a moment. A patch of 'memory' will allow the brain to make recognition matches and flesh out an experience with a halo of associative meaning, but an animal has no independent mechanism for returning to moments in its past.

Language acts in several ways to make recollection possible in humans. At the most obvious level, we can use self-questioning to steer our minds back to some occasion. For instance, we just have to ask ourselves what we had for breakfast or what it was like back in our school days for the words immediately to open up an angle into an area of past experience. The words will stir an anticipatory sense of what it would be like to be back there, living those same moments again. But we do not need to use such overt questioning for memories to flow. A lot of the time images will be jogged free simply as a result of a train of thought. A moment from our past will come to mind because some association has been struck. Thinking about breakfast may remind us of the time when we ate out under an awning at a hotel while on holiday. But the point is that words are needed for our brains to wander away from the moment in the first place. And if we wanted to, the potential for control is always there. We could say to ourselves: 'Stop day-dreaming. It's this morning's breakfast we want to get back to.'

Naturally, the accuracy and vividness of our recollections would depend on how crisply the moments had been trapped by the hippocampus at the time. Some episodes from our past, like accidents and embarrassing situations, are caught with a flashbulb crispness because of their emotion-laden impact. But psychologists analysing eyewitness accounts of staged dramas like bank robberies find that much of any remembered scene tends to be an invention. When asked to report details of a robbery, subjects inserted facts that they felt ought to be in their memories. And leading questions about what one of the robbers might have been wearing frequently implanted the conviction that such clothing was actually witnessed. In other words, there is not much difference between our imaginations and our memories except that with one, we know when we are inventing. With our recollections, the gist is usually accurate enough to serve our purposes, but the background details will be mostly a generalisation – what we might reasonably expect the scene to have looked like based on the blurred recall of many similar experiences.

If recollection is based on the power of language, then that other hallmark of the human mind, self-awareness, is based in turn on the ability to recall past states of awareness. Animals live with their noses pressed hard up against life. They are always in the thick of the moment and so have little chance to contemplate the fact of their own existence. But with words comes the ability to step back and start appreciating ourselves as beings enjoying states of representation.

Broadly speaking, the self-awareness of humans has two elements: the first is the act of being self-aware, of adopting a retrospective stance to each moment and taking notice of the fact that things are happening in our heads; the second is having knowledge of being a self. We learn to form a detached view of ourselves as a mental being.

Neither of these is natural to the brain. Even having an introspective slant to awareness is probably absent in animals. As argued, the brain has no evolutionary use for contemplation. It exists for the representation of reactions rather than sensations. Even the brain's fixing of memories is really a prospective rather than a retrospective step. The sole reason for the brain trapping the significant aspect of a moment is so as to save a viewpoint or habit which might improve its processing of future moments. The hippocampus is not taking snapshots to create a diary of where consciousness has been, it is merely hanging on to a potentially useful pattern of information for the minutes, hours, and days it takes for it to become built into the processing circuitry of the brain.

It may be a subtle point, but animals do not see into a moment, rather they look out from it. Subjectively, the animal brain would always be facing forward, focused not on where the latest shift in viewpoint has come from, but where it is heading. Rather than feeling like an observer or a passenger, an animal would have a feeling of simply being the vehicle, of doing the journey. This suggests that even our feeling of being there during a moment, observing, supervising, and taking decisions, is a habit grafted onto consciousness.

Language does several things to foster the human trick of self-consciousness. Simply being forced to speak grammatically – to break the world down into a story of actors and their actions – would have nurtured the idea of being a self in early humans. Sentences like 'I did this' or 'You must do that' would focus attention on the fact of there being an I and a you. The power that self-directed speech gives over thought and memory could then be used to think about the experiences of this newly discovered self. But the habit of being self-aware, of looking inwards and noticing, would not have been just a happy accident. Instead, it would have been developed to serve a social purpose. It was not that our hunter-gatherer ancestors wanted to create a tribe of moody, philosophising individualists; their lifestyle depended upon an ability to share and co-operate. So the main reason for instilling a habit of watching the self would be to make members of the group self-policing.

Animals are creatures of impulse. Even for a highly intelligent, highly social species such as the chimpanzee, self-restraint is difficult. Any co-operation or sharing of food is a fragile affair. When a group of chimps catches a colobus monkey, most of the spoils go to the strongest. And, certainly, it is hard to imagine a chimp stumbling on a bunch of bananas only to scoop it up and rush off excitedly to divide it with the rest of its troop. But language allowed humans to internalise a socially developed framework of control. Words stand for knots of ideas, and the ability to encode abstract social qualities such as duty, patience, kinship, fairness, and conscience would have been even more valuable to early humans than having names for talking about common objects like deer, stone axes, or fires. All the virtues needed by a society could be represented in the vocabulary passed from one generation to the next, so that children would learn the words, then learn the complex social attitudes that went with the words.

The next step after knowing what to do has to be the ability to tell ourselves to do it. By encouraging young children to get into the habit of looking inwards, they can be taught to guard against their anti-social impulses. The temptation to sneak food, be disrespectful, show fear in hunting, or leave the collecting of firewood until another day, can be resisted. The mind would become a supervised place, with the self acting as an outpost for social thinking. The teaching of this kind of self-awareness is obvious whenever parents are heard telling their children to think about what they are doing, or asking how they would like it if the same thing were done to them.

Of course, the social control over behaviour would never be perfect. However, anthropologists who have studied the few remaining modern-day hunter-gatherer tribes often remark how harmoniously they live. Rules are bent and voices often raised, yet self-awareness allows for a skilfully negotiated trade-off between the needs of the individual and the needs of the group. The socialisation of the brain was not about imposing a rigid, unquestioning pattern of behaviour, but taking the opportunity presented by language to make humans even more socially attuned.

So society educates us to look backwards through the moment and take responsibility for what we see. Even the normal definition of the word 'consciousness' is synonymous with introspection and control, with being in charge as things are happening. The idea of degrees of awareness or delays in brain processing makes us uncomfortable because we are supposed to be the 'I' that watches the tennis ball on to the strings, or makes the spontaneous decision to flex a finger. It is important that we believe our consciousness to be instant and all-seeing so that society can hold us accountable for our slips. But it is only once we have gone inside the making of a single moment of consciousness and seen how much is involved – how much planning, sorting, escalating, and reacting – that our conscious selves can really begin to come into some kind of focus.

CHAPTER FOURTEEN

Answering the Hard Question

At consciousness conferences in the 1990s, one speaker stuck out for his dress sense alone. David Chalmers, a boyish Australian mathematician turned philosopher, looked like his day job was as a singer in a heavy metal rock band. With a fluffed-out mane of hair, T-shirt, scuffed tennis shoes and calculatingly unmatched towelling socks – one red, one green – Chalmers seemed to have made a wrong turning on his way to the auditorium. But he was always applauded because there was a refreshing directness to what he had to say.

Chalmers stood for philosophy's backlash. In the late 1980s, the trend within philosophy had been to start taking neurology seriously. Inspired by the very successful examples of Daniel Dennett and Patricia Churchland, a gang of younger philosophers was anxious to position themselves as commentators ready to ride the wave of scientific discovery promised by the 1990s. If neuroscientists proved hesitant about drawing conclusions from their work, then philosophers could make a name by stating the obvious for them.

However, Chalmers would have none of this cosy camp following. Yes, he said, science was making steady progress on understanding all sorts of brain processes such as memory, imagery, and attention, but these were just the easy questions about consciousness. Nothing that was happening in science – or was even likely to happen – had made a dent on the hard question, the basic problem of how a lump of flesh like the brain could light up with the inner glow of subjective experience. All the neurological detail in the world would not give a reason for consciousness having the feeling of being a something rather than a nothing.

Chalmers argued that a robot could be programmed to have reactions to the world – to go through the whole rigmarole of representing, attending, and responding – but the calculations would be performed blindly. It could map the co-ordinates of the colour red, the smell of coffee, or a pang of longing, without ever needing to experience them in a conscious way. But for us, red has an inescapable and apparently irreducible mental quality about it – the sense of redness which makes it quite distinct from our experience of blue or yellow. If we happened to be unlucky enough to be born with only a black-and-white sense of vision, then no amount of careful description could convey what the mysterious quality of redness was like. For this reason, consciousness as it is experienced from the inside looks to lie forever beyond the grasp of scientific theory.

There was certainly nothing original about this point. However, the youthful vigour with which Chalmers set about reaffirming the special status of human consciousness struck a chord. As one of his fellow philosophers exclaimed joyfully during a conference break, just as neuroscientists were exhibiting signs of a stirring confidence – thinking that perhaps with scanners, neural networks, and everything else, a theory of mind might finally be in sight – along comes Chalmers to slap them back down again, reminding everyone that the winning post still looks impossibly far off. Sure, given another couple of centuries, science might understand the brain down to its very last synapse and oscillation, but this would not bring us any closer to being able to answer the hard question of why red feels red, and not blue.

In the end, Chalmers is probably right about the ineffable nature of conscious experience – except for the wrong reasons. As Chalmers makes clear, philosophers see the problem in terms of a causal gap. What they are looking for science to provide is the crucial step that turns a physical state into a mental state. Furthermore, this explanation should not be just a dry tale about some brain process. A successful theory of consciousness would have to be able to tell us exactly why things feel the way they do. The explanation has to get us inside the experience.

Yet posed in this manner, Chalmers's hard question is riddled with reductionist assumptions. First of all, the way that the hard question is phrased makes the assumption that consciousness is a bounded state, that there is one definable 'thing' sitting in the brain waiting to be explained. The transition from the realm of the physical to the mental is seen as being digitally crisp – there is either a state of subjective experience or there is not – instead of accepting that consciousness might be the gradual and variable result of brain activity. There is something we can call consciousness, but it has a dynamic continuity with things that we think of as being non-conscious, such as pre-conscious habits, the sprouting of new memory connections, and even the development of the brain over evolutionary timescales.

And then there is the matching assumption that a successful answer should feel instantly and completely understood. It might take us a while to 'get' the explanation – to turn things around in our heads until the detail snapped into focus – but when we do, the whole story would be told. The hard question implies a one-step answer and so, equally, our state of comprehension should click digitally from a state of not knowing to knowing. But really, we do not expect a theory about anything – whether it is car engines or the weather – to take us inside the experience of being those systems. When mechanics say they understand a motor car, what they mean is that they feel comfortably placed to start thinking about any problem connected with its running. They feel oriented with an angle in on an immense storehouse of personal knowledge, not that they feel immediately conscious of every detail they have ever learnt. The way that the human mind 'does' focal awareness means that we can never have the whole of anything burning brightly in our heads. What feels sharply understood is how to proceed if we want to explore our knowledge.

The same has to be true of an explanation of consciousness itself. What counts is the sense of being well positioned if we want to move about the subject, rather than having the whole story of the mind filling our heads at once. This does not mean that science cannot begin an explanation about the nature of subjective experience – of why things feel the way they do – just that it cannot finish it.

Reductionism puts all the emphasis on getting to the end of the journey, on getting past a lot of minor detail before breaking through to make the final spectacular causal connection. It wants to discover the brain's neural code or the prefrontal area that harbours the sense of self. The belief is that there is a single gimmick, or no more than a few gimmicks, waiting at the conclusion of the hunt. But the dynamic view is that it is how you start an explanation that matters most. You have to get yourself to a vantage point that then lets you look back across the grand sweep of everything. So it is not really so much a question of explaining consciousness as becoming familiar with its ways. As something that the brain creates dynamically afresh during every moment, consciousness comes in many shades and varieties; it is whatever the brain has managed or needed to do during the past half-second or so. There are an infinity of types of conscious moments that a brain can have, so it is getting things the wrong way round to pick an example of a canonical experience – such as the raw feeling of seeing red – and then ask science to build a bridge of theory to reach it.

The reason why the philosopher's favourite example, the colour red, seems to stand as the ultimate challenge to a theory of consciousness is that it is as near to a structureless and discrete brain state as we can imagine. With other mental activities, such as picturing a rhinoceros or hitting a tennis ball, we can at least see a link between the kind of activity going on in our brains and the form of subjective experience we are having. Knowing that our image of a rhinoceros is based upon a state of anticipation – an attempt to predict what it would be like to see a rhinoceros during the next moment of processing, with all the restless mix of concrete detail and open-ended association that goes with such a state of sensory priming – does seem to begin to explain something about the actual experience of having such a mental image. However, the experience of redness appears to have no features, no time-course, no sense of evolving structure. It just has a phenomenal quality – the feeling of there being something which it is like to be seeing red.

And yet any claims for the purity of this experience should immediately be questioned. To start with, there is the trivial point that there are an endless variety of shades of red that we can experience, with red shading into other colours like orange, brown, and purple at its margins. The kind of red we think we are seeing can also be affected by the texture of the surface, the evenness of the illumination, and many other physical factors. So exactly the same wavelength of light can appear harsh or soft, rich or flat, depending on the viewing conditions. But even if we were talking about an experimentally controlled experience in which all the natural variability of life was removed by seating subjects in front of a screen lit by a single specified frequency of light, there could never be a canonical act of seeing the colour red. No two moments of consciousness will ever be precisely the same because, for a start, the history of events leading up to them will always be different.

However, more tellingly we have to ask what it is we are actually doing when we try to fix our minds on an experience of redness. The brain is designed to handle as much of a moment as it can at a habitual, pre-conscious level, then orient itself to whatever aspect of the moment has turned out to be the most novel or significant. So simply asking ourselves to focus on the product of a low-level sensory habit – the mapping of a colour response – is a rather unnatural thing to do. Of course, we can do it because we have the words to organise our actions. We can tell ourselves to sit in a chair and concentrate on paying introspective attention to an experience of redness. An animal would look through the redness of an illuminated screen – its brain circuits would map the colour, but the focus of its thoughts would be on escape, boredom, the reward it should get for staring at the screen like the experimenter asked, or anything else which might present some degree of novelty about the moment. Only human beings would be able to organise themselves so as to be looking at an experience of pure colour.

Even then, while we can force ourselves to attend to our performance of a sensory habit, this does not lead to a digitally isolated state of awareness. We can push all other thoughts and sensations into the background for a moment, suppressing everything except for a highlighted sense of redness, yet this is only ever a pseudo state of isolation. Even suppressed, the activity is still there, giving shape to our overall state of representation. So seeing red is not an elemental experience. It is, in fact, a highly skilled and artificial use of the brain. We have to make an advance decision to ignore everything else that would naturally seem more important during a moment, such as an itch on the neck or a whole host of stray thoughts. It is the same as with Libet's spontaneous finger-flexing experiment. What happens during any one instant cannot be disconnected from all the planning that went before. If we manage to make an action or sensation appear nakedly simple, it is just because we are concealing all the effort that went into setting up the moment. The shape of the vessel is still expressing itself even if we are doing our best to make it appear hidden.

Finally, of course, as anyone who has ever tried to focus on a pure sensory experience will know, it is actually impossible to keep the background traffic of thought at bay for long. The inherent restlessness of the brain means that it will want to shift the viewpoint at least every half-second. So while our eyes may remain glued to the red screen, we will soon find ourselves looking through the experience. Our attention will be wandering away, and we may have to take a moment to regroup, to focus our minds again with the necessary purity. Maybe after a lifetime of training in Zen meditation, we might develop the ability to be able to focus on a patch of colour for lengthy periods without any other thought appearing to cross our minds, every threatened intrusion being stifled at a pre-conscious level by well-drilled habit. But the very fact that it would take so much time and practice to develop such a mental skill is simply further proof that there is much more to the act of seeing red than meets the philosopher's eye.

This does not solve the problem of redness. Even if our focus on a colour is fleeting and surrounded by an insistent buzz of thoughts, the experience still remains subjectively distinctive. Redness is not like greenness or blueness, just as sweetness is not a kind of sourness, or the ring of a bell like a kind of growl. But again, the point is that although we can talk about red as a category of experience – we can use words to snip the idea of redness free of its context – in real life, the experiencing of red is always part of a greater field of activity. The act of seeing red results from having taken a particular angle back through a wider state of representation, so it is continuous with everything else being mapped in the brain at the same moment, with any preparations for seeing red which were made in the time leading up to the moment, and even with the history of childhood development and species evolution that built the circuits used to map the experience of red.

It might seem a bit extreme to start throwing even our evolutionary history into the mix, but logically, the connection is there. Reductionism seeks some ultimate bedrock on which to build a theory of the mind. The point about the dynamic approach is that there is no final stopping place. All that happens as you chase the story outwards from the focus of the moment is that the influences which created the focus become more and more diffuse. So how far you pursue an explanation is a practical decision rather than a theoretical one. It might be enough just to consider the mapping activity taking place over the course of half a second or so of neural competition. This would cover most of what seems important in understanding how the brain represents redness, if not why this activity actually feels red.

However, philosophers like Chalmers are questioning what science can achieve in principle. And in principle, the forces that whip up a particular whorl of mental activity dissolve right back into the fabric of the universe itself. Where does neural activity or evolutionary history end and chemistry or the history of the Earth take over? There can be no winning post for a theory which attempts to explain the nature of subjective experience, or even just a dry theory about brain processing. Instead, for either kind of theory, science has to organise itself to find the best starting point. The aim must be to choose an angle in that explains as much as possible, because explaining everything is, in principle, impossible. The hard question can be partially answered, but never completely answered.

In fact, turning the reductionist position completely on its head, a successful theory of brain processing should lead the way towards a greater intensity of subjective experience. If introspection is a learnt skill, a language-based habit rather than an innate ability, then we can be pretty sure we have only a relatively poor focus on our interior states. The self-awareness granted us by 100,000 years of cultural evolution has been shaped to serve a social purpose. It is a socially useful view rather than a complete view. So good science should be able to arm us with a more objective framework of concepts and vocabulary with which to observe ourselves. Equipped with a better theoretical understanding, we will be able to focus more sharply on our mental states – for example, noticing how much is being pushed away into the background as we try to isolate a pure experience of redness. Instead of science being a passive tool of explanation, it would become the active vehicle of our own subjective explorations, deepening our sense of what it means to be aware.

This book has tried to track what would be a fundamental change in the science of the mind: a shift from reductionism to dynamism. In keeping with the spirit of this change, the aim has been to evolve a view. By sticking close to two key themes – the revolution promised by scanning technology and the puzzle of Libet's half-second – the hope was that the underlying issues would be made clear. Brain scanning takes us to the current front-line of mind science, revealing what has been happening as the confident assertions of cognitive science have come up against the harsh realities of neurology. The treatment of Libet's half-second results likewise tells us much about both the cultural setting in which current research is taking place and the basic facts with which a theory of consciousness must deal. It is plain from what has been seen that for science to take a dynamic approach to the mind the changes will have to be wholesale, not just at the level of theory-building, but with data collection and the very organisation of the field as well. Everything about the mind sciences must change to cut with the grain.

The difference in the style of theorising required is radical enough. Reductionism wants to identify an object and then break it down into its components. But there is no bounded object in a dynamic system. Like a whorl in a stream, consciousness is continuous with its background. There is a structure to be explained, but the background, the context, has to be included in the explanation. This need to be able to account for both the structure and the background means that some careful choices have to be made over the 'aperture' of a theory. Theories have to span sufficient space and time to catch the evolutionary logic of the brain at work.

They also have to be able to talk about the top-down needs shaping the system, as well as providing the bottom-up story of what the 'components' are doing. In practice, this means there must be a micro-structure explanation of consciousness – one that goes inside the moment to track the unfolding of a single processing cycle. But this must be balanced by a macro-structure tale. While the events within a moment represent the peak of the action, the most volatile and creative level of adjustment in the brain's state of organisation, there is also the story of how the brain builds its habits, memories, and processing pathways.

Finally, no matter what aspect of the brain is being considered, there will need to be a single common entry point for all theorising. If the human mind is culturally, as well as biologically, evolved, then this is the position from which any explanation must start. Even if we are talking about something as apparently elemental as the sight of redness or a spontaneous urge to flex a finger, the first thing to be considered is the social context. We have to ask how much is the mental activity a skill buttressed by language and cultural habit? The same goes for discussions about supposed faculties of the mind such as memory, attention, and imagery. The story has to start with a recognition that two kinds of shaping influences are at work when we are talking about the abilities of the human brain. This particularly affects extrapolations made from the study of the animal mind. Memory, attention, and everything else can be studied in laboratory animals, but only once researchers have made it clear what part of these abilities they feel to be strictly biological.

If theories have to have a different look, so does data. With data, there is again the question of finding the appropriate aperture. The design of experiments, the design of the recording technology, and even the chosen units of measurement have to be able to cope with the fluid dynamism of the information flows they are hoping to sample. And then the actual organisation of the mind sciences has to reflect a move away from reductionism. The traditional approach has been to slice the cake finely, creating hundreds of compartmentalised specialities in the belief that by getting lots of isolated bits of science right, the parts would automatically fall into place. However, what the field needs is some broader sense of direction, and all the academic machinery of conferences, journals, and research postings to foster such a change. Scanning technology has already had the beneficial effect of forcing the beginnings of a marriage between psychology and neurology. Computer science, through the interest in neural networks, has also grown much closer to the brain. But if the human mind is a social as well as a biological phenomenon, then yet further marriages are required with the 'soft' sciences of sociology and anthropology, and their many sub-disciplines.

Dynamism is not something which can simply be tacked on to the existing reductionist structure of mind science, it must replace it. The scale of this change is such that it makes it difficult to predict how fast things might actually move. Certainly, as has been seen, change has already been taking place in pockets of research and with individual researchers. But the call for a properly dynamic approach has also been made frequently enough in the past, only to be ignored. Psychology started with Helmholtz and Wundt attempting to measure the time-course of mental events. In the 1940s, Hebb was particularly clear in his descriptions of self-organising nerve networks and how they might lead to mental states. At the same time the Gestalt psychologist Wolfgang Köhler was making his attack on machine theory – the forerunner of cognitive science – and writing about the principles of a dynamic approach in astonishingly modern terms. In more recent times, thinkers like Neisser and Edelman have only had modest impact.

However, again the difference may be scanner technology. Never before has there been such a direct method for studying the human brain and its states. And not only does imaging make the dynamism of the brain nakedly apparent, but because of the money and hopes invested in the technique, people are having to pay attention to the results. The single-cell recording and brain lesioning work that dominated neuroscience during the 1980s had a low profile even within science itself. No one wanted to dwell too much on what was involved in experimenting on animals, and besides, few could make much sense of the results anyway. But brain scanning is exactly the kind of technically dazzling gee-whizzery that science likes to put on show. Having splashed out on the machines and hyped up the expectations, the neuroscience community will feel it has to deliver something.

To date, neuro-imaging may have disappointed many. However, the great advantage of scanning is that it does show everything. Even in the first fuzzy images of brains searching for a word association, picturing letters, or thinking about tapping out a sequence of finger movements, the evidence was there for those able to read it. And as the resolution improves and analysis methods are refined, the dynamic nature of brain processing can only become more obvious. It might be true when philosophers like Chalmers argue that the quest for an explanation of human consciousness may never have a final ending because there will always be more detail to discover and parts of the story which can be drawn into better focus, but at least soon there may be a generally agreed beginning – a way in to the last mystery.

Notes

General Reading

If there is a historical trajectory to the ideas outlined in this book, it runs through the writings of Thomas Hobbes, Wilhelm Wundt, William James, Lev Vygotsky, Wolfgang Köhler, Donald Hebb, Gilbert Ryle, Evgeny Sokolov and Ulric Neisser. It is hard to say anything that is not a repetition of what they have already understood. Only the detail is clearer.

Chapter 2: Disturbing the Surface

11 Pebbles in pond: Kosslyn expands on the analogy in an epilogue to *Wet Mind: The New Cognitive Neuroscience* by Stephen Kosslyn and Olivier Koenig (New York: Free Press, 1995).

14 The cognitive science movement: Good introductions are *The Mind's New Science: A History of the Cognitive Revolution* by Howard Gardner (New York: Basic Books, 1987) and *Cognitive Psychology: A Student's Handbook* by Michael Eysenck and Mark Keane (Hove, England: Lawrence Erlbaum Associates, 1990).

15 James on imagery as weak sensation: James talked about the top-down stimulation of sensory pathways and offered evidence such as the fact that people with damage to sensory cortex areas could no longer imagine those sensations either. See *The Principles of Psychology* by William James (Cambridge, Massachusetts: Harvard Univesity Press, 1981). A modern version of the same argument can be found in 'Loss of visual imagery and loss of visual knowledge – a case study', G Goldenberg, *Neuropsychologia 30*, 1081–99 (1992).

16 Ryle on resembling being a spectator: Ryle brilliantly dissects the experience of having images and other forms of mental activity in *The Concept of Mind* by Gilbert Ryle (London: Hutchinson, 1949).

– Timing mental rotation: 'Mental rotation of three-dimensional objects', R. N. Shepard and J. Metzler, *Science 191*, 701–3 (1971). For a review, see 'The mental image', R. N. Shepard, *The American Psychologist 33*, 125–37 (1978).

17 Kosslyn's treasure map experiment: 'Visual images preserve metric spatial

information: evidence from studies of image scanning', S. M. Kosslyn, T. M. Ball, B. J. Reiser, *Journal of Experimental Psychology: Human Perception and Performance 4*, 47–60 (1978).

– Kosslyn's rabbit next to elephant test: 'Measuring the visual angle of the mind's eye', S. M. Kosslyn, *Cognitive Psychology 10*, 356–89 (1978). For a review, see *Ghosts in the Mind's Machine: Creating and Using Images in the Brain* by Stephen Kosslyn (New York: Norton, 1983).

18 Kosslyn's computer display analogy: *Image and Mind* by Stephen Kosslyn (Cambridge, Massachusetts: Harvard University Press, 1980).

– Odd conjunctions take more time: *Ghosts in the Mind's Machine* (Kosslyn, Norton).

19 Pylyshyn counters: 'The imagery debate: analogue media versus tacit knowledge", Z. W. Pylyshyn, *Psychological Review 88*, 16–45 (1981). Note that Pylyshyn and others did make some valid points about the artificiality of Kosslyn's experiments, such as his requirement that subjects mentally 'pace' themselves by following an imaginary black dot when tracking across a map. In *Computation and Cognition* by Zenon Pylyshyn (Cambridge, Massachusetts: MIT Press, 1984), Pylyshyn reports on his own experiments in which subjects could eliminate delays by 'glancing across' rather than scanning across a map. See also commentaries in 'On the demystification of mental imagery', S. M. Kosslyn et al, *Behavioral and Brain Sciences 2*, 535–81 (1979).

20 Kosslyn says subjects not faking delays: 'Is time to scan visual images due to demand characteristics?' P. Jolicoeur and S. M. Kosslyn, *Memory and Cognition 13*, 320–32 (1985). See also a discussion in *Image and Brain: The Resolution of the Imagery Debate* by Stephen Kosslyn (Cambridge, Massachusetts: MIT Press, 1994).

– Tootell's autoradiography experiment: 'Deoxyglucose analysis of retinotopic organization in primate striate cortex', R. B. H. Tootell et al, *Science 218*, 902–4 (1982).

– The autoradiography method: 'The [^{14}C] deoxyglucose method for the measurement of local cerebral glucose utilization', L. Sokoloff et al, *Journal of Neurochemistry 28*, 897–916 (1977).

22 Topographical mapping in V1 not surprising: See 'The discovery of the visual cortex', M. Glickstein, *Scientific American* (September 1988), 84–91, for how war wounds led some to the same conclusion even before 1918.

23 Turing's famous proof: 'On computable numbers, with an application to the *Entscheidungs* problem', A. M. Turing, *Proceedings of the London Mathematics Society 2:42*, 230–65 (1936). Turing also claimed that machine intelligence would be indistinguishable from human intelligence by the year 2000 – 'Computing machinery and intelligence', A. M. Turing, *Mind 59*, 434–60 (1950).

- Putnam seized on the proof: 'Minds and machines', H. Putnam, in *Dimensions of Mind,* edited by Sidney Hook (New York: New York University Press, 1960). Another famous assertion that machines can be made intelligent came in 'Computer science as empirical enquiry: symbols and search', A. Newell and H. Simon, *Communications of the Association for Computing Machinery 19*, 113–26 (1976).

24 Searle lampoons strong AI: 'Minds, brains and programs', J. Searle, *Behavioral and Brain Sciences 3*, 417–57 (1980). 'Is the brain's mind a computer program?' J. Searle, *Scientific American* (January 1990), 19–25.

- Artificial intelligence only a matter of decades: See *AI: The Tumultuous History of the Search for Artificial Intelligence* by Daniel Crevier (New York: Basic Books, 1993). And the belief still lives – see *Mind Children: The Future of Robot and Human Intelligence* by Hans Moravec (Cambridge, Massachusetts: Harvard University Press, 1988) and *Impossible Minds: My Neurons, My Consciousness* by Igor Aleksander (London: Imperial College Press, 1996).

25 Brain's consumption of energy: 'The determination of cerebral blood flow in man by the use of nitrous oxide in low concentrations', S. S. Kety and C. E. Schmidt, *American Journal of Physiology 143*, 53–6 (1946).

- Broca put thermometers on the skull: 'Sur la thermométrie cérébrale', P. Broca, *Revue Scientifique 13*, 257 (1877). For more of the story, see *A Vision of the Brain* by Semir Zeki (Oxford: Blackwell Scientific Publications, 1993).

- Fulton's famous case: 'Observations upon the vascularity of the human occipital lobe during visual activity', J. F. Fulton, *Brain 51*, 310–20 (1928).

- Mosso's measurements from a peasant: *Ueber deb Kreislauf des Blutes in Menschlichen Gehirn* by Angelo Mosso (Leipzig: Verlag von Viet, 1881). This and other stories of the prehistory of scanning can be found in *Images of Mind* by Michael Posner and Marcus Raichle (New York: W. H. Freeman, 1994).

- Caton made first electrode recordings: 'The electric currents of the brain', R. Caton, *British Medical Journal 2*, 278 (1875). For a history of EEG, see 'Rise of neurophysiology in the 19th Century', M. A. B. Brazier, *Journal of Neurophysiology 20*, 212–26 (1957).

26 Alpha and beta brain waves: 'The origin of the Berger rhythm', E. D. Adrian and K. Yamagiwa, *Brain 58*, 323–51 (1935).

- The evoked response potential: For review, see *Psychophysiology: Human Behavior and Physiological Response* by John Andreassi (Hove, England: Lawrence Erlbaum Associates, 1989). Note, they can also be called event related potentials.

28 History of PET scanning: *Images of Mind* (Posner and Raichle, W. H. Freeman), *Exploring Brain Functional Anatomy with Positron Tomography: CIBA Foundation Symposium 163*, edited by Derek Chadwick and Julie

Whelan (Chichester, England: John Wiley, 1991), 'Positron-emission tomography', M. M. Ter-Pogossian, M. E. Raichle and B. E. Sobel, *Scientific American* (October 1980), 141–55, and 'Visualizing the mind', ME Raichle, *Scientific American* (April 1994), 36–42. For general principles of scanning, see *Human Brain Function*, edited by Richard Frackowiak et al (New York: Academic Press, 1997).

– The xenon technique: 'Brain function and blood flow', N. A. Lassen, D. H. Ingvar and E. Skinøj, *Scientific American* (October 1978), 50–9.

29 PET was big science: For a glimpse of the politics in other fields, see *Big Science: The Growth of Large-Scale Research*, edited by Peter Galison and Bruce Hevly (Stanford, California: Stanford University Press, 1992).

31 Kosslyn's PET experiment: 'Visual mental imagery activates topographically organized visual cortex: PET investigations', S. M. Kosslyn et al, *Journal of Cognitive Neuroscience 5*, 263–87 (1993). For other subsequent studies see 'Topographical representations of mental images in primary visual cortex', S. M. Kosslyn et al, *Nature 378*, 496–8 (1995), 'Individual differences in cerebral blood flow in area 17 predict the time to evaluate visualized letters', S. M. Kosslyn et al, *Journal of Cognitive Neuroscience 8*, 78–82 (1996), and also 'How the brain creates imagery: projection to primary visual cortex', Y. Miyashita, *Science 268*, 1719–20 (1995).

34 Kosslyn the first to scan subjective state: Note that the Hammersmith MRC Unit claim that the first truly ambitious bit of PET scanning was their own 'Willed action and the prefrontal cortex in man: a study with PET', C. D. Frith et al, *Proceedings of the Royal Society of London B 244*, 241–6 (1991), which attempted to find a locus for the conscious willing of movement.

35 Invention of MRI scanning: 'New imaging methods provide a better view into the brain', M. Barinaga, *Science 276*, 1974–6 (1997), and 'Dynamic magnetic resonance imaging of human brain activity during primary sensory stimulation', K. K. Kwong et al, *Proceedings of the National Academy of Sciences 89*, 5675–9 (1992).

36 Invention of MEG: 'How surgeons could navigate the brain', G. Harding, *New Scientist* (11 December 1993), 28–31.

Chapter 3: Ugly Questions about Chaos

41 Discovery that brain is made of neurons: An excellent account of Ramón y Cajal's pioneering work – carried out in his attic – can be found in *The Dreaming Brain* by J. Allan Hobson (New York: Basic Books, 1988).

42 Sherrington likened it to an enchanted loom: *Man On His Nature* by Charles Sherrington (Cambridge, England: Cambridge University Press, 1951).

– Hebb's ideas: Hebb's classic was *The Organization of Behavior: A*

Neuropsychological Theory by Donald Hebb (New York: Wiley, 1949). See also *Essay on Mind* by Donald Hebb (Hillsdale, New Jersey: Lawrence Erlbaum Associates, 1980), and for a biographical history, 'The mind and Donald O. Hebb', P. M. Milner, *Scientific American* (January 1993), 104–9. Hebb's original ideas were based on purely excitatory synapses because inhibitory connections had yet to be discovered, and so his student Peter Milner later refined the model – see 'The cell assembly: Mark II', P. Milner, *Psychological Review 64*, 242–52 (1957). Note that Hebb was certainly not the originator of the idea of neural networks as memory landscapes. The idea was surprisingly standard in early literature. See for example *The Physiology of Mind* by Henry Maudsley (London: Macmillan, 1876).

44 Hebb's own research on developing animals: *The Organization of Behavior: A Neuropsychological Theory* (Hebb, Wiley).

– Babies' brains tuned by experience: Recent research shows how this happens as a population voting change – 'Neural noise limitations on infant visual sensitivity', A. M. Skoczenski and A. M. Norcia, *Nature 391*, 698–700 (1998). For a more general treatment, see *Brain Development and Cognition: A Reader*, edited by Mark Johnson (Cambridge, Massachusetts: Blackwell, 1993).

45 McCulloch and Pitts classic paper: 'A logical calculus of the ideas immanent in nervous activity', W. S. McCulloch and W. H. Pitts, *Bulletin of Mathematical Biophysics 5*, 115–33 (1943). See also *Embodiments of Mind* by Warren McCulloch (Cambridge, Massachusetts: MIT Press, 1965). For a history of Hebb and McCulloch, see *Stairway to the Mind: The Controversial New Science of Consciousness* by Alwyn Scott (New York: Copernicus, 1995).

47 Move to neural networks: See *Artificial Minds* by Stan Franklin (Cambridge, Massachusetts: MIT Press, 1995), and *Simple Minds* by Dan Lloyd (Cambridge, Massachusetts: MIT Press, 1989).

– The backprop algorithm: This was first proposed by Paul Werbos as a student – 'Beyond regression: new tools for prediction and analysis in the behavioral sciences', P. Werbos (PhD thesis, Harvard University, 1974) – and then rediscovered and implemented by others. See *Parallel Distributed Processing: Explorations in the Microstructure of Cognition* (Volumes 1 and 2), edited by David Rumelhart, James McClelland and the PDP Research Group (Cambridge, Massachusetts: MIT Press, 1986).

49 Disillusionment with symbolic AI: *What Computers Can't Do: A Critique of Artificial Reason* by Hubert Dreyfus (New York: Harper and Row, 1972), *Computer Power and Human Reason* by Joseph Weizenbaum (San Fransciso: W. H. Freeman, 1976), *Artificial Intelligence: A Paper Symposium*, edited by J. Lighthill et al (London: Science Research Council of Great Britain, 1973).

– People should make the hardware smart: Note that most neural networks are actually implemented in software and run on ordinary computers. Special purpose hardware is expensive to manufacture – although evolving hardware circuitry is on the horizon. See 'Computer design meets Darwin', G. Taubes, *Science 277*, 1931–2 (1997). And one spooky result of such work is that evolving hardware appears to develop its own patches of analog circuitry! See 'Creatures from primordial silicon', C. Davidson, *New Scientist* (15 November 1997), 30–34.

– Field of artificial neural networks: The coming together of computer science and neurology can be seen in *The Computational Brain* by Patricia Churchland and Terrence Sejnowski (Cambridge, Massachusetts: MIT Press, 1992), *Computational Neuroscience*, edited by Eric Schwartz (Cambridge, Massachusetts: MIT Press, 1990) and *The Neurobiology of Neural Networks*, edited by Daniel Gardner (Cambridge, Massachusetts: MIT Press, 1994). See also 'A romance blossoms between gray matter and silicon', D. H. Freedman, *Science 265*, 889–90 (1994).

50 The principles of the digital computer: Building on the theories laid down by Alan Turing, John von Neumann established the basic design of the digital computer. See *Collected Works* (6 volumes) by John von Neumann, edited by A. H. Taube (Oxford: Pergamon Press, 1961–3). For an excellent review of the seductive simplicity of the digital view, see *Turing's Man: Western Culture in the Computer Age* by David Bolter (Harmondsworth, England: Pelican, 1986).

– Belief that neurons rise above noise: See *Spikes: Exploring the Neural Code* by Fred Rieke, David Warland, Rob de Ruyter van Steveninck and William Bialek (Cambridge, Massachusetts: MIT Press, 1997), 'Noise, neural codes and cortical organization', M. N. Shadlen and W. T. Newsome, *Current Opinion in Neurobiology 4*, 569–79 (1994), 'Reliability of spike timing in neocortical neurons', Z. F. Mainen and T. J. Sejnowski, *Science 268*, 1503–6 (1995), 'Simple codes versus efficient codes', W. R. Softky, *Current Opinion in Neurobiology 5*, 239–47 (1995), and 'Is neural noise just a nuisance?', D Ferster, *Science 273*, 1812 (1996). An alternative strategy is to argue that a measure of noise (chaotic 1/f flicker noise, to be precise) is positively useful in extracting a neural signal – see 'Noises on', K. S. Brown, *New Scientist* (1 June 1996), 28–31.

51 Axon depolarisation is black and white: Well, even this is not quite true. Jerome Lettvin of MIT showed that signals from neurons in a cat's eye do not automatically reach every branch at the end of the line. See 'Multiple meaning in single visual units', S. H. Chung, S. A. Raymond and J. Y. Lettvin, *Brain, Behavior and Evolution 3*, 72–101 (1970), and *Stairway to the Mind: The Controversial New Science of Consciousness* (Scott: Copernicus) for a review.

- Are neurons exclusively digital?: See 'Computation and the single neuron', C. Koch, *Nature 385*, 207–10 (1997).

52 Firing of neuron is an electro-chemical process: *The Neuron: Cell and Molecular Biology* by Irwin Levitan and Leonard Kaczmarek (Oxford: Oxford University Press, 1991).

55 Just one in ten spikes cause release of transmitter: 'An evaluation of causes for unreliability of synaptic transmission', C. Allen and C. F. Stevens, *Proceedings of the National Academy of Sciences 91*, 10380–83 (1994), and 'Noisy synapses and noisy neurons', D. K. S. Smetters and A. Zador, *Current Biology 6*, 1217–18 (1996).

56 Spike runs both ways through neuron: 'Active propagation of somatic action potentials into neocortical pyramidal cell dendrites', G. J. Stuart and B. Sakmann, *Nature 367*, 69–72 (1994), and 'Regulation of synaptic efficacy by coincidence of postsynaptic APs and EPSPs', H. Markram et al, *Science 275*, 213–15 (1997).

- Nitric oxide is a neurotransmitter: 'Nitric oxide: a novel neuronal messenger', D. S. Bredt and S. H. Snyder, *Neuron 8*, 8–11 (1992).

57 Synapses even on axon: See for example: 'A specific "axo-axonal" interneuron in the visual cortex of the rat', P. Somogyi, *Brain Research 136*, 345–50 (1977).

58 Computers specify a limited number of decimal places: Note that this round-off problem means that, strictly speaking, computers cannot simulate true chaos because they offer only a finite set of potential values for any simulated system to 'inhabit'. In effect, the simulation is put on to a new orbit with every iteration. However, there are tricks such as numerical shadowing which can be used to make sure that a simulation is not straying too far off track even with the constant rounding up of figures. See 'Simulating chaotic behavior with finite state machines', P. M. Binder and R. V. Jensen, *Physics Review 34A*, 4460–63 (1986). Or *Searching for Certainty* by John Casti (New York: William Morrow, 1991) for a more general discussion.

59 Chaos theory: A still unbeatable introduction to the history of the field is *Chaos: Making a New Science* by James Gleick (London: William Heinemann, 1988).

63 Lorenz and the butterfly effect: The tale of his work is well told in *Chaos: Making a New Science* (Gleick, William Heinemann). Lorenz's key paper was 'Deterministic nonperiodic flow', E. N. Lorenz, *Journal of the Atmospheric Sciences 20*, 130–41 (1963). Lorenz first used the famous butterfly example in 'Predictability: does the flap of a butterfly's wings in Brazil set off a tornado in Texas?', an address to the annual meeting of the American Association for the Advancement of Science in Washington, 29 December 1979.

67 Fractal geometry: The principles are explained in *The Fractal Geometry of Nature* by Benoit Mandelbrot (New York: W. H. Freeman, 1977). For its application to natural structure, see 'Fractal growth processes', L. M. Sander, *Nature 322*, 789–93 (1986) and *Fractals Everywhere* by Michael Barnsley (Orlando, Florida: Academic Press, 1988).

69 Attractor-based network designs: John Hopfield popularised this approach although others, like Stephen Grossberg, had been working along similar lines. See 'Neural networks and physical systems with emergent collective computational abilities', J. J. Hopfield, *Proceedings of the National Academy of Sciences 79*, 2554–8 (1982), and 'Absolute stability of global pattern formation and parallel memory storage by competitive neural networks', M. Cohen and S. Grossberg, *IEEE Transactions on Systems, Man and Cybernetics 13*, 815–26 (1983). For a review, see *The Computational Brain* (Churchland and Sejnowski, MIT Press).

– Freeman finds attractors in olfactory bulb: The work of Walter Freeman is difficult to place because it is clearly important yet also apparently deeply at odds with conventional neuroscience. For example, Freeman argues vehemently against the idea that cortex areas are topographically arranged to form maps. However many of the differences probably boil down to the fact that Freeman records EEGs, not the firing of individual cells. His electrode arrays sum the activity of hundreds of thousands of cells – or rather the dendritic currents of apical dendrites lying near the cortex surface. So where mainstream neuroscience takes measurements of brain activity at its most digital point – the depolarisation of a neuron – Freeman goes to the other extreme and samples the arriving wash of input and modulating feedback. Naturally, one approach encourages a pixel view of neuron function, the other suggests that the brain is purely chaotic. As argued here, the truth has to be some blend of this digitalism and dynamism.

Lack of space meant that I did not enter into a detailed discussion of Freeman's groundbreaking work, but see *Societies of Brains: A Study in the Neuuroscience of Love and Hate* by Walter Freeman (Hove, England: Lawrence Erlbaum Associates, 1995), 'How brains make chaos in order to make sense of the world', C. A. Skarda and W. J. Freeman, *Behavioral and Brain Sciences 10*, 161–95 (1987), 'The physiology of perception', W. J. Freeman, *Scientific American* (February 1991), 34–41, and 'Chaotic oscillations and the genesis of meaning in cerebral cortex', W. J. Freeman, in *Temporal Coding in the Brain*, edited by György Buzsáki et al (Berlin: Springer, 1994).

Another pioneer in the application of chaos theory to the brain is J. A. Scott Kelso of Florida Atlantic University, who has demonstrated that much of the processing order of the brain may come 'for free', rather than being an uphill struggle involving a lot of noise-defeating computational tricks. See *Dynamic Patterns: The Self-Organization of Brain and Behavior* by J. A. Scott Kelso (Cambridge, Massachusetts: MIT Press, 1995). Other attempts to apply 'raw' chaos theory to mind science include *Neocortical Dynamics and Human EEG Rhythms* by Paul Nunez (New York: Oxford University Press, 1995), *Mind as Motion: Explorations in the Dynamics of Cognition*, edited by Robert Port and Tim van Gelder (Cambridge, Massachusetts: MIT Press, 1995), and 'Dynamics of the brain at global and microscopic scales: neural networks and the EEG', J. J. Wright and D. T. J. Liley, *Behavioral and Brain Sciences 19*, 285–320 (1996). See also 'Mastering the nonlinear brain', J. Glanz, *Science 277*, 1758–60 (1997).

– Distinction between chaos and complexity: The obvious introductions to this area are the excellent *Complexity: Life on the Edge of Chaos* by Roger Lewin (London: J. M. Dent, 1993), *Complexity: The Emerging Science at the Edge of Order and Chaos* by Mitchell Waldrop (New York: Simon and Schuster, 1992), *Frontiers of Complexity: The Search for Order in a Chaotic World* by Peter Coveney and Roger Highfield (New York: Ballantine Books, 1995) and *Emergence: From Chaos to Order* by John Holland (Reading, Massachusetts: Addison Wesley, 1998).

71 Evolution as push of selection and pull of chaos: See *At Home in the Universe: The Search for Laws of Complexity* by Stuart Kauffman (New York: Oxford University Press, 1995), and *How the Leopard Changed its Spots: The Evolution of Complexity* by Brian Goodwin (London: Weidenfeld and Nicolson, 1995).

73 Talk about complexity as mystical force: See, for example, *Are We Unique?* by James Trefil (New York: Wiley, 1997), 'A framework for higher order cognition and consciousness', N. A. Baas, in *Toward a Science of Consciousness*, edited by Stuart Hameroff et al (Cambridge, Massachusetts: MIT Press, 1995), and *Stairway to the Mind: The Controversial New Science of Consciousness* (Scott: Copernicus).

Chapter 4: The Hunt for the Neural Code

75 Desimone's disconcerting result: 'Selective attention gates visual processing in the extrastriate cortex', J. Moran and R. Desimone, *Science 229*, 782–4 (1985).

– V4 as the brain's colour centre: Note that while V4 is central to colour perception, it is not a stand-alone colour processing and colour experiencing module as some more extreme theorists have implied (see *A Vision of the Brain* by Semir Zeki (Oxford, Blackwell Scientific Publications, 1993) for the contary position). As part of a dynamic hierarchy, V4 would merely play a central role in bringing colour representing activity into a contextualised focus. And there would be no reason why V4 should not also contribute to form perception. Indeed, recent research has revealed depth-coding cells in V4 – see 'Distance modulation of neural activity in the visual cortex', A. C. Dobbins et al, *Science 281*, 552–4 (1998). So to call V4 the colour centre is simply to be more right than wrong.

Another point is that, strictly speaking, regions such as V1, V4, MT, etc, refer to the macaque monkey brain – the standard subject of single cell experiments. It is only with scanning that their homologues are being identified in the human brain. See 'Borders of multiple visual areas in humans revealed by functional magnetic resonance imaging', M. I. Sereno et al, *Science 268*, 889–93 (1995).

– A hierarchy of visual processing: Several decades of neuroscience were summarised in two landmark articles, 'Distributed hierarchical processing in the primate cerebral cortex', D. J. Felleman and D. C. van Essen, *Cerebral Cortex 1*, 1–47 (1991), and 'Information processing in the primate visual system: an integrated systems perspective', D. C. van Essen, C. H. Anderson and D. J. Felleman, *Science 255*, 419–23 (1992). But while the van Essen-Felleman hierarchical chart of visual processing areas quickly took on an almost talismanic quality for neuroscientists, a group of neural network simulators cheekily suggested the actual hierarchy proposed was probably only in the top 100 million of the 10^{37} alternatives that were theoretically possible given 30 mapping areas and 318 reciprocal pathways between these areas – see 'Indeterminate organisation of the visual system', C. C. Hilgetag, M. A. O'Neill, M. P. Young, *Science 271*, 776–77 (1996). Also 'Constraints on cortical and thalamic projections: the no-strong-loops hypothesis', F. Crick and C. Koch, *Nature 391*, 245–50 (1998).

77 Idea of the receptive field: The term was coined in classic work on the frog's eye: 'The response of single optic nerve fibers of the vertebrate eye to illumination of the retina', H. K. Hartline, *American Journal of Physiology 121*, 400–15 (1938).

78 Neurons as pixel-like feature coding devices: Horace Barlow has been the most tireless champion of this view – see 'Single units of sensation: a neuron doctrine for perceptual psychology?', H. B. Barlow, *Perception 1*, 371–95 (1972), and also 'The neuron doctrine in perception', H. B. Barlow, in *The Cognitive Neurosciences*, edited by Michael Gazzaniga (Cambridge, Massachusetts: MIT Press, 1995). For a general dicussion, see *The Computational Brain* by Patricia Churchland and Terrence Sejnowski (Cambridge, Massachusetts: MIT Press, 1992). But perhaps the earliest clear exposition of the view of the cortex as a hierarchical analyser based on feature detecting cells comes in *Conditioned Reflexes and Neuron Organisation* by Jerzy Konorski (Cambridge: Cambridge University Press, 1948). See also *The Integrative Activity of the Brain: An Interdisciplinary*

Approach by Jerzy Konorski (Chicago: University of Chicago Press, 1967). Early criticism of the pixel model can be found in *Languages of the Brain* by Karl Pribram (Englewood Cliffs, New Jersey: Prentice Hall, 1971).

80 Integrate and fire model could predict the same: See 'In search of common foundations for cortical computation', W. A. Phillips and W. Singer, *Behavioral and Brain Sciences 20*, 657–722 (1997).

83 Modern neuroscience began with a famous bit of luck: The story is told in *Eye, Brain and Vision* by David Hubel (New York: W. H. Freeman, 1988). Luck features a lot in single cell experiments – a sign of how much researchers have generally been groping in the dark with no clear theories to guide them. Richard Jung, a contemporary of Hubel and Wiesel with a much better equipped lab, later complained that his own thoroughness in designing a mechanised stimulus display apparatus prevented him from making the kind of 'sloppy' mistake that proved so fruitful for the Harvard pair. See *The Neurosciences: Paths of Discovery*, edited by Frederic Worden, Judith Swazey and George Adelman (Cambridge, Massachusetts: MIT Press, 1975).

– Dynamicists felt brain wiring acted as a stage: The Gestalt school of psychology pushed the most extreme version of this view with their field theory of nerve energy. See *Principles of Gestalt Psychology* by Kurt Koffka (New York: Harcourt Brace, 1935) and *Dynamics in Psychology* by Wolfgang Köhler (New York: Grove Press, 1940). This suggested that consciousness was a free-flowing field of electrical potential. But their critics stuck shards of mica or silver needles – disrupting insulators or conductors – into the brains of animals to show such fields could not exist. See 'Physiological plasticity and brain circuit theory', R. W. Sperry, in *Biological and Biochemical Bases of Behavior*, edited by Harry Harlow and Clinton Woolsey (Madison, Wisconsin: University of Wisconsin Press, 1958).

Another important figure often mistakenly placed on the side of the naked dynamicists is Karl Lashley, but Lashley had a sophisticated understanding of the issues and was always wrestling with the problem of how brain circuitry could be both plastic and structured. For a discussion, see *Memory and Brain* by Larry Squire (New York: Oxford University Press, 1987).

84 Hubel and Wiesel's Nobel prize: For their work in mapping the visual cortex, they shared the Nobel Prize in Physiology or Medicine in 1981 with Roger Sperry.

– Hubel's discovery of colour cells in V1: Reported in brief as 'Colour vision cells found in visual cortex', G. Kolata, *Science 218*, 457–8 (1982), and in detail in 'Anatomy and physiology of a color system in the primate visual cortex', M. S. Livingstone and D. H. Hubel, *Journal of Neuroscience 4*, 309–56 (1984). Semir Zeki – an arch-rival of Hubel – argues that wavelength-responsive cells were actually found in V1 by others as early as

1974, but 'such reports made little impression on most, and made no impression at all on Hubel and Wiesel'. See *A Vision of the Brain*, 172 (Zeki, Blackwell Scientific Publications).

85 Representation of touch on cortex: For a general review, see 'Somatosensory cortex', J. H. Kass, in *Encyclopedia of Neuroscience* (volume 2), edited by George Adelman (Boston, Massachusetts: Birkhäuser, 1987). The idea that the somatosensory cortex mapping followed the topology of the body first arose in electrical stimulation experiments on humans undergoing brain surgery. See *The Cerebral Cortex of Man: A Clinical Study of Localization of Function* by Wilder Penfield and Theodore Rasmussen, (New York: Macmillan, 1950).

– Auditory cortex as map of frequencies: In fact, the mapping principles of the auditory hierarchy are poorly understood when compared to what is known about the visual cortex. This has much to do with sound being primarily a temporal phenomenon – it unfolds over time. Plainly, it is difficult for researchers with a rate-coding view of neurons to see how an auditory cell can stand for a pixel of aural activity in quite the same way that a visual neuron can code for a line orientation or a wavelength intensity at a certain location.

It has been shown that overall the primary auditory cortex maps a spectrum of frequencies, reflecting the way sound is filtered out by the tapering coil of the cochlea. But many cells actually respond best to swooping frequency changes – a more dynamic property – or to ecologically significant sounds such as cackles and trills. As might be expected, a simple form of tonotopy only shows up in anaesthetised animals. See 'Specialization of the auditory system for reception and processing of species-specific sounds', N. Suga, *Federal Proceedings 37*, 2342–54 (1978) for tonotopy findings, 'Auditory cortex of squirrel monkey: response patterns of single cells to species-specific vocalizations', Z. Wollberg and J. D. Newman, *Science 175*, 212–14 (1972) for complex responses, and *Perception* (third edition) by Robert Sekuler and Randolph Blake (New York: McGraw-Hill, 1994) for a review.

– Mapping of olfaction: See 'The molecular logic of smell', R. Axel, *Scientific American* (October 1995), 130–7, and 'Contributions of topography and parallel processing to odor coding in the vertebrate olfactory pathway', J. S. Kauer, *Trends in Neuroscience 14*, 79–85 (1991).

86 Cortex also has motor mappings: The existence of motor maps was suspected from electrode stimulation of the animal brain as early as the 1870s. See *Pioneers of Psychology* by Raymond Fancher (New York: Norton, 1990). A general review of motor areas can be found in *The Frontal Lobes and Voluntary Action* by Richard Passingham (Oxford: Oxford University Press, 1993). But a warning that coding of M1 neurons is actually a complex

business – with the topographical match to parts of the body only being approximate, the cells being grouped more in terms of action primitives – comes in 'Remapping the motor cortex', M. Barinaga, *Science 268*, 1696–8 (1995).

88 Six-layer design of cortex: See *Cerebral Cortex, Volume 1: Cellular Components of the Cerebral Cortex*, edited by Alan Peters and Edward Jones (New York: Plenum, 1984), 'Laminar origins and terminations of cortical connections of the occipital lobe in the rhesus monkey', K. S. Rockland and D. N. Pandya, *Brain Research 179*, 3–20 (1979), and 'A cannonical microcircuit for neocortex', R. J. Douglas, K. A. C. Martin and D. Whitteridge, *Neural Computation 1*, 480–8 (1989).

– Brodmann identified over 50 cortex divisions: *Vergleichende Lokalisationlehre der Grosshirnrinde in ihren Prinzipien dargestellt auf Grund des Zellenbaues* by Korbinian Brodmann (Leipzig: Barth, 1909).

89 V4 firing matches subjective experience: 'Colour coding in the cerebral cortex: the reaction of cells in monkey visual cortex to wavelengths and colours', S. Zeki, *Neuroscience 9*, 741–56 (1983). See also *A Vision of the Brain* (Zeki, Blackwell Scientific Publications) and *Colour Vision: A Study in Cognitive Science and the Philosophy of Perception* by Evan Thompson (London: Routledge, 1995) for discussion.

90 Workings of inferotemporal unlocked by another fluke: 'Visual properties of cells in inferotemporal cortex of the macaque', C. G. Gross, C. E. Rocha-Miranda and D. B. Bender, *Journal of Neurophysiology 35*, 96–111 (1972).

– Cells kept going as the memory trace: 'Neuron activity related to short-term memory', J. M. Fuster and G. E. Alexander, *Science 173*, 652–4 (1971), and 'Prefrontal cortical unit activity and delayed alternation performance in monkeys', K. Kubota and H. Niki, *Journal of Neurophysiology 34*, 337–47 (1971).

– Neurons stir in anticipation of events: 'Activity of superior colliculus in behaving monkey, II: effect of attention on neuronal responses', M. E. Goldberg and R. H. Wurtz, *Journal of Neurophysiology 35*, 560–74 (1972).

92 Grandmother cell conundrum: The problem of single cells standing for single events or memories – so-called pontifical cells or gnostic units – was widely discussed by pre-war neuroscientists like Sherrington and Konorski. See *Man On His Nature* by Charles Sherrington (Cambridge, England: Cambridge University Press, 1951) and *The Integrative Activity of the Brain: An Interdisciplinary Approach* (Konorski, University of Chicago Press). Then single cell recording brought the issue back into the limelight. Barlow suggested the brain might use small numbers of 'cardinal' cells so that the visual scene might be encoded by perhaps a thousand neurons. See 'Single units of sensation: a neuron doctrine for perceptual psychology?', H. B. Barlow, *Perception 1*, 371–95 (1972).

93 Brain better at detecting than receptive fields allow: This problem is known as hyperacuity. For a discussion, see *The Computational Brain* (Churchland and Sejnowski, MIT Press) and 'The grain of visual space', G. Westheimer, *Cold Spring Harbor Symposia on Quantitative Biology: The Brain 55*, 759–64 (1990).

95 Population coding around for years: See *Neurophilosophy: Toward a Unified Science of the Mind/Brain* by Patricia Smith Churchland (Cambridge, Massachusetts: MIT Press, 1986) for a history. Tensor networks advanced in 'Tensor network theory of the metaorganization of functional geometries in the central nervous system', A. Pellionisz and R. R. Llinás, *Neuroscience 16*, 245–73 (1985). Of course, there are differences between the various vector-type and ensemble theories – see *The Computational Brain* (Churchland and Sejnowski, MIT Press) and *Large-Scale Neuronal Theories of the Brain*, edited by Christof Koch and Joel Davis (Cambridge, Massachusetts: MIT Press, 1994) for a discussion.

96 Proof for population coding came quickly: 'Neuronal population coding of movement direction', A. P. Georgopoulos, A. B. Schwartz and R. E. Kettner, *Science 233*, 1416–19 (1986), 'Population coding of saccadic eye movements by neurons in the superior colliculus', C. Lee, W. H. Rohrer and D. L. Sparks, *Nature 332*, 357–60 (1988), 'Mental rotation of the neuronal population vector', A. Georgopoulos et al, *Science 243*, 234–6 (1989), and 'Direct cortical representation of drawing', A. B. Schwartz, *Science 265*, 540–2 (1994).

Proof that population coding is used in the sensory cortex has been harder to come by, but there are many experiments that are suggestive. See for example 'Dynamics of the hippocampal ensemble code for space', M. A. Wilson and B. L. McNaughton, *Science 261*, 1055–8 (1993), 'Sparse population coding of faces in the inferotemporal cortex', M. P. Young and S. Yamane, *Science 256*, 1327–31 (1992), and 'Columns for visual features of objects in monkey inferotemporal cortex', I. Fujita et al, *Nature 360*, 343–6 (1992).

97 The binding problem: Francis Crick pushed the issue of binding to prominence, championing a succession of possible solutions. In 'Functions of the thalamic reticular complex: the searchlight hypothesis', F. Crick, *Proceedings of the National Academy of Sciences 81*, 4586–93 (1984), he suggested the reticular nucleus of the thalamus might serve to select the 'in focus' percept. In 'Toward a neurobiological theory of consciousness', F. Crick and C. Koch, *Seminars in the Neurosciences 2*, 263–75 (1990), Crick suggested that synchronous oscillations were involved.

For other discussions, see 'Binding in models of perception and brain function', C. von der Malsburg, *Current Opinion in Neurobiology 5*, 520–6 (1995), 'A feature integration theory of attention', A. Treisman, *Cognitive Psychology 12*, 97–136 (1980), 'Principles of feature integration in visual perception', W. Prinzmetal, *Perception and Psychophysics 30*, 330–40 (1981), and 'Psychology's "binding problem" and possible neurobiological solutions', V Hardcastle, *Journal of Consciousness Studies 1*, 66–90 (1994).

98 Colour of object misbound to another: 'Illusory conjunctions in the perceptions of objects', A. M. Triesman and H. Schmidt, *Cognitive Psychology 14*, 107–41 (1982).

– Singer and Gray's binding experiment: 'Oscillatory responses in cat visual cortex exhibit inter-columnar synchronization which reflects global stimulus properties', C. M. Gray et al, *Nature 338*, 334–7 (1989). See also 'Stimulus-specific neuronal oscillations in orientation columns of cat visual cortex', C. M. Gray and W. Singer, *Proceedings of the National Academy of Sciences 86*, 1698–1702 (1989). And for work by a rival group, 'Coherent oscillations: a mechanism for feature linking in the visual cortex', R. Eckhorn et al, *Biological Cybernetics 60*, 121–30 (1988).

99 Other synchrony experiments followed: 'Interhemispheric synchronisation of oscillatory neuronal responses in cat visual cortex', A. K. Engel et al, *Science 252*, 1177–9 (1991), 'Relation between oscillatory activity and long-range synchronization in cat visual cortex', P. Konig, A. K. Engel and W. Singer, *Proceedings of the National Academy of Sciences 92*, 290–4 (1995), 'Sensorimotor encoding by synchronous neural ensemble activity at multiple levels of the somatosensory system', M. A. L. Nicolelis et al, *Science 268*, 1353–8 (1995), 'Dynamics of neuronal interactions in monkey cortex in relation to behavioural events', E. Vaadia et al, *Nature 373*, 515–18 (1995), and 'Visuomotor integration is associated with zero time-lag synchronisation among cortical areas', P. R. Roelfsema et al, *Nature 385*, 157–61 (1997).

100 Cortex cells connect more to each other than outside world: See *The Synaptic Organisation of the Brain*, edited by Gordon Shepherd (New York: Oxford University Press, 1990). For a model of cortex circuitry based on feedback dynamics, see 'Recurrent excitation in neocortical circuits', R. J. Douglas et al, *Science 269*, 981–5 (1995).

Chapter 5: A Dynamical Computation

102 Maunsell shows movement cells doubling their rate: 'Attentional modulation of visual motion processing in cortical areas MT and MST', S. Treue and J. R. Maunsell, *Nature 382*, 539–41 (1996). For reviews of attention effect findings, see 'The brain's visual world: representation of visual targets in cerebral cortex', J. R. Maunsell, *Science 270*, 764–9 (1995), 'Neural mechanisms of selective visual attention', R. Desimone and J. Duncan, *Annual Review of Neuroscience 18*, 193–222 (1995), and 'Neural mechanisms for visual memory and their role in attention', R. Desimone, *Proceedings of the National Academy of Sciences 93*, 13494–9 (1996).

102 Crick's Nobel: Crick shared the 1962 Nobel Prize for Physiology or Medicine with James Watson and Maurice Wilkins for his 1953 paper on the structure of DNA.

– Crick found viscosity a dull problem: See 'A mind's eye view', N. Hawkes, *The Times Magazine* (30 April 1994), 21–2.

103 Crick's belief that V1 is hardwired: 'Are we aware of neural activity in primary visual cortex?' F. Crick and C. Koch, *Nature 375*, 121–3 (1995), and 'Visual perception: rivalry and consciousness', F. Crick, *Nature 379*, 485–6 (1996).

– Logothetis finds even V1 not sacrosanct: 'Activity changes in early visual cortex reflect monkeys' percepts during binocular rivalry', D. A. Leopold and N. K. Logotheis, *Nature 379*, 549–53 (1996). For other evidence of attention effects reaching all the way down the hierarchy, see 'Focal attention produces spatially selective processing in visual cortex areas V1, V2 and V4 in the presence of competing stimuli', B. C. Motter, *Journal of Neurophysiology 70*, 909–19 (1993), and 'The neural mechanisms of spatial selective attention in areas V1, V2 and V4 of macaque visual cortex', S. J. Luck, L. Chelazzi, S. A. Hillyard and R. Desimone, *Journal of Neurophsyiology 77*, 24–42 (1997).

104 Merzenich shows plasticity of finger representations: 'Variability in hand surface representations in area 3b and area A in adult owl and squirrel monkeys', M. M. Merzenich et al, *Journal of Comparative Neurology 258*, 281–96 (1987), 'Cortical representation plasticity', M. M. Merzenich et al, in *Neurobiology of the Neocortex*, edited by Patricia Rakic and Wolf Singer (New York: Wiley, 1988), 'Topographical reorganization of the hand representation in cortical area 3b of owl monkeys trained in a frequency discrimination task', G. H. Recanzone et al, *Journal of Neurophysiology 67*, 1031–56 (1992). See also 'Massive cortical reorganization after sensory deafferentation in adult macaques', T. P. Pons et al, *Science 252*, 1857–60 (1991), and an fMRI study showing changes in human musicians: 'Increased cortical representation of the fingers of the left hand in string players', T. Elbert et al, *Science 270*, 305–7 (1995).

105 Tendency was to downplay Merzenich's results: Many researchers quite correctly pointed out that Merzenich's findings could be explained by the re-surfacing of existing but weak or disused connections rather than axonal growth changes as such. A hand cell might have a few latent connections with more distant parts of the motor map and these would come to drive the cell as its normal input stream dried up. Such a mechanism seemed enough to explain the 2 millimetre radius shifts seen by Merzenich or the 10 millimetre longer-term changes reported by Pons. But while there may be little evidence that input-hungry neurons in a mature brain can do more than turn to existing faint connections, the point is that they still show

marked plasticity in their receptive field properties. See 'The brain remaps its own contours', M. Barinaga, *Science 258*, 216–18 (1992), for a review. Also 'Perceptual correlates of massive cortical reorganization', V. S. Ramachandran et al, *Science 258*, 1159–60 (1992), 'Phantoms of the brain', J. H. Kaas, *Nature 391*, 331–3 (1998), and 'Capacity for Plasticity in the Adult Owl Auditory System Expanded by Juvenile Experience', E. I. Knudsen, *Science 279*, 1531–3 (1998).

106 Cat's eyeball shows plasticity in minutes: See 'Spatial integration and cortical dynamics', C. D. Gilbert et al, *Proceedings of the National Academy of Sciences 93*, 615–22 (1996), for a review and 'How cortex reorganizes', J. H. Kaas, *Nature 375*, 735–6 (1995), for comment.

112 Ethics of animal research: See *The Monkey Wars* by Deborah Blum (Oxford: Oxford University Press, 1994), *The Unheeded Cry: Animal Consciousness, Animal Pain and Science* by Bernard Rollin (Oxford: Oxford University Press, 1989), and 'The benefits and ethics of animal research', a forum in *Scientific American* (February 1997), 63–77. A standard guideline covering procedures such as the puncturing of lungs in single cell work is *Preparation and Maintenance of Higher Mammals During Neuroscience Experiments*, NIH publication no. 91-3207 (Bethesda, Maryland: National Institutes of Health, 1991).

115 Friston says measurements should not make assumptions: Reverse correlation is being used increasingly to allow neurons to define their own feature set. For a review, see 'What's the best sound?', E. D. Young, *Science 280*, 1402–3 (1998). Evidence that cells may take part in population votes merely by shifting the timing of their spikes, rather than changing their firing rates, comes from 'Dynamics of neuronal interactions in monkey cortex in relation to behavioural events', E. Vaadia et al, *Nature 373*, 515–18 (1995). See also *Information Processing in the Cortex: Experiments and Theory*, edited by Ad Aertsen and Valentino Braitenberg (Heidelberg: Springer, 1992).

116 Synchrony a loose phenomenon: Many researchers argue that timing is in fact incredibly precise. Research into the way that the barn owl uses slight differences in the arrival time of spikes from either ear to calculate the location of a noise has been taken as good evidence that, at least in some cases, timing is all. A dynamic view would say that the brain would be flexible in its coding strategy, taking more notice of relative timing when dealing with time-sensitive information, but probably making a more 'relaxed' use of synchrony in other situations.

For review of barn owl findings, see 'Computational maps in the brain', E. I. Knudsen, S. du Lac and S. D. Esterly, *Annual Review of Neuroscience 10*, 41–65 (1987). See also 'Reliability of spike timing in neocortical neurons', Z. F. Mainen and T. J. Sejnowski, *Science 268*, 1503–6 (1995). For models which show how the brain might exploit precise timing, see 'Pattern recognition computation using action potential timing for stimulus representation', J. J. Hopfield, *Nature 376*, 33–6 (1995), and 'Storage of 7 plus or minus 2 short-term memories in oscillatory subcycles', J. E. Lisman and M. A. P. Idiart, *Science 267*, 1512–15 (1995). For general reviews, see 'Temporal coding in neural populations?', E. E. Fetz, *Science 278*, 1901–2 (1997), 'Neural coding: the enigma of the brain', C. F. Stevens and A. Zador, *Current Biology 5*, 1370–1 (1995), and *Spikes: Exploring the Neural Code* by Fred Rieke, David Warland, Rob de Ruyter van Steveninck and William Bialek (Cambridge, Massachusetts: MIT Press, 1997). For evidence against precise timing, see 'Noise, neural codes and cortical organization', M. N. Shadlen and W. T. Newsome, *Current Opinion in Neurobiology 4*, 569–79 (1994).

– Friston's solution was the neural transient: 'Neuronal transients' by K. J. Friston, *Proceedings of the Royal Society London B 261*, 401–5 (1995), 'Another neural code?', K. J. Friston, *NeuroImage 5*, 213–20 (1997), 'Transients metastability and neuronal dynamics', K. J. Friston, *NeuroImage 5*, 164–71 (1997), and 'Transient phase-locking and dynamic correlations: Are they the same thing?', K. J. Friston, K.-M. Stephan and R. S. J. Frackowiak, *Human Brain Mapping 5*, 48–57 (1997).

Chapter 6: Benjamin Libet's Half-Second

120 Research by Helmholz and Wundt: For a general account, see *Pioneers of Psychology* by Raymond Fancher (New York: Norton, 1990). See also *Selected Writings of Hermann von Helmholtz*, edited by Russell Kahl (Middletown, Connecticut: Wesleyan University Press, 1971), *Wilhelm Wundt and the Making of a Scientific Psychology*, edited by Robert Rieber (New York: Plenum, 1980), and 'On the speed of mental processes', F. C. Donders, *Acta Psychologia 30*, 412–31 (1969).

– Nerves conduct at varying speeds: For a theoretical analysis, see 'Effect of geometrical irregularities on propagation delay in axonal trees', Y. Manor, C. Koch and I. Segev, *Biophysical Journal 60*, 1424–37 (1991).

122 The phi effect: Originally noted by Sigmund Exner in Wundt's laboratory, phi or the apparent motion effect was explored systematically by the founder of Gestalt psychology, Max Wertheimer – 'Experimentelle Studien über das Sehen von Bewegung', M. Wertheimer, *Zeitschrift für Psychologie 61*, 161 (1912). For review, see 'Apparent movement', S. M. Antis, *Handbook of Sensory Physiology 8*, 655–73 (1978).

– Fusing also with touch: *Sensory Saltation* by Frank Gelard and Carl Sherrick (New York: Lawrence Erlbaum Associates, 1975). Also 'Anticipated stimuli across skin', M. P. Kilgard and M. M. Merzenich, *Nature 373*, 663 (1995).

– Lights change colour halfway across: 'Shape and colour in apparent motion', P. A. Kolers and M. von Grünau, *Vision Research 16*, 329–35 (1976).

124 Libet's puzzling results: For a comprehensive collection of Libet's timing research, see *Neurophysiology of Consciousness: Selected Papers and New Essays* by Benjamin Libet (Boston, Massachusetts: Birkhäuser, 1993).

- Penfield made name stimulating the brain: *The Cerebral Cortex of Man: A Clinical Study of Localization of Function* by Wilder Penfield and Theodore Rasmussen (New York: Macmillan, 1950) and *The Excitable Cortex in Conscious Man* by Wilder Penfield (Springfield, Illinois: Thomas, 1958).

128 Libet's first results: 'Production of threshold levels of conscious sensation by electrical stimulation of human somatosensory cortex', B. Libet et al, *Journal of Neurophysiology 27*, 546–78 (1964).

129 Eccles a controversial figure: Eccles eventually came to feel that quantum effects might rule synapse activity – with the soul acting at each nerve ending to tip the balance of quantum probabilities in the direction it wanted. See *The Principles of Design and Operation of the Brain*, edited by John Eccles and Otto Creutzfeldt (Berlin: Springer, 1990), and *Evolution of the Brain: Creation of the Self* by John Eccles (London: Routledge, 1989).

- Libet invited to speak at Vatican conference: Reported in *Brain and Conscious Experience*, edited by John Eccles (New York: Springer, 1966).

- Libet seemed like Eccles's protégé: Libet's actual feelings about the meaning of his own work were always rather cryptic. His standard position was that he talked experimental fact and did not indulge in empty theorising. Yet Libet often seemed to be lending support to an Eccles-like interpretation of the half-second lag by insisting that the onset of awareness was digitally sudden. Despite the results of his own subliminal experiments, Libet did not believe in preconsciousness or slow degrees of awareness. Consciousness was something that – when it finally arrived – was absolute. All neural activity beforehand was unconscious and qualia-free (personal communication).

It was only with the publication of his collected papers in 1993 – and at seventy-seven, too old to do further experiments himself – that Libet ventured into print to outline his own guess. Aligning himself more with Sperry ('Neurology and the mind-brain problem', R. W. Sperry, *American Scientist 40*, 291–312, 1952) than Eccles, Libet put forward a 'mental force field' type theory. In an epilogue written for *Neurophysiology of Consciousness: Selected Papers and New Essays* (Libet, Birkhäuser), he suggested that brain circuits worked in an unconscious fashion. But through some unknown mechanism, their mass activity must lead to the emergence of a mental field—a unified subjective state. This mental field would appear after half a second and then could act in a general way to will the brain circuitry into carrying out specified actions. There would be an interaction at a global level, but not at the level of individual synapses as argued by Eccles

True to form, Libet said he was only willing to put forward this theory because he also had a test that could disprove it. The experiment would involve isolating a chunk of cortex (using a fine wire to undercut the white matter connections while leaving the blood supply intact), then stimulating the chunk with an electrode to see if it could 'broadcast' the resulting sensation it was feeling to the rest of the brain. This idea also published as 'A testable field theory of mind-brain interaction', B. Libet, *Journal of Consciousness Studies 1*, 119–26 (1994).

– Phenomenon of backward masking: For review, see *Visual Masking: An Integrative Approach* by Bruno Breitmeyer (Oxford: Oxford University Press, 1984), 'Backward masking', D. Raab, *Psychological Bulletin 60*, 118–29 (1963), and 'Semantic activation without conscious identification in dichotic listening, parafoveal vision and visual masking: a survey and appraisal', D. Holender, *Behavioral and Brain Sciences 9*, 1–66 (1986).

130 Third stimulus could mask the masker: 'Recovery of masked visual targets by inhibition of the masking stimulus', W. N. Dember and D. G. Purcell, *Science 157*, 1335–7 (1967)

– Libet's backward masking experiment: 'Cortical and thalamic activation in conscious sensory experience', B. Libet et al, in *Neurophysiology Studied in Man*, edited by George Somjen (Amsterdam: Excerpta Medica, 1972).

131 Libet's enhancement experiment: 'Neuronal vs subjective timing for a conscious sensory experience', B. Libet, in *Cerebral Correlates of Conscious Experience*, edited by Pierre Buser and Arlette Rougeul-Buser (Amsterdam: Elsevier/North-Holland Biomedical Press, 1978). These results were mentioned only in passing here because Libet felt he had not used quite enough subjects for them to be statistically bullet-proof. His intention was to add more subjects, but then his collaborator, Bertram Feinstein, died. Eventually Libet was persuaded the results could be published in full. See 'Retroactive enhancement of a skin sensation by a delayed cortical stimulus in man: evidence for delay of a conscious sensory experience', Libet et al, *Consciousness and Cognition 1*, 367–75 (1992).

132 Kornhuber's readiness potential experiments: 'Hirnpotentialänderungen bie Wilkürbewegungen und passiven Bewegungen des Menschen: Beritschaftspotential und reafferente Potentiale', H. H. Kornhuber and L. Deeke, *Pfügers Archiv für Gesamte Physiologie 284*, 1–17 (1965). See also *Psychophysiology: Human Behavior and Physiological Response* by John Andreassi (Hove, England: Lawrence Erlbaum Associates, 1989).

– Libet's freewill experiment: 'Time of conscious intention to act in relation to onset of cerebral activity (readiness-potential): the unconscious initiation of a freely voluntary act', B. Libet et al, *Brain 106*, 623–42 (1983). Experiment described again and its implications debated in: 'Unconscious cerebral initiative and the role of conscious will in voluntary action', B. Libet, *Behavioral and Brain Sciences 8*, 529–66 (1985).

134 Critics felt timing an urge created an artifact: See for example 'Toward a psychophysics of intention', L. E. Marks, *Behavioral and Brain Sciences 8*, 547 (1985) and 'Timing volition: questions of what and when about W', J. L. Ringo, *Behavioral and Brain Sciences 8*, 550 (1985).

135 Preconscious and automatic modes of processing: The term preconscious is chosen in preference to others, such as subconscious, subliminal, implicit – or worst of all, unconscious – because it has the useful connotation of something that is on its way to being conscious, or which could become escalated to consciousness in the right circumstances. The implication is that all processing shares a common path and what differs is the depth of processing. Preattentive processing is perhaps a more standard term, but sounds as if it applies only to the sensory half of the equation and not to the full arc of processing that allows us to execute acts like changing gears, or opening doors, habitually.

For an excellent review of how the concept of preconsciousness has been dealt with in mind science, see 'Is human information processing conscious?', M. Velmans, *Behavioral and Brain Sciences 14*, 651–725 (1991). See also *Preconscious Processing* by Norman Dixon (Chichester, England: John Wiley, 1981), 'The psychological unconscious and the self', J. F. Kihlstrom, in *Experimental and Theoretical Studies of Consciousness: CIBA Foundation Symposium 174*, edited by Gregory Bock and Joan March (Chichester, England: John Wiley, 1993), 'Discrimination and learning without awareness: a methodological survey and evaluation', C. W. Eriksen, *Psychological Review 67*, 279–300 (1960), and *Perception Without Awareness: Cognitive, Clinical and Social Perspectives*, edited by Robert Bornstein and Thane Pittman (New York: Guilford Press, 1992).

– Natural to see conscious and preconscious as separate modes: The reductionist hunt for some sharp dividing line between conscious and non-conscious brain processes has muddied the water tremendously, leading to much confusion in the literature. For recent examples of experiments that 'prove' a brain mechanism difference, see 'Classical conditioning and brain systems: the role of awareness', R. E. Clark and L. R. Squire, *Science 280*, 77–81 (1998), 'Brain regions responsive to novelty in the absence of awareness', G. S. Berns, J. D. Cohen and M. A. Mintun, *Science 276*, 1272–5 (1997), and 'Dissociation of the neural correlates of implicit and explicit memory', M. D. Rugg et al, *Nature 392*, 595–8 (1998). For a review, see 'Characteristics of dissociable human learning systems', D. R. Shanks and M. F. St John, *Behavioral and Brain Sciences 17*, 367–447 (1994).

136 Baars's global workspace model: See *A Cognitive Theory of Consciousness* by

Bernard Baars (Cambridge: Cambridge University Press, 1988) and *In The Theater of Consciousness* by Bernard Baars (New York: Oxford University Press, 1998). For others with similar models of an escalation of initially preconscious processing, see *Cognition and Reality: Principles and Implications of Cognitive Psychology* by Ulric Neisser (New York: W. H. Freeman, 1976), *Chronometric Explorations of Mind* by Michael Posner (Hillsdale, New Jersey: Lawrence Erlbaum Associates, 1978), and 'Integrated cortical field model of consciousness', M. Kinsbourne, in *Experimental and Theoretical Studies of Consciousness: CIBA Foundation Symposium 174* (Bock and March, John Wiley).

– Libet's subjects knew what action to produce: The point that the students went into the freewill experiment with a consciously-held context was obvious to many commentators. See, for example, 'Problems with the psychophysics of intention', B. G. Breitimeyer, *Behavioral and Brain Sciences 8*, 539–40 (1985), and 'Conscious intention is a mental fiat', E. Scheerer, *Behavioral and Brain Sciences 8*, 552–3 (1985).

137 Libet's experiment showing subconscious reactions: 'Control of the transition from sensory detection to sensory awareness in man by the duration of a thalamic stimulus: the cerebral 'time-on' factor', B. Libet et al, *Brain 114*, 1731–57 (1991).

138 Variation on a standard subliminal test: 'Conscious and unconscious perception: experiments on visual masking and word recognition', A. J. Marcel, *Cognitive Psychology 15*, 197–237 (1983).

Chapter 7: A Moment of Anticipation

140 Keep your eye on the ball: Research shows that players have to predict where to focus their eyes, moving them into place well ahead of time. See 'Why can't batters keep their eyes on the ball', A. T. Bahill and T. LaRitz, *American Scientist 72*, 249–53 (May-June, 1984).

– The gifted athlete seems able to conjure with time: See 'Good Timing', J. McCrone, in *The Science of Sport*, a supplement to the *New Scientist* (9 October 1993), 10–12.

– Top athletes score averagely in reaction tests: *Acquiring Ball Skill: A Psychological Interpretation* by Harold Whiting (London: Bell and Sons, 1969).

141 McLeod's cricket player study: 'Visual reaction time and high-speed ball games', P. McLeod, *Perception 16*, 49–59 (1987). For details of the time constraints in cricket, see 'Mechanisms of skill in cricket batting', B. Abernethy, *Australian Journal of Sports Medicine 13*, 3–10 (1981).

142 Cricket balls can develop a late swing: For the physics of spinning balls, see 'The seamy side of swing bowling', W. Brown and R. Mehta, *New Scientist*

(21 August 1993), 21–4, and 'Working knowlege: baseball pitches', A. M. Nathan, *Scientific American* (September 1997), 83–4.

144 Electrical recording of the muscles: For review, see *Psychophysiology: Human Behavior and Physiological Response* by John Andreassi (Hove, England: Lawrence Erlbaum Associates, 1989).

145 Film clips demonstrate anticipation: 'Anticipation in sport: a review', B. Abernethy, *Physical Education Review 10*, 5–16 (1987), and 'Visual search strategies and decision-making in sport', B. Abernethy, *International Journal of Sport Psychology 22*, 189–210 (1991).

147 Brain predicts when reaching: *The Neural and Behavioural Organization of Goal-Directed Movements* by Marc Jeannerod (Oxford: Oxford University Press, 1988). The psychophysics literature is generally full of evidence for anticipation. One broad area of research concerns what is known as reafference messages or corollary discharges – the idea that the motor areas of the brain need to tell the sensory areas about planned actions so that this self-generated movement can be subtracted from the conscious experience. See for example 'An internal model for sensorimotor integration', D. M. Wolpert, Z. Ghahramani and M. I. Jordan, *Science 269*, 1880–2 (1995), and 'Visual decomposition of colour through motion extrapolation', R. Nijhawan, *Nature 386*, 66–9 (1997). Another telling line of research is work on priming – particularly the distinction made between conscious priming leading to a narrow state of expectation, while subconscious priming produces a more general, open-ended, associative state. See 'Is human information processing conscious?', M. Velmans, *Behavioral and Brain Sciences 14*, 651–725 (1991).

149 Exceptions who took anticipation seriously: See *A Cognitive Theory of Consciousness* by Bernard Baars (Cambridge: Cambridge University Press, 1988), and *Cognition and Reality: Principles and Implications of Cognitive Psychology* by Ulric Neisser (New York: W. H. Freeman, 1976).

Of course, many other individuals have put anticipation centre-stage. Both Wundt and James considered the issue in depth – see James's discussion of the experiments of Wundt and others in *The Principles of Psychology* by William James (Cambridge, Massachusetts: Harvard Univesity Press, 1981). See also 'William James symposium: attention', D. L. LaBerge, *Psychological Science 1*, 156–62 (1990), for a review of how modern James's ideas still sound. Other more recent instances can be found in *Attentional Processing: The Brain's Art of Mindfulness* by David LaBerge (Cambridge, Massachusetts: Harvard University Press, 1995), 'The attentive brain', S. Grossberg, *American Scientist* (September-October 1995), 438–49, 'Memory of the future: an essay on the temporal organization of conscious awareness', D. H. Ingvar, *Human Neurobiology 4*, 127–36 (1985), *Attention and Effort* by Daniel Kahneman (Englewood Cliffs, New Jersey: Prentice-

Hall, 1973), and *Preparatory States and Processes*, edited by Sylvan Kornblum and Jean Requin (Hillsdale, New Jersey: Lawrence Erlbaum Associates, 1984).

150 Desimone's search image experiment: 'A neural basis for visual search in inferior temporal cortex', L. Chelazzi, E. K. Miller, J. Duncan and R. Desimone, *Nature 363*, 345–47 (1993). For an early hint of the same finding, see 'Activity of superior colliculus in behaving monkey, II: effect of attention on neuronal responses', M. E. Goldberg and R. H. Wurtz, *Journal of Neurophysiology 35*, 560–74 (1972). For review, see 'Seeing the tree for the woods', A. Cowey, *Nature 363*, 298 (1993), and 'Neural mechanisms of selective visual attention', R. Desimone and J. Duncan, *Annual Review of Neuroscience 18*, 193–222 (1995).

Changes in rates of firing are one way to prime an area of circuitry. A second way would be an anticipatory shift in the level of synchrony – and, indeed, recent research has suggested this happens. See 'Spike synchronisation and rate modulation differentially involved in motor cortical function', A. Riehle, S. Grün, M. Diesmann and A. Aertsen, *Science 278*, 1950–3 (1997).

– Tanaka's study of coding in IT: See 'Coding visual images of objects in the inferotemporal cortex of the macaque monkey', K. Tanaka et al, *Journal of Neurophysiology 66*, 170–189 (1991), 'Neuronal mechanisms of object recognition', K. Tanaka, *Science 262*, 685–8 (1993), and 'Optical imaging of functional organization in the monkey inferotemporal cortex', G. Wang, K. Tanaka and M. Tanifuji, *Science 272*, 1665–8 (1996).

151 Experiment reminiscent of Merzenich's finger studies: 'Long-term learning changes the stimulus selectivity of cells in the inferotemporal cortex of adult monkeys', E. Kobatake, K. Tanaka and Y. Tamori, *Neuroscience Research 17*, 237 (1992).

153 Neurons are always firing: It is rare to find any textbook that makes play of this fact, even though it is obvious in every recording of a neuron. The computational view is that firing is stimulus-driven and so a cell's 'baseline' firing – its spontaneous or random activity – is essentially meaningless. One of the few to take a continual state of representation as a starting point for theorising is Walter Freeman – see *Societies of Brains: A Study in the Neuroscience of Love and Hate* by Walter Freeman (Hove, England: Lawrence Erlbaum Associates, 1995). Although, of course, Freeman dislikes the term 'representation' because of its static overtones and prefers instead to speak of a state of neural intention.

A further point not made often enough is that neurons really seem designed to communicate news about significant changes in their input rather than report raw values. A cell will quickly adapt to the constant sight of a red light or whatever else it is supposed to be tuned to detecting. So the idea of a fixed feature-coding device is even more mythical. For review, see 'More than just frequency detectors?', A. M. Thomson, *Science 275*, 180 (1997), and 'Computation and the single neuron', C. Koch, *Nature 385*, 207–10 (1997).

154 We don't see black when eyes are closed: Some argue that the spontaneous rustle is merely the stray pop of retinal cells, others that it is the chatter of cortex pathways. Most likely it is both. Any pressure on the eyeballs certainly sparks a flood of lights, suggesting we are seeing 'real' retinal input – at least while the cortex itself is still in a state of taut alertness. But as the cortex circuits themselves become relaxed and decoupled, as in the hypnagogic state on the edge of sleep, the spontaneous activity we see becomes more vivid. There are sudden floods of colour and usually crawling or spiralling patterns – and even fleeting, ghostly faces – will be seen. Then when we enter the sleep state proper, as the brain is shut off from external stimulation by a gating of the thalamus, our minds erupt into full-colour, dream-like imagery. The stray firing appears to self-organise to produce fully-fledged pseudo-experiences.

See *The Perception of Brightness and Darkness* by Leo Hurvich and Dorothea Jameson (Boston: Allyn and Bacon, 1966). And for a discussion of hypnagogia and dream states, see *The Myth of Irrationality: The Science of the Mind From Plato to Star Trek* by John McCrone (London: Macmillan, 1993), *Dying to Live: Science and the Near-Death Experience* by Susan Blackmore (London: Grafton, 1993) and *Hypnagogia: The Unique State of Consciousness Between Wakefulness and Sleep* by Andreas Mavromatis (London: Routledge and Kegan Paul, 1987).

156 Anticipations flow from whatever has just been escalated: A point well made in *A Cognitive Theory of Consciousness* (Baars, Cambridge University Press).

158 Activity would stir other sensory modalities: There is no definite story of how modalities connect. There appears to be a convergence on the hippocampal formation but multimodal overlap occurs in the prefrontal cortex and also lower motor areas. Some have even singled out sub-cortical structures like the superior colliculus, claustrum and cerebellum. In truth, cross-modal convergence probably happens at many levels of the processing hierarchy. The phenomenon of synesthesia suggests that lines may even connect unimodal mapping areas like V4 and the auditory cortex. For one view, see *The Merging of the Senses* by Barry Stein and Alex Meredith (Cambridge, Massachusetts: MIT Press, 1993).

159 Grand stack of mapping can be turned on its head: Again, even at the turn of the century, such speculation was common. See *The Principles of Psychology* by William James (Cambridge, Massachusetts: Harvard Univesity Press, 1981). For modern examples, see 'Adaptive resonance

theory: self-organizing networks for stable learning, recognition, and prediction', S. Grossberg and G. A. Carpenter, in *The Handbook of Neural Computation*, edited by Emile Fiesler and Russell Beale (New York: Oxford University Press, 1997), 'The role of attention in auditory information processing as revealed by event-related potentials and other brain measures of cognitive function', R. Näätänen, *Behavioral and Brain Sciences 13*, p201–88 (1990), and 'Visual search and stimulus similarity', J. Duncan and G. W. Humphreys, *Psychological Review 96*, 433–58 (1989).

– Motor areas transmit intention to move: The best proof comes from eye movements and the illusory stability of our visual experience. See 'False perception of motion in a patient who cannot compensate for eye movements', T. Haarmeier et al, *Nature 389*, 849–52 (1997), and 'A theory of visual stability across saccadic eye movements', B. Bridgeman, A. H. C. van der Heijden and B. M. Velichkovsky, *Behavioral and Brain Sciences 17*, 247–92 (1994).

163 Neisser wrote imagery was first half of perceptual cycle: *Cognition and Reality: Principles and Implications of Cognitive Psychology* (Neisser, W. H. Freeman). See also 'Theories relating mental imagery to perception', R. Finke, *Psychological Bulletin 98*, 236–59 (1985), and 'The nature of imagery', P. V. Horne, *Consciousness and Cognition 2*, 58–82 (1993).

Chapter 8: The Needs that Shape the Brain

166 Effects of sensory deprivation: *Sensory Deprivation: Fifteen Years of Research*, edited by John Zubek (New York: Appleton-Century-Crofts, 1969). Note that this was a key prediction to emerge from Donald Hebb's theories – see *Essay on Mind* by Donald Hebb (Hillsdale, New Jersey: Lawrence Erlbaum Associates, 1980).

161 Edelman's Nobel prize: Edelman shared the Nobel Prize in Physiology or Medicine in 1972 with Rodney Porter for his work on immunology.

167 Edelman's monastery of science: 'Center for the mind pleases the senses', J. Cohen, *Science 270*, 572 (1995).

– Interviewed in *The New Yorker:* 'Dr Edelman's brain', S. Levy, *The New Yorker* (2 May 1994), 62–73.

– Edelman's trilogy of books: *Neural Darwinism: The Theory of Neuronal Group Selection* by Gerald Edelman (New York: Basic Books, 1987), *Topobiology: An Introduction to Molecular Embryology* by Gerald Edelman (New York: Basic Books, 1988), and *The Remembered Present: A Biological Theory of Consciousness* by Gerald Edelman (New York: Basic Books, 1989). His theory was first sketched out in *The Mindful Brain*, edited by Gerald Edelman and Vernon Mountcastle (Cambridge, Massachusetts: MIT Press, 1978), and a popular account can be found in *Bright Air, Bright Fire: On the*

Matter of the Mind by Gerald Edelman (New York: Basic Books, 1992).

169 In a snub that typified the general response: See p.284 of *The Astonishing Hypothesis: The Scientific Search for the Soul* by Francis Crick (New York: Simon and Schuster, 1994). See also 'Neural Edelmanism', F. Crick, *Trends in Neuroscience 12*, 240–8 (1989), 'A global brain theory', W. Calvin, *Science 240*, 1802–3 (1989), and 'Neuroscience: a new era?', H. B. Barlow, *Nature 331*, 571 (1988).

– Edelman made presentational mistakes: In a strange twist, Edelman was the dynamicist with a reductionist approach to his social style while Crick was the very opposite. As the *New Yorker* interview indicated, many people objected to Edelman because it sounded as if he believed that he was the one person with the one essential insight who was going to solve the problem of consciousness. Instead of being dynamic in his interaction with the rest of the field – that is, emphasing the continuity of his ideas with the theories of others like Hebb and Jean-Pierre Changeux, so making clear the long history of similar thinking – he seemed intent on being digital – to offer the world a unique, contextless and descrete theory. Crick, on the other hand, was the unrepentant reductionist (see for example p.207 of *The Astonishing Hypothesis: The Scientific Search for the Soul,* Crick, Simon and Schuster) who only ever claimed to be drawing existing thinking into sharper focus. The result was that Crick was embraced while fellow academics delighted in presenting Edelman with lists of all the people who had thought of Neural Edelmanism long before he did. See for instance p.184 of *Consciousness Explained* by Daniel Dennett (London: Penguin, 1993), in which Dennett even manages to cite his own doctoral dissertation in 1965.

171 The 'what' and 'where' streams of processing: See 'Two cortical visual systems', L. G. Ungerleider and M. Mishkin, in *Analysis of Visual Behavior*, edited by David Ingle, Melvyn Goodale and Richard Mansfield (Cambridge, Massachusetts: MIT Press, 1982). Note that there are undoubtedly more accurate ways of characterising the difference. Calling them object-location and object-identity pathways is a rather static, sensory processing-biased, way of looking at the split. Others have put a more active slant on the distinction. For example, Melvyn Goodale thinks of it as perceptual identification versus sensorimotor transformation – the temporal lobe path asks 'What is it?' in a general, decontexualised, way that allows an object to be recognised in any situation. The parietal lobe on the other hand asks 'What can I do about it?', pinning the object to a spatial context and supporting the frontal cortex in its generation of a response plan. But Goodale probably does go too far in suggesting that the temporal stream is the conscious aspect of experience. See *The Visual Brain in Action* by David Milner and Melvyn Goodale (Oxford: Oxford University Press, 1995).

Another suggestion is that the split is between a temporal pathway specialising in far space and a parietal pathway dealing in near space. Because V1 is mapped upside down on the cortex, the upper half of the visual scene would feed naturally into the temporal path which would become specialised for the search and recognition of objects seen off towards the horizon, while the parietal lobe would feed off the lower hemifield and so become specialised for the near space representations needed to control the accurate fall of our feet or manipulations by our hands. See 'Functional specialization in the lower and upper visual fields in humans: its ecological origins and neurophysiological implications', F. H. Previc, *Behavioral and Brain Sciences 13*, 519–75 (1990).

172 Magnocellular and parvocellular outputs: 'Psychophysical evidence for separate channels for the perception of form, color, movement and depth', M. S. Livingstone and D. H. Hubel, *Journal of Neuroscience 7*, 3416–68 (1987).

173 What-where division rules frontal cortex as well: This has been the subject of some controversy and certainly a what-where split would characterise the prefrontal lobes in only a general way. Patricia Goldman-Rakic argues for a prefrontal cortex that is strongly divided, while others find evidence for zones of integrated activity. See 'Integration of what and where in the primate prefrontal cortex', S. Chenchal Rao, G. Rainer and E. K. Miller, *Science 276*, 821–24 (1997), and for review, 'The machinery of thought', T. Beardsley, *Scientific American* (August 1997), 58–63.

174 Difference between two sides of brain is one of processing style: See 'Hemispheric specialization and the growth of human understanding', M. Kinsbourne, *American Psychologist 37*, p411-420 (1982), and 'Rethinking the right hemisphere', J. W. Brown, in *Cognitive Processing in the Right Hemisphere*, edited by Ellen Perecman (New York: Academic Press, 1983). For recent experimental evidence, see 'Where in the brain does visual attention select the forest and the trees?', G. R. Fink et al, *Nature 382*, 626–8 (1996), and 'Neural mechanisms of global and local processing: a combined PET and ERP study', H. J. Heinze et al, *Journal of Cognitive Neuroscience 10*, 485–98 (1998).

Then for reviews of the relative role of the two hemispheres in language and spatial processing, see *Hemispheric Asymmetry: What's Right and What's Left* by Joseph Hellige (Cambridge, Massachusetts: Harvard University Press, 1993) and *Left Brain, Right Brain* by Sally Springer and Georg Deutsch (New York: W. H. Freeman, 1993). Note that there is evidence that any brain lateralisation is more extreme in men than women, so adding yet another level of complexity to the discussion. See 'Sex differences in functional organization of the brain for language', B. A. Shaywitz et al, *Nature 373*, 607–9 (1995), and *Sex Differences in Cognitive*

Abilities by Diane Halpern (Hillsdale, New Jersey: Lawrence Erlbaum Associates, 1992).

175 People with right brain damage cannot judge emotion: See *Human Cognitive Neuropsychology* by Andrew Ellis and Andrew Young (Hove, England: Lawrence Erlbaum Associates, 1988).

– Classic test of attentional style: 'Forest before trees: the precedence of global features in visual perception', D. Navon, *Cognitive Psychology 9*, 353–83 (1977). For performance following brain damage, see 'Neuropsychological contributions to theories of part/whole organization', L. C. Robertson and M. R. Lamb, *Cognition 23*, 299–330 (1991).

176 Lateralisation began with tool use: For review, see *The Lopsided Ape: Evolution of the Generative Mind* by Michael Corballis (New York: Oxford University Press, 1991), and 'Language, tools and brain: the ontogeny and phylogeny of hierarchically organized sequential behavior', P. M. Greenfield, *Behavioral and Brain Sciences 14*, 531–95 (1991).

– Receptive fields may be different in left and right brain: For review, see *Left Brain, Right Brain* (Springer and Deutsch, W. H. Freeman), and *Image and Brain: The Resolution of the Imagery Debate* by Stephen Kosslyn (Cambridge, Massachusetts: MIT Press, 1994).

178 The emotional jolt of an 'aha!': These feelings were dealt with at length in my two earlier books, *The Myth of Irrationality: The Science of the Mind From Plato to Star Trek* by John McCrone (London: Macmillan, 1993) and *The Ape That Spoke: Language and the Evolution of the Human Mind* by John McCrone (London: Macmillan, 1990). For a review of experiments and history of research in this area, see *Metacognition: Knowing about Knowing*, edited by Janet Metcalfe and Arthur Shimamura (Cambridge, Massachusetts: MIT Press, 1994). Note that a valuing step has become central to many recent brain-based theories of consciousness. See for example *The Remembered Present: A Biological Theory of Consciousness* (Edelman, Basic Books), and *Descartes' Error: Emotion, Reason and the Human Brain* by Antonio Damasio (New York: Putnam, 1994).

179 Sokolov's orientation response: *Perception and the Conditioned Reflex* by Evgeny Sokolov (New York: Macmillan, 1963), *Neuronal Mechanisms of the Orienting Reflex*, edited by Evgeny Sokolov and O. S. Vinogradova (Hillsdale, New Jersey: Lawrence Erlbaum Associates, 1975), and *Attention, Arousal and the Orientation Response* by Richard Lynn (Oxford: Pergamon Press, 1966).

181 Students respond to slides of hobbies: 'Interest as a predeterminer of the GSR index of the orienting reflex', J. A. Wingard and I. Maltzman, *Acta Psychologica 40*, 153–160 (1980).

182 The P300 response: First noted in 'Evoked potential correlates of stimulus uncertainty', S. Sutton, M. Braren and J. Zubin, *Science 150*, 1187–8 (1965). For

review, see *Psychophysiology: Human Behavior and Physiological Response* by John Andreassi (Hove, England: Lawrence Erlbaum Associates, 1989).

183 Some felt P300 due to inhibition wave: See, for example, commentaries by Eric Halgren and Niels Birbaumer in 'Event-related potentials and cognition: a critique of the context updating hypothesis and an alternative interpretation of P3', R. Verleger, *Behavioral and Brain Sciences* 11, 343–427 (1988). The general idea that a wave of suppression is needed to make the focus of attention stand out has a long history. See for example 'The physiological basis of perception', E. D. Adrian, in *Brain Mechanisms and Consciousness*, edited by J. F. Delafresnay (Oxford: Blackwell, 1954), or 'Lateral inhibition and cognitive masking: a neuropsychological theory of attention', R. E. Walley and T. D. Weiden, *Psychological Review 80*, 284–302 (1973). For a discussion of what various sub-components of the P300 might represent, see 'The role of attention in auditory information processing as revealed by event-related potentials and other brain measures of cognitive function', R. Näätänen, *Behavioral and Brain Sciences 13*, 201–88 (1990)

– Preattentive processing: 'A feature integration theory of attention', A. Triesman and G. Gelade, *Cognitive Psychology 12*, 97–136 (1980). The phrase was first coined in *Cognitive Psychology* by Ulric Neisser (New York: Appleton-Century-Crofts, 1967). The question of exactly how much is mapped before attentional focusing begins has become a central issue for attention researchers. Some believe in early selection – attention is needed to organise the perceptual field. Others believe in late selection – the spread of sensation is mapped to an advance state of recognition before a choice is made. The dynamic view taken here makes this a non-issue. A skilled brain maps as much as it can, to as high a level as it can, and the spotlight of attention is 'filled' either by what is not fitting into place or what has been actively sought. This means that many factors, such as the level of preparedness, will affect how the brain seems to perform during a moment. See 'Competitive brain activity in visual attention', J. Duncan, R. Ward and G. Humphreys, *Current Opinion in Neurobiology 7*, 255–61 (1997).

184 P300 also cleans up the memory trace: See 'P3: byproduct of a byproduct', N. Birbaumer and T. Elbert, *Behavioral and Brain Sciences 11*, 375–7 (1988).

186 Self-organisation and plasticity of the cortex: See *Brain Development and Cognition: A Reader*, edited by Mark Johnson (Cambridge, Massachusetts: Blackwell, 1993), 'The control of neuron number', R. W. Williams and K. Herrup, *Annual Review of Neuroscience 11*, 423–53 (1988), and 'Synaptic activity and the construction of cortical circuits', L. C. Katz and C. J. Shatz, *Science 274*, 1133–9 (1996). For Hebb's work, see *The Organization of Behavior: A Neuropsychological Theory* (New York: Wiley, 1949). Note that there is considerable debate over which counts more, the pruning of unused connections or the growth of new ones. See 'Selective stabilisation of

developing synapses as a mechanism for the specification of neuronal networks', J. P. Changeux and A. Danchin, *Nature 264*, 705–12 (1976), and 'The neural basis of cognitive development: a constructivist manifesto', S. R. Quartz and T. J. Sejnowski, *Behavioral and Brain Sciences 20*, 537–96 (1997).

188 Speech areas of Broca and Wernicke: For the history of early localisation work, see *Pioneers of Psychology* by Raymond Fancher (New York: Norton, 1990). The standard view of language production before PET scanning can be seen in 'Specializations of the human brain', N. Geschwind, *Scientific American* (September 1979), 158–68.

189 Washington PET group's first results: 'Positron emission tomographic studies of the cortical anatomy of single-word processing', S. E. Petersen et al, *Nature 331*, 585–9 (1988). Also 'Positron emission tomographic studies of the processing of single words', S. E. Petersen et al, *Journal of Cognitive Neuroscience 1*, 153–70 (1989).

190 Discovery of the practice effect: For an account, see *Images of Mind* by Michael Posner and Marcus Raichle (New York: W. H. Freeman, 1994). Results were eventually published as 'Practice-related changes in human brain functional anatomy during non-motor learning', M. E. Raichle et al, *Cerebral Cortex 4*, 8–26 (1994). Of course, ERP research had already suggested much the same principle. Recording from a cat showed that with habituation to a stimulus, EEG activity shrunk until only small areas of sensory mapping would respond. See *Foundations of Cognitive Processes*, R. W. Thatcher and E. R. John (New York: Harper and Row, 1977). The Hammersmith MRC PET unit also beat their Washington rivals into print with reports of a practice effect in finger-tapping exercises. See 'Motor practice and neurophysiological adaptation in the cerebellum: a positron emission tomography study', K. J. Friston et al, *Proceedings of the Royal Society London B 248*, 223–8 (1992).

192 Shrinkage in Haier's IQ studies: 'Regional glucose metabolic changes after learning a complex visuospatial/motor task: a positron emission tomographic study', R. J. Haier et al, *Brain Research 570*, 134–43 (1992), and 'Intelligence and changes in regional cerebral glucose metabolic rate following learning', R. J. Haier et al, *Intelligence 16*, 415–26 (1992).

– Brain shrinks activity as hard as possible: Single cell recording evidence has supported the idea that cells on the edge of the population vote needed to mount a response are weeded out with learning, so creating a skeletal template. See 'The representation of stimulus familiarity in anterior inferior temporal cortex', L. Li, E. K. Miller and R. Desimone, *Journal of Neurophysiology 69*, 1918–29 (1993).

193 Downloading of new skills takes time: Evidence for just how long comes from 'Neural correlates of motor memory consolidation', R. Shadmehr and H. H. Holcomb, *Science 277*, 821–5 (1997), which showed that the 'internal

model' of a newly learnt action shifted steadily down the brain output hierarchy over the course of six hours, becoming more stable over that time. The tracking of movements by sensory memories in chick imprinting work suggests a similar story. See *The Making of Memory* by Steven Rose (London: Bantam Press, 1992).

Chapter 9: Consciousness's Twin Peaks

The frontal cortex hierarchy: For a general account, see *The Frontal Lobes and Voluntary Action* by Richard Passingham (Oxford: Oxford University Press, 1993).

194 Scanning reveals roles of SMA and premotor cortex: 'Activation of the supplementary motor area during voluntary movement in man suggests that it works as a supramotor area', J. M. Orgogozo and B. Larsen, *Science* 206, 847–50 (1979), and 'Supplementary motor area and other cortical areas in organization of voluntary movements in man', P. E. Roland et al, *Journal of Neurophysiology* 43, 118–36 (1980). Note that the original idea of a 'master-slave' relationship between the SMA and premotor cortex has given way to a more sophisticated reading. For example, Passingham sees the premotor as being for guiding action in the presence of clear cues while the SMA is for guiding action when memories have to be used to read the situation. To emphasise the complementary roles, Passingham prefers to call the SMA the medial premotor cortex. See *The Frontal Lobes and Voluntary Action* (Passingham, Oxford University Press).

195 The cerebellum smooths execution: The cerebellum was originally felt to be involved in simply the fine-grain execution of motor acts. But with the evidence of scanning, it has become obvious that the cerebellum helps with the fine-grain anticipating and output control of all mental activity. This is, of course, no surprise if human speech and language-driven thinking are seen as merely a third flank to the frontal cortex output hierarchy. For discussion of the changing view, see 'Prediction and preparation: fundamental functions of the cerebellum', E. Courchesne and G. Allen, *Learning and Memory 4*, 1–35 (1997), and 'On the specific role of the cerebellum in motor learning and cognition: clues from PET activation and lesion studies in man', W. T. Thach, *Behavioral and Brain Sciences 19*, 411–31 (1996).

196 Gibson's theory of affordances: See *The Ecological Approach to Visual Perception* by James Gibson (Boston: Houghton-Mifflin, 1979). Gibson's 'direct perception' approach to awareness has become a rallying point for many psychologists disaffected with the overly-modular view of cognitive psychology. But Gibson's approach is too extreme when it is taken to deny any form of extended processing in order to extract information about a scene. For reactions to Gibson, see *The Mind's New Science: A History of the*

Cognitive Revolution by Howard Gardner (New York: Basic Books, 1987).

– Planning of eye shifts: *Eye Movements and Psychological Processes*, edited by Richard Monty and John Senders (Hillsdale, New Jersey: Lawrence Erlbaum Associates, 1976). Note that as well as saccades, we also have systems to handle steady tracking eye movements and adjustments to compensate for the movement of our own head and body. Evidence of just how much needs to be learnt comes from studies of the lack of precision and prediction in the eye movements of young children – 'Eye movements of preschool children', E. Knowler and A. J. Martins, *Science 215*, 997–9 (1982).

198 Tootell shows central vision is exaggerated: 'Borders of multiple visual areas in humans revealed by functional magnetic resonance imaging', M. I. Sereno et al, *Science 268*, 889–93 (1995).

199 Language hierarchy parallels the other two output hierarchies: For review of the evidence, see *The Frontal Lobes and Voluntary Action* (Passingham, Oxford University Press).

– Complex organisation of the prefrontal cortex: See *The Prefrontal Cortex: Anatomy, Physiology and Neuropsychology of the Frontal Lobe* by Joaquin Fuster (New York: Raven Press, 1989).

204 Hippocampus as a memory trace trap: Historically, there are many views about what the hippocampus is for, including that it is merely a navigational mapping of surrounding space – see *The Hippocampus as a Cognitive Map* by John O'Keefe and Lynn Nadel (Oxford: Oxford University Press, 1978). A champion of the memory trap story has been Howard Eichenbaum. See 'Thinking about brain cell assemblies', H. Eichenbaum, *Science 261*, 993–4 (1993), and *Memory, Amnesia and the Hippocampal System* by Neal Cohen and Howard Eichenbaum (Cambridge, Massachusetts: MIT Press, 1993). See also 'The medial temporal lobe memory system', L. R. Squire and S. Zola-Morgan, *Science 253*, 1380–6 (1991), and *The Neurobiology of Memory: Concepts, Findings, Trends* by Yadin Dudai (New York: Oxford University Press, 1989).

– LTP and memory cascade: How the brain builds memory traces into its circuitry is still very poorly understood. The discovery of the LTP reaction – 'Long-lasting potentiation of synaptic transmission in the dentate area of the anesthetized rabbit following stimulation of the perforant path', T. V. P. Bliss and T. Lomo, *Journal of Physiology 232*, 331–56 (1973) – gave researchers a vital first landmark. But LTP is undoubtedly just part of a complex chain of events. LTP may also underpin other kinds of functions, like anticipatory priming. See 'Long-term potentiation: what's learning got to do with it?', T. J. Shors and L. D. Matzel, *Behavioral and Brain Sciences 20*, 597–655 (1997), for a critical review. And note that LTP has attracted so much attention precisely because it seems to be a digitally crisp response – it is the kind of memory mechanism expected by the computationally-minded. The full story of memory trapping is bound to be include some more subtle forms of adaptation. See for instance 'Homeostasis or synaptic plasticity?', Y. Frégnac, *Nature 391*, 845–6 (1998).

For a more generally dynamic account of the story of memory trace formation, see *The Making of Memory* by Steven Rose (London: Bantam Press, 1992). For swelling of dendrites, see 'Stimulation induced changes in the dimensions of stalks of dendritic spines in the dentate molecular layer', E. Fifkova and C. L. Anderson, *Experimental Neurology 74*, 621–7 (1981) – although again the relevance of such changes is the subject of continued debate. For the jangling of stimulated nerve networks, see 'Lasting changes in spontaneous multiunit activity in the chick forebrain following passive avoidance training', R. Mason and S. P. R. Rose, *Neuroscience 21*, 931–41 (1987), and 'Reactivation of hippocampal ensemble memories during sleep', M. A. Wilson and B. L. McNaughton, *Science 265*, 676–9 (1994).

– The NMDA receptor: Note that that this is really a subtype of a glutamate receptor. Its name derives from N-methyl-D-aspartate, a drug used to mimic glutamate's excitatory effect in the lab.

205 Bower says acetycholine cleans up memory trace: 'Acetylcholine and memory', M. E. Hasselmo and J. M. Bower, *Trends in Neuroscience 16*, 218–22 (1993).

206 Flashbulb sharp memories: For review of the veridity of such memories, see *Autobiographical Memory: An Introduction* by Martin Conway (Milton Keynes, England: Open University Press, 1990).

208 Capacity limits of working memory: 'The magical number seven, plus or minus two: some limits on our capacity for processing information', G. A. Miller, *Psychological Review 63*, 81–97 (1956). A more dynamic view of working memory suggests that capacity is limited by perceptual binding mechanisms. See 'The capacity of visual working memory for features and conjunctions', S. J. Luck and E. K. Vogel, *Nature 390*, 279–81 (1997).

209 Working memory as maintained firing: For the experiments from Desimone's lab, see 'A neural mechanism for working and recognition memory in inferior temporal cortex', E. K. Miller, L. Li and R. Desimone, *Science 254*, 1377–9 (1991), and 'Dual mechanisms of short-term memory: ventral prefrontal cortex', L. Chelazzi, E. K. Miller, A. Lueschow and R. Desimone, *Society for Neurosciences Abstracts 19*, 975 (1993). For early evidence of the principle, see 'Neuron activity related to short-term memory', J. M. Fuster and G. E. Alexander, *Science 173*, 652–4 (1971), and 'Prefrontal cortical unit activity and delayed alternation performance in monkeys', K. Kubota and H. Niki, *Journal of Neurophysiology 34*, 337–47 (1971). For review, see *Memory in the Cerebral Cortex: An Emperical Approach to Neural Networks in the Human and Nonhuman Primate* by Joaquin Fuster (Cambridge, Massachusetts: MIT Press, 1995).

- Ungerleider's f-MRI study: See 'Functional brain imaging studies of cortical mechanisms for memory', L. G. Ungerleider, *Science 270*, 769–75 (1995), and 'Transient and sustained activity in a distributed neural system for human working memory', S. M. Courtney, L. G. Ungerleider, K. Keil and J. V. Haxby, *Nature 386*, 608–12 (1997).

210 Lesioning studies of hippocampus and prefrontal: For review, see *Memory and Brain* by Larry Squire (New York: Oxford University Press, 1987).

211 The case of HM in the 1950s: 'Loss of recent memory after bilateral hippocampal lesions', W. B. Scoville and B. Milner, *Journal of Neurology, Neurosurgery and Psychiatry 20*, 11–21 (1957), and 'Further analysis of the hippocampal amnesic syndrome: 14-year follow-up study of HM', B. Milner, S. Corkin and H. L. Teuber, *Neuropsychologia 6*, 215–34 (1968).

212 The more recent case of CW: 'Dense amnesia in a professional musician following herpes simplex virus encephalitis', B. A. Wilson, A. D. Baddeley and N. Kapur, *Journal of Clinical and Experimental Neuropsychology 17*, 1–14 (1995). The documentary about CW was 'Prisoner of Consciousness', *Equinox* (London: Channel Four, 1987).

212 Effects of prefrontal lobe damage: See *The Frontal Granular Cortex and Behavior*, edited by J. M. Warren and Konrad Akert (New York: McGraw-Hill, 1964), and *The Working Brain: An Introduction to Neuropsychology* by Alexander Luria (New York: Basic Books, 1973).

213 Motor cortex riddled with sensory neurons: For review, see 'Neuronal networks for movement preparation', J. Requin, A. Riehle and J. Seal, in *Attention and Performance XIV*, edited by David Meyer and Sylvan Kornblum (Cambridge, Massachusetts: MIT Press, 1992). For Graziano's premotor work, see 'Coding of visual space by premotor neurons', M. S. A. Graziano, G. S. Yap and C. G. Gross, *Science 266*, 1054–7 (1994), and 'Coding the locations of objects in the dark', M. S. A. Graziano, X. T. Hu and C. G. Gross, *Science 277*, 239–41 (1997). Of course, the opposite is also true – sensory areas like the PPC have neurons that appear to code for motor acts. See 'Coding of intention in the posterior parietal cortex', L. H. Snyder, A. P. Batista and R. A. Andersen, *Nature 386*, 167–70 (1997).

214 Dr Strangelove syndrome: The medical name for the condition is alien hand sign. See 'The alien hand sign: localization, lateralization and recovery', G. Goldberg and K. K. Bloom, *American Journal of Physical Medicine and Rehabilitation 69*, 228–38 (1990).

Chapter 10: Of Sub-cortical Bottlenecks

218 People can react faster with anticipation: For a review of LaBerge's work, see *Attentional Processing: The Brain's Art of Mindfulness* by David LaBerge (Cambridge, Massachusetts: Harvard University Press, 1995). For example

of reaction time work, see 'Presentation probability and choice time', D. L. LaBerge and J. R. Tweedy, *Journal of Experimental Psychology 67*, 71–9 (1961). See also *Response Times: Their Role in Inferring Elementary Mental Organization* by R. Duncan Luce (New York: Oxford University Press, 1986).

220 White matter takes up much of brain: Estimates actually vary between 40 and 60 per cent. See *The Human Brain in Figures and Tables: A Quantitative Handbook*, S. M. Blinkov and I. I. Glezer (New York: Plenum Press, 1968). The issue of direct cortico-cortico connections being unmanaged has been largely ignored, but has recently been picked up by Francis Crick. See 'Constraints on cortical and thalamic projections: the no-strong-loops hypothesis', F. Crick and C. Koch, *Nature 391*, 245–50 (1998).

221 Divisions of the thalamus: The standard reference is *The Thalamus* by Edward Jones (New York: Plenum Press, 1985). For a review of its topographic organisation, see *Attentional Processing: The Brain's Art of Mindfulness* (LaBerge, Harvard University Press).

222 Theories that the thalamus might focus consciousness: The central position of the thalamus and its role in wakefulness has led a great many to suggest it has some sort of gateway or bottleneck role. An example of the kind of finding that fueled early speculation was the fact that electrical stimulation of the thalamus could produce a psychological 'arrest'. See 'Highest level seizures', W. Penfield and H. Jasper, *Research Publications of the Association for Research in Nervous and Mental Disease 26*, 252–71 (1947).

Modern theorising began with revelations about feedback connections. See 'Gating of thalamic input to cerebral cortex by nucleus reticularis thalami', C. D. Yingling and W. Singer, in *Attention, Voluntary Contraction and Event-related Cerebral Potentials*, edited by John Desmedt (Basel, Switzerland: Karger, 1977), 'The function of the thalamic reticular complex: the searchlight hypothesis', F. Crick, *Proceedings of the National Academy of Sciences 81*, 4586–90 (1984), 'The control of retinogeniculate transmission in the mammalian lateral geniculate nucleus', S. M. Sherman and C. Koch, *Experimental Brain Research 63*, 1–20 (1986), *A Cognitive Theory of Consciousness* by Bernard Baars (Cambridge: Cambridge University Press, 1988), and 'Towards a neural network model of the mind', J. G. Taylor, *Neural Network World 2*, 797–812 (1992).

– Thalamus might act as the pacemaker: 'Towards a neurobiological theory of consciousness', F. Crick and C. Koch, *Seminars in the Neurosciences 2*, 263–75 (1990), 'Search for coherence: a basic principle of cortical self-organization', W. Singer, *Concepts in Neuroscience 1*, 1–26 (1990), 'Of dreaming and wakefulness', R. R. Llinás and D. Pare, *Neuroscience 44*, 521–35 (1991). For review, see 'Thalamic contributions to attention and consciousness', J. Newman, *Consciousness and Cognition 4*, 172–93 (1995).

- Reverberatory loop theory: See *The Astonishing Hypothesis: The Scientific Search for the Soul* by Francis Crick, (New York: Simon and Schuster, 1994). For evidence of an amplifying circuit, see *Thalamic Oscillations and Signaling* by Mircea Steriade, Edward Jones and Rodolfo Llinás (New York: Wiley, 1990).

223 LaBerge's theory and hi-fi analogy: See *Attentional Processing: The Brain's Art of Mindfulness* (LaBerge, Harvard University Press), and 'Attention, awareness and the triangular circuit', D. L. LaBerge, *Consciousness and Cognition 6*, 149–81 (1997).

225 LaBerge's two PET experiments: 'Positron emission tomographic measurements of pulvinar activity during an attention task', D. L. LaBerge and M. S. Buchsbaum, *Journal of Neuroscience 10*, p613-619 (1990), and 'PET measurements of attention to closely spaced visual shapes', M. Liotti, P. T. Fox and D. L. LaBerge, *Society for Neurosciences Abstracts 20*, 354 (1994).

228 Passingham's research involved mostly lesion studies: See *The Frontal Lobes and Voluntary Action* by Richard Passingham (Oxford: Oxford University Press, 1993).

229 Passingham's PET experiments: 'Regional cerebral blood flow during voluntary hand and arm movements in human subjects', J. G. Colebatch et al, *Journal of Neurophysiology 65*, 1392–1401 (1991), 'Motor practice and neurophysiological adaptation in the cerebellum: a positron tomographic study', K. J. Friston et al, *Proceedings of the Royal Society of London B 248*, 223–8 (1992), 'Motor sequence learning: a study with positron emission tomography', I. H. Jenkins et al, *Journal of Neuroscience 14*, 3775–90 (1994), 'Anatomy of motor learning II: subcortical structures and learning by trial and error', M. Jueptner et al, *Journal of Neurophysiology 77*, 1325–37 (1997), and 'The time course of changes during motor sequence learning: a whole-brain fMRI study', I. Toni et al, *NeuroImage 8*, 50–61 (1998).

   For other similar scanning studies of motor learning and skill 'consolidation', see 'Motor learning in man: a positron emission tomographic study', Seitz et al, *Neuroreport 1*, 57–60 (1990), 'Functional MRI evidence for adult motor cortex plasticity during motor skill learning', A. Karni et al, *Nature 377*, 155–8 (1995), and 'Neural correlates of motor memory consolidation', R. Shadmehr and H. H. Holcomb, *Science 277*, 821–5 (1997).

231 Standard plan of basal ganglia loops: 'Parallel organisation of functionally segregated circuits linking basal ganglia and cortex', G. E. Alexander, M. R. DeLong and P. L. Strick, *Annual Review of Neuroscience 9*, 357–81 (1986). See also *Models of Information Processing in the Basal Ganglia*, edited by James Houk, Joel Davis and David Beiser (Cambridge, Massachusetts: MIT Press, 1995).

232 Neuromodulators have complex effects: See *Brain Biochemistry and Brain Disorders* by Philip Strange (Oxford: Oxford University Press, 1992).

233 Goldman-Rakic dopamine experiments: 'Modulation of memory fields by dopamine D1 receptors in prefrontal cortex', G. V. Williams and P. S. Goldman-Rakic, *Nature 376*, 572–5 (1995).

234 Cognitive symptoms of Parkinson's disease: 'Cognitive function in Parkinson's disease: from description to theory', R. G. Brown and C. D. Marsden, *Trends in Neurosciences 13*, 21–9 (1990).

235 Meck's time estimation studies: 'Peak-interval timing in humans activates frontal-striatal loops', S. C. Hinton, W. H. Meck and J. R. MacFall, *NeuroImage 3*, 224–8 (1996), and 'Scalar expectancy theory and peak-interval timing in humans', B. C. Rakitin et al, *Journal of Experimental Psychology: Animal Behavior Processes 24*, 15–33 (1998).

 – Subthalamic damage leads to ballism: 'Chorea and myoclonus in the monkey induced by gamma aminobutyric acid antagonism in the lentiform complex', A. R. Crossman et al, *Brain 111*, 1211–33 (1988).

 – Graybiel's basal ganglia study: 'The basal ganglia and adaptive motor control', A. M. Graybiel et al, *Science 265*, 1826–30 (1994). For experiment in which dopamine supply was lesioned, see 'Effect of the nigrostriatal dopamine system on acquired neural responses in the striatum of behaving monkeys', T. Aosaki, A. M. Graybiel and M. Kimura, *Science 265*, 412–15 (1994).

237 Dopamine not just joy juice: 'Getting the brain's attention', I. Wickelgren, *Science 278*, 35–7 (1997).

 – Schultz's dopamine research: 'Neuronal activity in monkey ventral striatum related to the expectation of reward', W. Schultz et al, *Journal of Neuroscience 12*, 4595–9 (1992), and 'A neural substrate of prediction and reward', W. Schultz, P. Dayan and P. R. Montague, *Science 275*, 1593–5 (1997). The lucky accident that led to results recounted in 'Getting the brain's attention', I. Wickelgren, *Science 278*, 35–7 (1997).

238 Brain set up for automating thinking and acting: Note that much of the basal ganglia research has been carried out against the background idea that the basal ganglia are part of an implicit learning circuit. As said above, many mind scientists have been seeking to make a reductionist split between conscious and unconsious activity in the brain. But the same data that appears to show two separate pathways can also be read as a shift from the global, top-heavy, style of processing needed to deal with novel or difficult tasks, to the local, low-level, style of processing suitable for well-learnt habits. As evidence that there is a true living continuum of response – with 'implicit learning modules' like the basal ganglia showing learning to events too dull, stealthy or regular to attract a full-blown shift of attention – see 'Brain regions responsive to novelty in the absence of awareness', G. S. Berns, J. D. Cohen and M. A. Mintun, *Science 276*, 1272–5 (1997).

Chapter 11: The Brain's Forking Pathway

242 Sokolov's orientation response: *Perception and the Conditioned Reflex* by Evgeny Sokolov (New York: Macmillan, 1963). For his anticipatory neuronal model, see 'The modeling properties of the nervous system', E. N. Sokolov, in *A Handbook of Contemporary Soviet Psychology*, edited by Michael Cole and Irving Maltzman (New York: Basic Books, 1973). Gray says his thinking was also shaped by the perceptual cycle of Ulric Neisser – see *Cognition and Reality: Principles and Implications of Cognitive Psychology* by Ulric Neisser (New York: W. H. Freeman, 1976)

Note that the noticing of negative events like the fridge going off becomes much easier to understand when it is seen that the brain's circuits are always firing and so always in a state of representative tension. Any cessation of a stimulus will cause an easily detectable dip in an already existing landscape of activity, so there is no need for an active preparation of the surface (although conscious priming would tweak the circuits still further, turning the attentional spotlight onto the auditory pathway and creating a state of tuned vigilance).

243 Gray's research on anxiety: For review, see 'The neuropsychology of anxiety: reprise', J. A. Gray and N. McNaughton, in *Nebraska Symposium on Motivation: Perspectives of Anxiety, Panic and Fear*, edited by Debra Hope (Lincoln, Nebraska: University of Nebraska Press, 1995).

246 LeDoux's work on the brainstem and amygdala: For review, see 'Emotion, memory and the brain', J. LeDoux, *Scientific American* (June 1994), 32–9, and *The Emotional Brain: The Mysterious Underpinnings of Emotional Life* by Joseph LeDoux (New York: Simon and Schuster, 1996). Note that other parts of the brain, such as the superior and inferior colliculi, would play important roles in any putative interrupt circuit. And like the amygdala, they could also be recruited for top-down use by the cortex planning stream, so would not be purely interrupt structures.

247 Stimulation of monkey amygdala: 'Electrical and chemical stimulation of frontotemporal portion of limbic system in the waking animal', P. D. Maclean and J. M. R. Delgado, *Electroencephalography and Clinical Neurophysiology 5*, 91–100 (1953). For classic ablation studies, see 'Psychic blindness and other symptoms following bilateral temporal lobectomy in rhesus monkeys', H . Klüver and P. Bucy, *American Journal of Physiology 119*, 352–3 (1937). For review, see *The Amygdala: Neurobiological Aspects of Emotion, Memory and Mental Dysfunction*, edited by John Aggleton (New York: Wiley-Liss, 1992).

249 Nucleus accumbens as a contention switch: For review, see 'The neuropsychology of schizophrenia', J. A. Gray et al, *Behavioral and Brain Sciences 14*, 1–84 (1991), and 'The neuropsychology of anxiety: reprise', J. A. Gray and N. McNaughton, in *Nebraska Symposium on Motivation:*

Perspectives of Anxiety, Panic and Fear (Hope, University of Nebraska Press).

– PET implicates amygdala in emotional assessment: For review, see 'How scary things get that way', M. Barinaga, *Science 258*, 887–8 (1992), and 'A new image of fear and emotion', S. E. Hyman, *Nature 393*, 417–18 (1998).

250 LeDoux finds LTP in amygdala: 'LTP is accompanied by commensurate enhancement of auditory-evoked responses in a fear conditioning circuit', M. T. Rogan and J. E. LeDoux, *Neuron 15*, 127–36 (1995).

252 Top-down bouncing of plans off amygdala: For relationship between ventromedial prefrontal cortex and amygdala, see *Descartes' Error: Emotion, Reason and the Human Brain* by Antonio Damasio (New York: Putnam, 1994).

253 Anatomy of cingulate cortex: *Neurobiology of Cingulate Cortex and Limbic Thalamus: A Comprehensive Handbook*, edited by Brent Vogt and Michael Gabriel, (Boston, Massachusetts: Birkhäuser, 1993), and 'Contributions of anterior cingulate cortex to behaviour', O. Devinsky, M. J. Morrell and B. A. Vogt, *Brain 118*, 279–306 (1995).

254 Cingulate damage led to empty awareness: 'Emotional disturbances associated with focal lesions of the limbic frontal lobe', A. R. Damasio and G. W. Van Hoesen, in *Neuropsychology of Human Emotion*, edited by Kenneth Heilman and Paul Satz (New York: Guilford Press, 1983). The story of Mrs T, a case of akinetic mutism, is also told in *Descartes' Error: Emotion, Reason and the Human Brain* (Damasio, Putnam).

– PET scanning of Stroop test: 'The anterior cingulate cortex mediates processing selection in the Stroop attentional conflict paradigm', J. V. Pardo et al, *Proceedings of the National Academy of Sciences 87*, 256–9 (1990). For a review of such experiments, see 'Isolation of specific interference processing in the Stroop task: PET activation studies', S. F. Taylor et al, *NeuroImage 6*, 81–92 (1997). See also 'Anterior cingulate cortex, error detection and the online monitoring of performance', C. S. Carter et al, *Science 280*, 747–9 (1998), for strong evidence of a scaffolding role.

255 PET scanning of pain sensation: 'Pain and Stroop interference tasks activate separate processing modules in anterior cingulate cortex', S. W. J. Derbyshire, B. A. Vogt and A. K. P. Jones, *Experimental Brain Research 118*, 52–60 (1998). See also 'Pain affect encoded in human anterior cingulate but not somatosensory cortex', P. Rainville et al, *Science 277*, 968–71 (1997).

256 Posner's interpretation of cingulate evidence: 'Attentional mechanisms and conscious experience', M. I. Posner and M. K. Rothbart, in *The Neuropsychology of Consciousness*, edited by David Milner and Michael Rugg (London: Academic Press, 1992), 'Attentional networks', M. I. Posner and S. Dehaene, *Trends in the Neurosciences 17*, 75–9 (1994), and 'Constructing neuronal theories of mind', M. I. Posner and M. K. Rothbart,

in *Large-Scale Neuronal Theories of the Brain*, edited by Christof Koch and Joel Davis (Cambridge, Massachusetts: MIT Press, 1994).

Note that Posner's model of the attention/escalation process, and the balancing of interruptions against plans, places the emphasis on rather different bits of neural machinery to, say, Gray, LeDoux, LaBerge or Baars. For example, as well as laying special stress on the anterior cingulate, Posner talks more about the role of the posterior parietal cortex in disengaging the brain from its current point of focus and the superior colliculus in executing the shift. The argument is that the focusing of consciousness during a cycle of processing is a whole brain exercise and so all these theorists are simply feeling different parts of the same elephant.

On the other hand, there are differences among these theorists as to where they lie on a computational-dynamic spectrum. Gray, who probably has the most thoroughly worked-out circuit model, is also probably the most computational in some parts of his thinking. For example, he suggests that the hippocampal theta rhythm – a seven-to-ten-beat per second EEG oscillation – may effectively be the sound of the comparator apparatus cranking out its results at the rate of one frame of settled awareness every 100 milliseconds or so. Posner, LeDoux and Baars have moved towards more dynamic descriptions of what began originally as strongly computational models, while LaBerge's work perhaps fits most naturally into an adaptive systems mould.

260 Gray's publication of theories: 'The contents of consciousness: a neuropsychological conjecture', J. A. Gray, *Behavioral and Brain Sciences 18*, 659–722 (1995), 'The neuropsychology of schizophrenia', J. A. Gray et al, *Behavioral and Brain Sciences 14*, 1–84 (1991), and 'Consciousness: what is the problem, and how should it be addressed?', J. A. Gray, *Journal of Consciousness Studies 2*, 5–9 (1995). For his early work on a behavioural inhibition pathway, see *The Neuropsychology of Anxiety: An Enquiry into the Functions of the Septo-Hippocampal System* by Jeffrey Gray (Oxford: Oxford University Press, 1982).

Chapter 12: Getting It Backwards

268 Tuscon conference at Woodstock: From an account of the conference in 'Can science explain consciousness?', J. Horgan, *Scientific American* (July, 1994), 88–94.

270 Mental responses of animals: See *Animal Thought* by Stephen Walker (London: Routledge and Kegan Paul, 1983).

272 Half second measures time to make a global shift: Two further lines of evidence for this come from the psychophysical phenomena of iconic memory and the attentional blink.

Iconic memory can be interpreted as a delayed 'shrink' of the perceptual field. Instead of creating an escalated focus as fast as possible, the escalation is deliberately delayed, so telling something about the natural cycle time of the process. In a typical experiment, a slide containing three rows of four letters is flashed up on a screen for just 50 milliseconds. Subjects have to avoid reading any of the letters until a tone sounds to tell them which column they are mentally supposed to turn to and report. As long as the signal is not delayed more than a third to half a second, their performance is almost perfect. They can inspect a still lingering iconic image and read off the chosen column. But this act of looking then wipes out all memory for the other two columns, as if the top-down act of focusing irrevocably sculpts what had been a flat and even spread of preconscious mapping. See 'The information available in brief visual presentations', G. Sperling, *Psychological Monographs 74:11* (1960). Note that it has been suggested that the cortex grabs the information direct from lingering retinal activity. See 'Locus of short-term visual storage', B. Sakitt, *Science 190*, 1318–19 (1975).

The attentional blink is a more recently discovered effect which reveals the brain needs about half a second to recover from being 'pinched up' to catch a perceptual event – it cannot focus sharply on a second event until it has had time to realign its anticipatory state. In a typical experiment, subjects are shown a series of alphabet letters and asked not only to report a sighting of any x's, but also the occasional insertion of the number 4. When asked to spot x's or 4's alone, people can catch them all even at a presentation rate of eight a second. But with two clashing targets to report, a half-second mental blindspot appears. This suggests that the more rapid attentional performance is only possible because a single anticipatory framework remains in place. Once subjects have to swap between goal states, the full cycle time is exposed.

See 'Temporary suppression of visual processing with an RSVP task: an attentional blink?', J. Raymond, K. Shapiro and K. M. Arnell, *Journal of Experimental Psychology: Human Perception and Performance 18*, 849–60 (1992). For evidence that information is still being handled at a preconscious level during the blink, see 'Word meanings can be accessed but not reported during the attentional blink', S. J. Luck, E. K. Vogel and K. L. Shapiro, *Nature 383*, 616–18 (1996). The dynamic nature of this effect is evident in the fact that lesser levels of goal conflict lead to apparently faster switching. See *Perception and Communication* by Donald Broadbent (London: Pergamon Press, 1958) for early discussions about attentional bottlenecks.

Note further that the question of attentional dwell time or the brain's processing frame rate has been a source of great confusion in cognitive psychology because it is assumed that as a computational device, the brain should have a single, fixed, processing cycle speed. Attention shifts should take a set time. The idea that the brain is always representing, and that sharp attention is the pinching up of this surface, helps explain the widely variable performance that is actually observed. For the evolutionary nature of attentional states, see 'Integrated field theory of consciousness', M. Kinsbourne, in *Consciousness in Contemporary Science*, edited by Anthony Marcel and Edoardo Bisiach (Oxford: Oxford University Press, 1988).

275 In shocking moments events seem frozen: See 'When a second lasts forever', J. McCrone, *New Scientist* (1 November 1997), 52–6.

Chapter 13: The Ape that Spoke

278 How the human mind is different: The story of human evolution and the learnt nature of our mental skills is dealt with in detail in *The Ape That Spoke: Language and the Evolution of the Human Mind* by John McCrone (London: Macmillan, 1990) and *The Myth of Irrationality: The Science of the Mind From Plato to Star Trek* by John McCrone (London: Macmillan, 1993).

– Rise of Homo sapiens: For a good general account, see *Lucy: The Beginnings of Humankind* by Donald Johanson and Maitland Edey (New York: Warner Books, 1982), and *The Origin of Modern Humans* by Roger Lewin (New York: W. H. Freeman, 1993).

279 Chimpanzee intelligence and tool-use: See *The Chimpanzees of Gombe* by Jane Goodall (Cambridge, Massachusetts: Harvard University Press, 1986), *Chimpanzee Politics: Power and Sex Among Apes* by Frans de Waal (London: Jonathan Cape, 1982), and *Chimpanzee Material Culture: Implications for Human Evolution* by William McGrew (New York: Cambridge University Press, 1992).

280 Art as sign of modern mind: *Becoming Human: Evolution and Human Uniqueness* by Ian Tattersall (New York: Harcourt Brace, 1998).

281 Tool use preadapted brain for speech: 'Neuromotor mechanisms in the evolution of human communication', D. Kimura, in *Neurobiology of Social Communication in Primates: An Evolutionary Perspective*, edited by Horst Steklis and Michael Raleigh (New York: Academic Press, 1979). Interestingly, another early proponent of this idea now believes that lateralisation exists in most animals and so tool use could only have further refined an existing feature of the brain. See 'Towards a unified view of cerebral hemispheric specializations in vertebrates', P. F. MacNeilage, in *Comparative Neuropsychology*, edited by David Milner (Oxford: Oxford University Press, 1998).

282 Johanson on change in parenting style: *Lucy: The Beginnings of Humankind* (Johanson and Edey, Warner Books).

283 Chimps can be taught words: For review of the vexed issue of ape language

competence, see *Aping Language* by Joel Wallman (New York: Cambridge University Press, 1992).

285 Over 6,000 dialects and 200 language families: This is a popularly quoted figure – see 'Hard Words', P. E. Ross, *Scientific American* (April 1991), 70–9 – but others suggest the true number may be double this. See 'Language diversity', J. A. Allan, P. Baker and M. Farmer, *New Scientist* (10 February 1996), 48.

– Variations in sentence order: *Universals of Language*, edited by Joseph Greenberg (Cambridge, Massachusetts: MIT Press, 1963), and *The Language Instinct: The New Science of Language and Mind* by Steven Pinker (New York: William Morrow, 1994).

286 Neanderthals probably not articulate: *Uniquely Human: The Evolution of Speech, Thought and Selfless Behavior* by Philip Lieberman (Cambridge, Massachusetts: Harvard University Press, 1991).

287 Animals locked into the present: See *The Ape That Spoke: Language and the Evolution of the Human Mind* (McCrone, Macmillan), *Animal Thought* by Stephen Walker (London: Routledge and Kegan Paul, 1983), *An Essay Concerning Human Understanding* by John Locke, edited by Peter Nidditch (Oxford: Clarendon Press, 1975), and *The World as Will and Idea* by Arthur Schopenhauer, translated by R Hackforth (Cambridge: Cambridge University Press, 1972).

288 Wittgenstein on dog not afraid of a beating tomorrow: *Philosophical Investigations* by Ludwig Wittgenstein (Oxford: Basil Blackwell, 1976).

– Idea speech responsible is old: See *The Myth of Irrationality: The Science of the Mind From Plato to Star Trek* (McCrone, Macmillan), and *Understanding Vygotsky: A Quest for Synthesis* by René van der Veer and Jaan Valsiner (Oxford: Blackwell, 1991).

– Our ancestors were thinkers before being speakers: This is the standard line taken by cognitive scientists. So, for example, Steven Pinker says: 'It seems uncontestable, even banal, to say that the language faculty was selected for its ability to communicate thought.' See 'Facts about human language relevant to its evolution', S. Pinker, in *Origins of the Human Brain*, edited by Jean-Pierre Changeux and Jean Chavaillon (Oxford: Oxford University Press, 1995). See also *How the Mind Works* by Steven Pinker (London: Allen Lane, 1998), and *Language, Learning and Thought*, edited by John MacNamara (New York: Academic Press, 1977).

290 PET studies of the mapping of vocabulary: 'Discrete cortical regions associated with knowledge of colour and knowledge of action', A. Martin et al, *Science 270*, 102–5 (1995), 'Neural correlates of category-specific knowledge', A. Martin, J. Haxby, F. Lalonde, C. Wiggs and L. Ungerleider, *Nature 379*, 649–52 (1996), and 'A neural basis for lexical retrieval', H. Damasio et al, *Nature 380*, 499–505 (1996).

291 Myelinisation is slow in humans: See *Brain Development and Cognition: A Reader*, edited by Mark Johnson (Cambridge, Massachusetts: Blackwell, 1993), 'Development of cortical circuitry and cognitive function', P. S. Goldman-Rakic, *Child Development 58*, 601–22 (1987), and 'Myelinisation of cortical-hippocampal relays during late adolescence', F. M. Benes, *Schizophrenia Bulletin 15*, 585–93 (1991).

292 Children pick up grammar from inference: Ever since Noam Chomsky drew a line in the sand with *Syntactic Structures* (The Hague: Mouton, 1957), there has been a bitter divide between those who believe in nature and those who argue for nurture. It has only been with the advent of neural networks and theories about self-organising systems that it has become possible to see a mutual resolution – see *Rethinking Innateness: A Connectionist Perspective on Development*, edited by Jeffrey Elman et al (Cambridge, Massachusetts: MIT Press, 1996), 'Innateness, autonomy, universality? Neurobiological approaches to language', R.-A. Mueller, *Behavioral and Brain Sciences 19*, 611–75 (1996), and 'Statistical learning by 8-month-old infants', J. R. Saffran, R. N. Aslin and E. L. Newport, *Science 274*, 1926–8 (1996). For predisposing instincts like turn-taking and gaze-following, see *Language Development: A Reader*, edited by Andrew Lock and Eunice Fisher (London: Croom Helm, 1984), and *Joint Attention: Its Origins and Role in Development*, edited by Chris Moore and Philip Dunham (Hillsdale, New Jersey: Lawrence Erlbaum Associates, 1995).

295 Often our inner dialogue seems sketchy: For evidence that it takes time mentally to unpack speech acts, see *Speaking: From Intention to Articulation*, edited by Willem Levelt (Cambridge, Massachusetts: MIT Press, 1989), and 'Brain activity during speaking: from syntax to phonology in 40 milliseconds', M. van Turennout, P. Hagoort and C. M. Brown, *Science 280*, 572–4 (1998).

297 Recollection and self-awareness are word-based: For review, see *The Myth of Irrationality: The Science of the Mind From Plato to Star Trek* (McCrone, Macmillan). Eyewitness experiments in *Eyewitness Testimony* by Elizabeth Loftus (Cambridge, Massachusetts: Harvard University Press, 1979). See also *Private Speech: From Social Interaction to Self-Regulation*, edited by Rafael Diaz and Laura Berk (Hillsdale, New Jersey: Lawrence Erlbaum Associates, 1992), and *The Disappearance of Introspection* by William Lyons (Cambridge, Massachusetts: MIT Press, 1986).

Chapter 14: Answering the Hard Question

303 Chalmers strikes a chord: 'The puzzle of conscious experience', D. J. Chalmers, *Scientific American* (December 1995), 62–8, *The Conscious Mind: In Search of a Fundamental Theory* by David Chalmers (Oxford: Oxford

University Press, 1996), and *Explaining Consciousness: The Hard Problem*, edited by Jonathan Shear (Cambridge, Massachusetts: MIT Press, 1997).

305 Colour as the philosopher's favourite challenge: There are a wealth of books on the subject, for example, *Colour Vision: A Study of Cognitive Science and the Philosophy of Perception* by Evan Thompson (London: Routledge, 1995), *Readings on Color, Volume 1: The Philosophy of Color*, edited by Alex Bryne and David Hilbert (Cambridge, Massachusetts: MIT Press, 1997), and *Color for Philosophers: Unweaving the Rainbow* by Clyde Hardin (Cambridge, Massachusetts: Hackett, 1988).

The difficulty of achieving a canonical experience of a colour has been widely discussed. See, for instance, 'Colors, normal observers and standard conditions', C. L. Hardin, *Philosophical Quarterly 21*, 125–33 (1983). But colour vision is actually such a dynamically-constructed experience that even at a 'raw' neural level – with the extra overlay of human thought discounted – it would be hard to talk about a baseline experience of colour. The perceived colour of one corner of the visual field is always affected by what surrounds it – the local colour response has to 'feel the shape of the vessel'. For the importance of the physical context, let alone the mental context, see 'The colors of things', P. Brou et al, *Scientific American* (September 1986), 80–7, and 'Color vision: a review from a neuro-physiological perspective', P. Gouras and E. Zrenner, *Progress in Sensory Physiology 1*, 139–79 (1981).

309 Shifting from reductionism to dynamism: Note that reductionism, *per se*, is not the problem but its indiscriminate use. The ability to isolate sharply defined cause and effect relationships is a powerful tool. But like a surgeon's scalpel, what counts is the skill with which it is used. Rather than hacking away at the body of the problem, the aim should be to make the single cut that reveals what lies within, then to move on to the next deeper cut. So with the problem of consciousness, for example, the first cut is to distinguish between the language-enhanced human mind and 'locked in the present' animal consciousness. Then would come some attempt to define a basic cycle of preparation, mapping and reaction.

The problem with unconstrained reductionism is that it creates a flat, featureless, field of knowledge fragments, in which no part seems to count more than any other, rather than an organised landscape with natural high points. It also fosters the belief that science can simply start its explorations at any point – so researchers can begin with something simple such as the 'neural correlates of visual experience' or the process of memory and then move on to a more general theory of mind. The dynamic view is that science has to be properly orientated to its subject – it has to begin with a broad perspective that makes it clear the route to take to travel smoothly into the detail of visual mapping or memory fixing. The fractalisation of much of current science is caught well in 'What cannot be said in science', M. T. Greene, *Nature 388*, 619–20 (1997). See also *Lifelines: Biology, Freedom, Determinism* by Steven Rose (London: Allen Lane, 1997).

311 Köhler's attack on machine theory: *Dynamics in Psychology* by Wolfgang Köhler (New York: Grove Press, 1940).

Index